Recognition in Microbe-Plant Symbiotic and Pathogenic Interactions

NATO ASI Series

Advanced Science Institutes Series

A series presenting the results of activities sponsored by the NATO Science Committee, which aims at the dissemination of advanced scientific and technological knowledge, with a view to strengthening links between scientific communities.

The Series is published by an international board of publishers in conjunction with the NATO Scientific Affairs Division

A Life Sciences **B Physics**	Plenum Publishing Corporation London and New York
C Mathematical and Physical Sciences	D. Reidel Publishing Company Dordrecht, Boston, Lancaster and Tokyo
D Behavioural and Social Sciences **E Applied Sciences**	Martinus Nijhoff Publishers Boston, The Hague, Dordrecht and Lancaster
F Computer and Systems Sciences **G Ecological Sciences** **H Cell Biology**	Springer-Verlag Berlin Heidelberg New York London Paris Tokyo

Series H: Cell Biology Vol. 4

Recognition in Microbe-Plant Symbiotic and Pathogenic Interactions

Edited by

Ben Lugtenberg

Department of Plant Molecular Biology
University of Leiden
Leiden, The Netherlands

Springer-Verlag
Berlin Heidelberg New York London Paris Tokyo
Published in cooperation with NATO Scientific Affairs Divison

Proceedings of the NATO Advanced Research Workshop on Recognition in Microbe-Plant Symbiotic and Pathogenic Systems held at Biddinghuizen, The Netherlands, May 11–16, 1986

ISBN-13: 978-3-642-71654-6 e-ISBN-13: 978-3-642-71652-2
DOI: 10.1007/978-3-642-71652-2

Softcover reprint of the hardcover 1st edition 1986

2131/3140-543210

PREFACE

This volume comprises the lectures of the speakers at the NATO Advanced Research Workshop held at the Congress Centre The Flevohof at Biddinghuizen, The Netherlands, May 11-16, 1986. The purpose of the workshop was to bring together experts in symbiosis, plant pathology and plant molecular biology in order to discuss recent progress in the field of microbe - plant recognition at the molecular level, to promote integration of various disciplines, and to define recommendations for future research and applications.

Plants have developed a variety of sophisticated defence mechanisms to cope with an environment in which many different microbes live. Most microbes which colonize plant tissues are harmless. Some microbes have developed ways to attack plants successfully, resulting in enormous losses of crop yields. Other microbes have reached an agreement with the host plant which is beneficial for both: these microbes live in symbiosis with the plant and provide their host plant for example with substantial amounts of atmospheric nitrogen. Chemical protection of crops is a necessity in modern crop management but this treatment has some negative effects as well. Therefore scientists are looking for alternative, biological, ways to control crop pests.

Against this background specialists from eleven countries discussed the results of their most recent work on the molecular background of microbe - plant interactions. It appeared that, in order to capitalize the recent rapid progress made in the molecular genetical studies on Rhizobium-legume and pathogen - host plant interactions, a multidisciplinary approach is required. Plant molecular biologists, phytopathologists, molecular microbiologists, plant physiologists, soil biologists and agronomists should collaborate closely and define recommendations for future research on the molecular basis of microbe - plant interactions. The ultimate goal is application of this knowledge for the benefit of mankind.

The atmosphere of the workshop was excellent. Several lectures and discussions resulted in new collaborations and opened links to other disciplines. The scientific success of the workshop was to a major extent the result of the excellent, critical and friendly advice of my colleagues Prof. Peter Albersheim, Prof. Klaus Hahlbrock and Prof. Robbert Schilperoort, to whom I am most grateful. Essentially all speakers and discussion leaders contributed their manuscripts in time. I thank all participants for a week of excellent science.

The meeting would not have been possible without the help of many people, most of whom I will not mention here. I want to express my admiration for Sonja Chang, who spent many hours at the organization of the workshop. I am also in debt to her for her careful and efficient handling of the manuscripts and for improving and preparing many of the excellent typescripts from which this book was produced. The selfless handling of logistic and financial aspects of the meeting by my colleagues Dr. Anton van Brussel and Dr. Robert Okker was warmly appreciated.

The financing of the workshop was helped greatly by contributions of several bodies, especially the Adviescommissie IOP-b, Gist-brocades, Mogen International B.V., Multiplant B.V. and Shell International Petroleum Company Limited, who paid part of the expenses of several participants. The Scientific Affairs Division of NATO was our major sponsor. On behalf of all participants I thank them for the very generous grant which enabled us to finance the scientific meeting. The contacts with and valuable advice from Dr. Truels Truelsen and Dr. Mario di Lullo was greatly appreciated. It was sad to learn that Dr. di Lullo died recently.

We hope that this volume will prove to be useful in sharpening our perspective of how microorganisms and plants communicate at the molecular level.

B. Lugtenberg

Leiden, September 1986

CONTENTS

I. RECOGNITION IN NODULATION BY RHIZOBIUM

IV. RECOGNITION IN BIOLOGICAL CROP PEST CONTROL

V. RECOMMENDATIONS FOR FUTURE RESEARCH AND APPLICATIONS

GENETIC ANALYSIS OF RHIZOBIUM-PLANT INTERACTIONS

Sharon R. Long, N. Kent Peters, John T. Mulligan, Mark E. Dudley and Robert F. Fisher
Department of Biological Sciences
Stanford University
Stanford, CA 94305
U.S.A.

INTRODUCTION

In approaching the question of plant-microbe recognition, we require studies at several levels. In addition, it is important to understand the many stages at which bacteria and plants interact. This is particularly relevant to the legume-Rhizobium symbiosis, which proceeds by a series of steps involving both the host plant and the bacterial symbiont. These steps, originally outlined by Vincent (1980), include attachment, root hair curling, meristem initiation, invasion of the plant, release of bacteria into plant cells, and successful expression of nitrogen fixation and other essential genes. At all of these stages, recognition events may occur.

For any one stage of nodule formation, various research groups are concerned with analysis at levels including genetics, biochemistry, and cell biology. Among the specific research goals are: identification of the genes necessary for nodulation, host range, release and function; analysis of gene regulation, as studied by gene fusions, promoter analysis and in vitro transcription; and study of gene products, whose nature is hinted at by sequence identification, of higher-level assemblies (polysaccharides, for example) and of other cell properties which are controlled by the bacterial genes.

Genetics has a unique and unifying role in the study of plant-microbe specificity (Long 1984). The construction of mutants and variants permits a test of phenotypes and biochemical properties which may be important in interacting systems; the identification of symbiotic genes and their transfer into other backgrounds makes it possible to highlight essential functions apart from background events. And finally, the identification, cloning, and molecular manipulation of

NATO ASI Series, Vol. H 4
Recognition in Microbe-Plant Symbiotic and Pathogenic Interactions. Edited by B. Lugtenberg

individual genes should permit us to work from the gene towards the protein (John et al 1985; Egelhoff and Long 1985).

Nodulation genetics has been studied primarily in three fast growing species, Rhizobium meliloti, R. leguminosarum, and R. trifolii. A consensus of gene maps and nomenclature has been arrived at by a combination of mutagenesis, gene transfer, and sequencing (figure 1)(Downie et al 1983, 1985; Rossen et al 1984; Djordevic et al 1984, 1985; Schofield and Watson 1986; Wijffelman et al 1985; Kondorosi et al 1984, 1985; Torok et al 1984; Long et al 1982; Jacobs et al 1985; Egelhoff et al 1985; and several articles in this volume). Recent work is also expanding our understanding of the slow-growing bacteria, e.g. Bradyrhizobium japonicum (Lamb and Hennecke 1986; Russell et al 1985), and of unique bacteria such as R. parasponiae and R. loti (Scott 1986; Scott et al 1985). Studies on gene maps and phenotypes in these various systems show both similarities and differences in their organization and apparent action, as will be apparent through this volume. In my presentation, I will begin with an overview of nodulation events, and will pose some questions which researchers in the field find intriguing. Then, focussing on a few early stages, I will present our work assigning particular genes to events in nodulation, and further studies on the molecular biology of these genes.

RHIZOBIUM-PLANT INTERACTION: PHENOMENA AND QUESTIONS

If we consider the primary root axis and free living bacterium before their physical contact, we now know about several features which may be important for interaction and recognition. On the part of the plant, it has been reported by Bauer and colleagues (Bauer 1981) that the most infectible root hairs are those which are still growing. This may indicate something important about the mechanism of invasion. On the bacteria's part, it now appears that a signal from the plant stimulates the expression of the nodulation (nod) gene (Mulligan and Long 1985; Rossen et al 1985; Innes et al 1985; van Brussel et al 1986; Canter-Cremers et al 1986). This signal has been characterized by several laboratories, including ours, to be a flavone (see below, and reports by Johnston et al , Redmond et al , and Wijffelman et al , in this volume).

The events of early nodulation have been characterized by

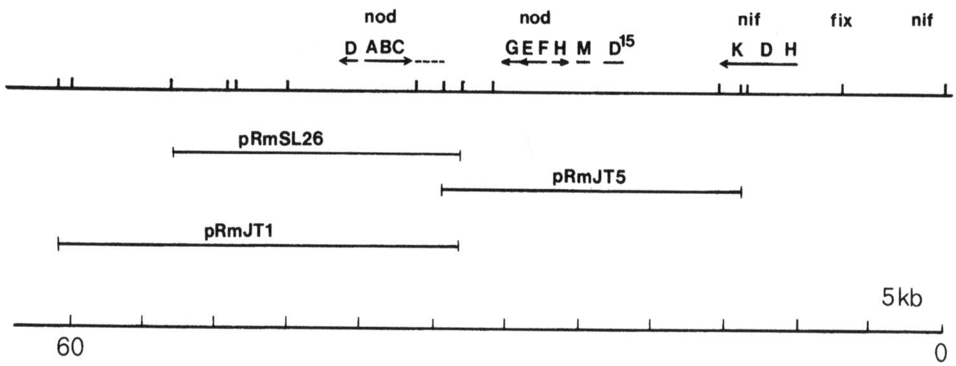

Figure 1. Map of the extended symbiotic region in R. meliloti 1021. Dotted lines near nodABC indicate the position predicted for nodI and nodJ (C. Wijffelman, personal communication; Shearman et al 1986; M. Djordevic and B. Rolfe, personal communication). Map for nodFEG and H, from data of R. Fisher, J. Swanson et al is consistent with that of Kondorosi et al (1985) and Dénarié et al (this volume). For discussion of common nod genes DABC, see text.

microscopy, but biochemically they are obscure. In the infectible zone, homologous bacteria will cause deformation and curling of the host epidermal root hairs, and this is one of the first specific events which shows recognition of the bacteria and plant (Vincent 1980). Among the questions which we might ask about this process are: How does the bacterium interact with the growing tip of the root hair? Also, what is the relationship of root hair branching and root hair deformation -- which can be stimulated by some heterologous Rhizobium species and by some mutants in nodulation genes -- to the kind of root hair curling or shepherd's-crook formation which is diagnostic for a homologous, successful interaction?

In the crook of curled root hairs, infection threads initiate and grow into the plant. Several things are not known about these structures. First, what is the composition of the infection thread? What determines its synthesis? And also, how is a dynamic equilibrium maintained between the growth of the infection thread and the deposition of wall around it? Several research groups have recently found that the bacterial extracellular polysaccharide, while not required for the bacterial induction of cell division (see below), appear to be essential for stimulating the formation of infection threads (for example, see Finan et al 1985; Leigh et al 1985 and Pühler et al , this volume). The exact role of the polysaccharide, however, is not known.

It has been known for some time that cell divisions in many species are formed in the inner cortex at a time when the penetrating infection thread was still in the outer cortex (see Newcomb 1981) or middle cortex (Libbenga and Harkes 1973). Pushing observations back one step earlier, Calvert et al (1984) reported that cell division of soybean is initiated prior to any infection thread penetration of the root by Bradyrhizobium japonicum. We have recently carried out similar studies in alfalfa, by inoculating bacteria onto roots and examining longitudinal root sections at 12, 18, 24 and 30 hours after inoculation. We also found that prior to the formation of any infection threads, alfalfa activates cells of the inner root cortex and these cells begin the cell division which create the nodule meristem (M. Dudley S. Long, unpublished observations).

Many mysteries remain concerning this stage. First, we do not know what the signal is which causes cell divisions in the plant, apparently acting across a distance of several cell layers. Whether the signal for cell division comes directly from the bacterium, or is produced in the outer plant cells and transduced into a secondary signal, is also not known. Bauer et al (1985) have reported that this signal is host specific, in that a Rhizobium will not stimulate significant cell division in a heterologous plant root even if it is stimulated by its own host plant. Thus, the stimulation of cell division also seems to be a stage at which recognition must take place. In addition to understanding what the signal is, we would also like to know which plant cells respond to this and how targeted reponse results in the morphogenesis of the nodule.

The penetration of infection threads into proliferating plant cells permits the release of bacteria from infection threads into cells of the host. During the release, bacteria emerge from the infection thread in unwalled droplets which are surrounded by plant plasma membrane (see Newcomb 1981, for review). This membrane continues to envelope the bacterial cell, and as the bacteria differentiate into bacteroids, the plant plasma membrane may also differentiate (see also Brewin, this volume). The mature nodule includes both infected and uninfected plant cells. In addition, the nodule displays a very distinctive morphology, in which the vascular strands are peripheral rather than central.

GENES USED IN NODULATION

In order to discover what genes are involved in causing nodule formation, we carried out transposon mutagenesis of R. meliloti. Complementation of Nod^- mutants led us to investigate an 8.7 kilobase EcoRl fragment, carried on plasmid pRmSL26 (Long et al 1982). Saturation transponon mutagenesis Tn5 of the 8.7 kb fragment showed that a 3.5 kb DNA segment contained genes for nodulation (Jacobs et al 1985; Kondorosi et al 1984). Sequence and complementation analysis showed that this segment contains four genes, nodD and nodABC (Torok et al 1984; Jacobs et al 1985; Egelhoff et al 1985). The nodD transcript reads in the opposite direction from the nodABC operon, and the start sites of the open reading frames are about 250 base pairs apart (Egelhoff et al 1985).

To investigate the function of these genes we have analyzed the phenotype of mutants in nodA and nodC. Alfalfa roots were inoculated with bacteria either wild type or a mutant of nodA or of nodC, and after 48 hours the inoculated plant segment was excised and sectioned for microscopic examination. In a wild type Rhizobium-plant interaction, extensive cell division has taken place at this point. By contrast, we found that mutants in nodA and nodC did not provoke any cell division in the plant. In addition, these bacterial mutants do not cause root hair curling on the plant hosts (Jacobs et al 1985). From this, we can conclude that the nod genes A and C are essential for stimulating not only root hair curling (Hirsch et al 1982; Jacobs et al 1985; Rossen et al 1984; Djordevic et al 1984), but also early cell divisions in the root's inner cortex.

The function of the nodABC and nodD genes are conserved in different Rhizobium species. This was shown by constructing strains in which mutants of these nod genes were complemented with cloned DNA segments from Rhizobium with other host ranges. We found, for example, that a nodA mutants of Rhizobium meliloti could be complemented and therefore become Nod^+ on alfalfa, if the strain contained cloned nodA genes of Rhizobium trifolii (Fisher et al 1985). Parallel experiments in many laboratories (Kondorosi et al 1984; Djordevic et al 1984; Downie et al 1983) have shown this phenomenon. For this reason, the nodDABC group are referred to as "common" nod genes.

In addition to the common nod genes, there are other genes required for, or involved in, alfalfa nodulation by R. meliloti. Such genes were first reported by Kondorosi et al (1984) in strain R. meliloti 41. Comparison of the restriction maps of strains 41 and strain 1021 shows that there are considerable differences in this region. Therefore we set out to see whether such extra nodulation genes were present in R. meliloti 1021. By complementation of a point mutant, WL131, with extended clone pRmJT5, we discovered that DNA in between the common nod genes and the nif genes encodes additional nodulation genes (Long et al 1985). This region should correspond to the reported hsn gene region of Kondorosi et al (1984, 1985). To investigate the map of possible nodulation genes in this segment, we have carried out transposon Tn5 mutagenesis, and have analyzed the nodulation phenotype of the resulting strains (Long et al 1985; J. Tu, J. Swanson, J. Ogawa, R. Fisher, R. Sanga and S. Long, manuscript in preparation). We have found several regions in which transposons altered the nodulation behavior of the bacterium. We conclude that while the physical maps are very different, there are probably similar genes in R. meliloti 1021 and in R. meliloti strain 41. An extensive study of this region in strain 2011 is also reported by Dénarié and colleagues, in this volume. In addition, we have found genes lying outside the probable hsn region which are apparently involved in regulation, as described below.

The transfer of host specific nod genes into other species of Rhizobium, or into Agrobacterium tumefaciens, confers the ability to nodulate alfalfa on those recipient strains (Kondorosi et al 1984; Downie et al 1983; Djordevic et al 1985; Dénarié et al , this volume; Tu et al , manuscript in preparation). This would indicate that host range is conferred by positive factors in Rhizobium. However, it has been reported that Rhizobium trifolii host specificity genes have the property that, when mutated, the strain acquires a broader host range (Djordevic et al 1985). This would be evidence of negatively acting host range genes, in addition.

REGULATION OF NOD GENES

We have analyzed the expression of the nodulation genes D and ABC by using translational gene fusions. The *E. coli* beta-galactosidase gene, *lacZ*, was fused in frame to *nodD* and also to *nodC*. Using these constructs we found that the *nodD* locus was expressed whether the cells were grown in culture, or in the presence of a host plant. However, the *nodABC* genes appeared not to be expressed in free-living bacterial cells, while the *nodABC* expression increased greatly in cells exposed to alfalfa plant exudates (Mulligan and Long 1985). These exudates were prepared both from alfalfa roots and alfalfa seeds, and involved simply placing plant material in water, and later decanting the water to use as inducer. Inducibility of *nod* genes was also reported by Rossen et al (1985); van Brussel et al (1986) and Spaink (this volume), and Innes et al (1985). These latter studies have shown that several loci, not just *nodABC*, are inducible by plant exudates.

The expression of *nodABC-lacZ* was considerably lower when it was born in its normal location on the megaplasmid than when carried on a vector. We found that expression of a megaplasmid (pSym)-borne *nodABC-lacZ* fusion was greatly enhanced if extra copies of the *nodD* gene were carried on vectors in the *Rhizobium* cells. Further analysis suggests strongly that the effect of the plant exudate on expression of *nodABC* requires the protein product from the *nodD* gene (Figure 2). It is not clear what the mechanism of the *nodD* effect is, but two general models can be proposed: first, the *nodD* gene product could be a protein which acts as a receptor for the plant signal, or as part of the signal transduction apparatus of the *Rhizobium* cell. The second general model would propose that the *nodD* gene product acts as a positive transcriptional activator for the *nodABC* genes, and that the plant inducer is required either directly or indirectly for the *nodD* gene product to be active. These models require testing by *in vitro* and genetic studies. We have begun preparation for such experiments by purification and characterization of RNA polymerase from *Rhizobium meliloti*. We have found that *R. meliloti* RNAP accurately transcribes *nodD* *in vitro*, but not *nodABC* (Fisher et al 1986). This is consistent with the model that *nodABC* transcription is regulated by a positive transcriptional activator, and suggest that *in vitro* expression analysis may be helpful in defining such positive factors.

The availability of *nodC-lacZ* fusions has also permitted

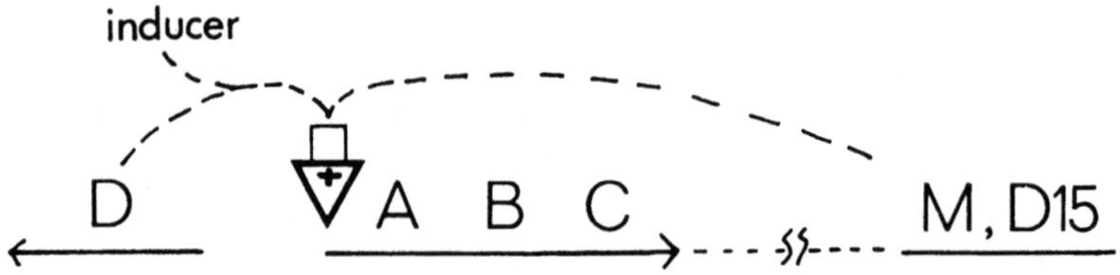

Figure 2. A tentative scheme for multiple genes influencing nodABC. A multiple copy clone bearing M and D15 causes nodABC to become constitutive.

us to look for other regulatory genes. We found that a clone, pRmJT5, had an interesting regulatory effect. When this clone was placed in a Rhizobium cell containing a nodC-lacZ fusion, it caused the nodC gene to be expressed constitutively at high level. Mutagenesis of clone pRmJT5 showed that this regulatory effect was partly due to a gene we have designated M. This gene is located between the host specific nod genes and the nif genes (see figure 1).

Parallel studies on nod gene regulation had revealed an interesting problem in Rhizobium meliloti. We found, using a DNA probe internal to nodD, that R. meliloti appears to have several genes with homology to nodD. This may explain why mutations in nodD in R. meliloti are not completely Nod$^-$, although nodD mutants of other species are completely Nod$^-$ (Downie et al 1985; Djordevic et al 1984)). We wanted to see whether these extra copies of nodD are functional. One of the extra copies of nodD, referred to here as nodD15, is located on the 15 kb EcoR1 fragment bearing the host-specific (hsn) nodulation genes and nodM. We found that nodD15 is also important, along with the nodM gene, in causing constitutive expression of nodABC. The nodM-nodD15 effect occurs at the same promoter (same transcription start site), as the nodD/plant inducer effect, as shown by RNA primer extension analysis. It therefore appears that the nodABC promoter can be affected by several regulatory loci (Figure 2). However, the effect of nodM-nodD15 is not a simple one. When nodD15 alone is placed in a cell it affects the level of uninduced nodABC expression, as well as permitting nodABC to be induced by plant exudate. The relative roles of nodM, nodD15 and nodD require definition by biochemical and genetic analysis.

IDENTIFICATION OF A NOD GENE INDUCING SUBSTANCE FROM ALFALFA

We had observed that alfalfa seed exudates or root exudates were capable of causing nodABC induction. In order to understand the molecular mechanism of activation, it is essential to know the identity of the plant signal molecule. Several lines of evidence indicated that this might be due to a small molecule with substantial aromatic character. We found that the inducing activity dialyzed through membrane with a molecular weight cut-off of 2,000, was removed from solution by charcoal, and was extracted into ether if the aqueous phase pH was less than 10. This suggested a somewhat hydrophobic molecule, bearing a titratable proton with a pK of approximately 10.

To purify the inducer we fractionated alfalfa seed exudate, which had been concentrated with ether extraction, by reverse phase HPLC chromatography. The C-18 column was eluted with a 0% to 100% methanol gradient, and 1 ml fractions of this column were collected and assayed for their ability to induce a nodC-lacZ fusion. We found that almost all of the inducing activity eluted in the most hydrophobic portion of the gradient. This inducing activity resolved into several peaks, of which the most active peak was collected, repurified and identified by several types of spectroscopy. In particular, UV-visible spectroscopy suggested that the molecule was a flavone, and analysis of spectra in acid, in base, and when complexed with aluminum, suggested that the structure was a tetrahydroxyflavone. Further analysis, including proton NMR and mass spectroscopy, confirmed our identification of the alfalfa inducer as 3',4',5,7- tetrahydroxyflavone, whose trivial name is luteolin (Figure 3) (Peters et al , in press). It is apparent that alfalfa seeds, and alfalfa roots, release a mixture of molecules which have inducing activity. Although the structures of these are not completely known, preliminary evidence suggests they are modified (e.g. methoxylated) relatives of luteolin.

We assayed exudates of a variety of other plants to

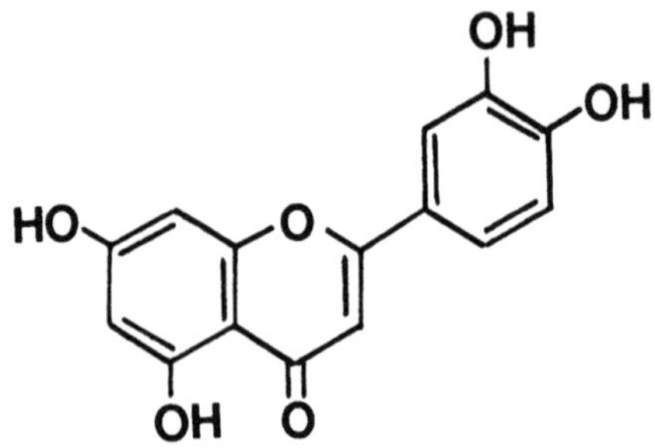

Luteolin

Figure 3. The structure of luteolin (3',4',5,7-tetrahydroxyflavone), a nod gene inducer purified from alfalfa exudates.

determine whether these could induce the R. meliloti nod gene. We found that in general, exudates from legumes are capable of inducing R. meliloti, even if those legumes are not hosts. Also in general, we found that non-legumes tend not to have inducing activity present in their exudate. Some of these plants, however, have inducing activity present if they were extracted. This indicates that the level of synthesis, or the export of compounds, may be important regulatory levels.

CONCLUSION

An intensive study of the common nod genes D and ABC from R. meliloti has shown that these genes are required both for root hair curling and for initiation of cell division in the plant host. The divergently transcribed promoters for nodD and nodA are closely spaced, and controlled in a distinctly different way. nodD is constitutive, while nodABC is induced by plant exudates. We have purified from alfalfa seed exudate the most active of several inducing compounds and have identified it as luteolin (3',4',5,7-tetrahydroxyflavone). In vitro transcription analysis of the nodD-nodABC is consistent with the model that nodABC is a positively regulated promoter. Reconsitution studies should make it possible to identify proteins and other factors required for nodABC promoter activation.

The ultimate goals of our studies are to answer the questions of, what is the mechanism of nodulation, and how is nodulation regulated. The study of bacterial genes and gene products will give direct answers for only part of these questions. Ultimately a study of the plant partner at the level of genetics, cell biology, and whole plant physiology, will be necessary to achieve a complete understanding of the interaction.

REFERENCES

Bauer WD (1981) Infection of legumes by Rhizobium. Ann Rev Plant Physiol 32:407-449

Bauer WD, Bhuvaneswari TV, Calvert HE, Law IJ, Malik NSA, Vesper SJ (1985) Recognition and infection by slow-growing Rhizobi. In:Evans HJ, Bottomley PJ, Newton WE (eds) Nitrogen Fixation Research Progress 6th International Symposium on Nitrogen Fixation 4-10 Aug 1985. Martinus Nijhoff, The Netherlands, p 247

van Brussel AAN, Zaat SAJ, Canter Cremers HCJ, Wijffelman CA, Pees E, Tak T, Lugtenberg BJJ (1986) Role of plant root exudate and sym plasmid-localized nodulation genes in the synthesis by Rhizobium leguminosarum of Tsr factor. J Bacteriology 165:517-522

Calvert HE, Pence MK, Pierce M, Malik NSA, Bauer WD (1984) Anatomical analysis of the development and distribution of Rhizobium infections in soybean roots. Can J Bot 62:2375-2384

Canter Cremers HCJ, van Brussel AAN, Plazinski J, Rolfe BG (1986) Sym plasmid and chromosomal gene products of Rhizobium trifolii elicit developmental responses on various legume roots. J Plant Physiol 122:25-40

Downie JA, Hombrecher G, Ma Q-S, Knight CD, Wells B, Johnston AWB (1983) Cloned nodulation genes of Rhizobium leguminosarum determine host-range specificity. Mol Gen Genet 190:359-365

Downie JA, Knight CD, Johnston AWB, Rossen L (1985) Identification of genes and gene products involved in the nodulation of peas by Rhizobium leguminosarum. Mol Gen

Genet 198:255-262

Djordjevic MA, Schofield PR, Rolfe BG (1985) Tn5 mutagenesis of Rhizobium trifolii host-specific nodulation genes result in mutants with altered host-range ability. Mol Gen Genet 200:463-471

Djordjevic MA, Schofield PR, Ridge RW, Morrison NA, Bassam BJ, Plazinski J, Watson JM, Rolfe BG (1984) Rhizobium nodulation genes involved in root hair curling (Hac) are functionally conserved. PMB 132:1-14

Egelhoff TT, Fisher RF, Jacobs TW, Mulligan JT, Long SR (1985) Nucleotide sequence of Rhizobium meliloti 1021 nodulation genes: nodD is read divergently from nodABC. DNA 4:241-248

Egelhoff TT, Long SR (1985) Rhizobium meliloti nodulation genes: identification of nodDABC gene products, purification of nodA protein and verification of nodA in Rhizobium. J Bacteriol 164:591-599

Finan TM, Hirsch AM, Leigh JA, Johansen E, Kuldau GA, Deegan S, Walker GC, Signer ER (1985) Symbiotic mutants of Rhizobium meliloti that uncouple plnat from bacterial differentiation. Cell 40:869-877

Fisher RF, Tu JK, Long SR (1985) Conserved nodulation genes in R. meliloti and R. trifolii. Appl Envir Microbiol 49:1432-1435

Fisher RF, Brierley HL, Long SR (1985) Transcription of Rhizobium meliloti nodulation genes: identification of a nodD transcription initiation site in vitro. J Biol Chem, in revision

Hirsch AM, Long SR, Bang M, Haskins N, Ausubel FM (1982) Structural studies of alfalfa roots infected with nodulation mutants of Rhizobium meliloti. J Bacteriol 151:411-419

Jacobs TW, Egelhoff TT, Long SR (1985) Physical and genetic map of a Rhizobium meliloti nodulation gene region and nucleotide sequence of nodC. J Bacteriol 162:469-476

John M, Schmidt J, Wieneke U, Kondorosi E, Kondorosi A, Schell J (1985) Expression of the nodulation gene nodC of

Rhizobium meliloti in Escherichia coli: role of the nodC gene product in nodulation.The EMBO Journal 4:2425-2430

Kondorosi A, Horvath B, Göttfert M, Putnoky P, Rostas K, Györgypal Z, Kondorosi E, Török I, Bachem C, John M, Schmidt J, Schell J (1985) Identification and organization of Rhizobium meliloti genes relevant to the initiation and development of nodules. In: Evans HJ, Bottomley PJ, Newton WE (eds) 6th Int Symposium on Nitrogen Fixation. Martinus Nijhoff, The Netherlands, p 73

Kondorosi E, Banfalvi Z, Kondorosi A (1984) Physical and gene analysis of a symbiotic region of Rhizobium meliloti: identification of nodulation genes. Mol Gen Genet 193:445-452

Lamb JW, Hennecke H (1986) In Bradyrhizobium japonicum the common nodulation genes, nodABC, are linked to nifA and fixA. Mol Gen Genetics 202:512-517

Leigh JA, Signer ER, Walker GC (1985) Exopolysaccharide-deficient mutants of Rhizobium meliloti that form ineffective nodules. Proc Natl Acad Sci USA 82:6231-6235

Libbenga KR, Harkes PAA (1973) Initial proliferation fo cortical cells in the formation of root nodules of Pisum sativum L. Planta 114:17-28

Long SR, Jacobs TW, Egelhoff TT, Tu J, Fisher RF (1985) Genes used in Rhizobium-plant interactions. In: Ludden PW, Burris JE (eds) Nitrogen fixation and CO_2 metabolism. Elsevier Science Publishing Co., Inc., p75-82

Long SR (1984) Nodulation genetics. In: Kosuge T, Nester EW (eds) Plant-microbe interactions. Macmillian Publishing Co., New York, p 265-306

Long SR, Buikema W, Ausubel FM (1982) Cloning of Rhizobium meliloti nodulation genes by direct complementation of Nod mutants. Nature 298:485-488

Mulligan JT, Long SR (1985) Induction of Rhizobium meliloti nodC by plant exudate requires nodD. Proc Natl Acad Sci USA 82:6609-6613

Newcomb W (1981) Nodule morphogenesis and differentiation.

In:Giles KL, Atherly AG (eds) Biology of the Rhizobiaceae. Academic Press, Inc., New York London, p247-297

Peters NK, Frost JW, Long SR (1986) A plant flavone, luteolin, induces expression of Rhizobium meliloti nodulation genes. Science p 233 (in press)

Török I, Kondorosi E, Stepkowski T, Pósfai J, Kondorosi A (1984) Nucleotide sequence of Rhizobium meliloti nodulation genes. Nucleic Acids Research 12:9509-9524

Rossen L, Johnston AWB, Downie JA (1984) DNA sequence of the Rhizobium leguminosarum nodulation genes nodAB and C required for root hair curling. Nucleic Acids Research 12:9497-9508

Rossen, L, Shearman CA, Johnston AWB, Downie JA (1985) The nodD gene of Rhizobium leguminosarum is autoregulatory and in the presence of plant exudate induces the nodA, B, C genes. The EMBO Journal 4:3369-3373

Russell P, Schell MG, Nelson KK, Halverson LJ, Sirotkin KM, Stacey G (1985) Isolation and characterization of the DNA region encoding nodulation functions in Bradyrhizobium japonicum. J Bacteriol 164:1301-1308

Schofield PR, Watson JM (1986) DNA sequence of Rhizobium trifolii nodulation genes reveals a reiterated and potentially regulatory sequence preceding nodABC and nodFE. Nucleic Acids Research 14:2891-2903

Scott DB, Chua K-Y, Jarvis BDW, Pankhurst CE (1985) Molecular cloning of a nodulation gene from fast- and slow-growing strains of Lotus rhizobia. Mol Gen Genetics 201:43-50

Scott KF (1986) Conserved nodulation genes from the non-legume symbiont Bradyrhizobium sp. (Parasponia). Nucleic Acids Research 14:29052919

Shearman CA, Rossen L, Johnston AWB, Downie JA (1986) The Rhizobium leguminosarum nodulation gene nodF encodes a polypeptide similar to acyl-carrier protein and is regulated by nodD plus a factor in pea root exudate. The EMBO Journal 5:647-652

Vincent J (1980) Factors controlling the legume-Rhizobium

symbiosis. In: Newtown WE, Orme-Johnson WH (eds) Nitrogen Fixation, volume II. University Park Press, 1980, pp 102-129

Wijffelman CA, Pees E, van Brussel AAN, Okker RJH, Lugtenberg BJJ (1985) Genetic and functional analysis of the nodulation region of the Rhizobium leguminosarum Sym plasmid pRL1JI. Arch Microbiol 143:225-232

Yelton MM, Mulligan JM, Long SR (submitted for publication) Rhizobium meliloti nodC is induced by extracts from several plants and is active in other bacterial backgrounds. J Bacteriology, in review

RESPECTIVE ROLES OF COMMON AND SPECIFIC *Rhizobium meliloti* *nod* GENES IN THE CONTROL OF LUCERNE INFECTION

F. Debellé, S.B. Sharma, C. Rosenberg, J. Vasse, F. Maillet, G. Truchet and J. Dénarié
Laboratoire de Biologie Moléculaire des Relations Plantes-Microorganismes, CNRS-INRA, B.P. 27, 31326 Castanet-Tolosan Cedex, France

INTRODUCTION

Rhizobia are soil bacteria which can infect specifically the roots of leguminous hosts and elicit the formation of root nodules in which they reduce dinitrogen into ammonia. The degree of specificity of this relationship is extremely varied. Some tropical strains can nodulate and fix nitrogen on a large number of genera belonging to various tribes of Leguminosae and even on the non-legume genus *Parasponia*. On the other hand *Rhizobium meliloti* strains are reported to have a very limited host range : each strain can nodulate effectively only some species of *Medicago*, *Melilotus* and *Trigonella* (see the review of Long 1984). *R. meliloti* is thus a suitable organism with which to investigate the genetic and molecular basis of a highly specific plant-bacteria interaction.

In *Medicago sativa*-*R. meliloti* symbiosis the infection process leading to nodule formation can be divided schematically as follows : attachment to root hairs, root hair curling, infection thread formation within root hairs, growth of the thread toward the inner root cortex, initiation of a nodule meristem and nodule organogenesis (Truchet et al 1984 ; Truchet et al 1985). Whether host specificity is involved in each step of the infection process is not known.

In *R. meliloti* RCR2011 (=SU47) nodulation (*nod*) genes required for root hair curling have been mapped on the pSym megaplasmid at around 25 kb from the nitrogenase genes (Rosenberg et al 1981 ; Long et al 1982). They have been called common *nod* genes since mutations in these genes can be complemented or replaced (for nodulation on the homologous host, lucerne) by nodulation genes from other *Rhizobium* species (Fischer et al 1985 ; Truchet et al 1985). *R. meliloti* common *nod* genes are organized in two divergent transcription units, one containing *nodABC* and the other the regulatory *nodD* gene (Egelhoff et al 1985 ; Jacobs et al 1985). Transcription of *nodABC* genes is stimulated by plant

NATO ASI Series, Vol. H 4
Recognition in Microbe-Plant Symbiotic and Pathogenic Interactions. Edited by B. Lugtenberg

root exudates of both homologous and heterologous hosts (Mulligan and Long 1985). Thus it is unlikely that nodABC genes determine host range specificity.

In this paper we report the cloning and sequencing of R. meliloti RCR2011 genes which control host range specificity (nodFE and nodH) and, after a detailed cytological study of the various nod mutant phenotypes, we propose a model describing the respective roles of R. meliloti common and specific nod genes in the control of root hair curling, infection thread formation and nodule induction.

CLONING OF GENES CONTROLLING HOST RANGE

Transfer of the pSym megaplasmid of R. meliloti RCR2011 into Agrobacterium tumefaciens (strains cured of the Ti plasmid) enables this bacterium to induce nodule formation on lucerne but not on clover, which shows that genes determining nodule initiation and host range specificity are located on pSym (Truchet et al 1984). A series of derivatives of the conjugative plasmid RP4, carrying large inserts of pSym, have been constructed (Batut et al 1985). One of them, pGMI71, carrying a pSym insert of around 70 kb, renders A. tumefaciens able to nodulate specifically lucerne (Truchet et al 1985). Plasmid pGMI71 was thus used as a source of DNA to clone the host range genes of R. meliloti. Partial EcoR1 digests were cloned in E. coli into the EcoRI site of pRK290, a broad host range cloning vehicle. The pRK290 random insert population was then mobilized into two R. meliloti Nod$^-$ recipients carrying large pSym deletions (namely deletions of the region cloned in pGMI71). R. meliloti exconjugant populations were inoculated "en masse" on lucerne seedlings to select for R. meliloti Nod$^+$ transconjugants by the procedure of Long et al (1982). Bacteria isolated from nodules carried pRK290 inserts of varying size, but these plasmids always included EcoRI fragments of 8.7, 1.8, 1.2, 2.1 and 15.3 kb. These fragments are adjacent on the EcoRI map of the corresponding pSym region (Long et al 1982 ; Batut et al 1985 ; Jacobs et al 1985). One of the smallest insert clones, pGMI149, was used in further experiments.

A. tumefaciens (pGMI149) transconjugants induced marked root hair curling (= shepherd's crooks) and nodule formation on lucerne and not on clover : genes determining nodule initiation and controlling the specificity of hair curling are therefore present in pGMI149. Plasmid pGMI149 was then transferred into

R. trifolii and R. leguminosarum strains : these pGMI149 transconjugants induced abundant nodulation on lucerne. Marked hair curling and formation of infection threads within root hairs were normal : therefore sequences required for infection thread specific development in lucerne root hairs are also present in pGMI149.

MUTAGENESIS OF NODULATION GENES

To define the nodulation regions, the 29.1-kb pSym fragment cloned in pGMI149 was mutagenized with transposon Tn5. Sixty one clones were found to be altered in the nodulation process, either completely defective (Nod$^-$) or delayed (Nod^d). Localization of nod::Tn5 inserts was achieved by digesting the mutated pGMI149nod::Tn5 plasmids with various combinations of PstI, EcoRI, BamHI and HindIII (see Figure 1). Tn5 inserts associated with nodulation defects are all located within approximately 16 kb and clustered in three regions.

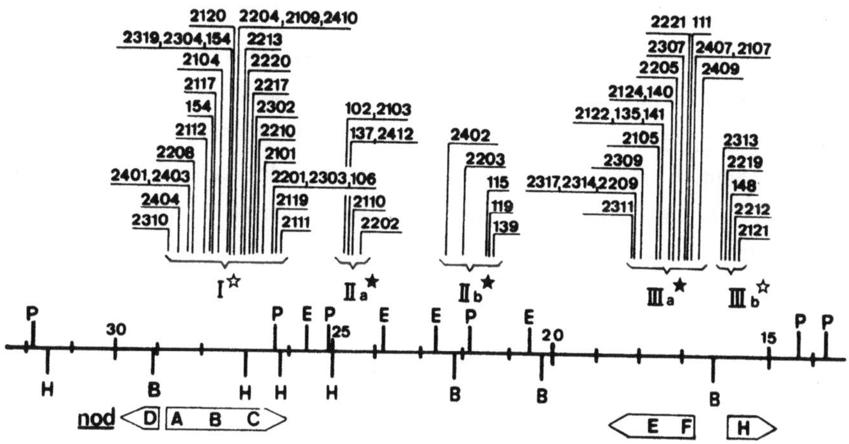

Figure 1. Physical and genetic map of the nodulation region of the R. meliloti RCR2011 pSym megaplasmid. Position of Tn5 inserts is shown by a vertical line. B = BamHI ; E = EcoRI ; P = PstI ; H = HindIII.
Symbiotic phenotypes : open stars = Nod$^-$; solid stars = Nod^d

Region I (3.5 kb) includes the common nodABC region distal to nidHDK already reported (Egelhoff et al 1985 ; Jacobs et al 1985) ; region II, including inserts in two clusters (IIa and IIb) over about 4 kb, is in the center of the cloned region ; region III, also including inserts in two clusters (IIIa and IIIb) over about 3.5 kb, is proximal to nifHDK. Region I mutants are all Nod^- ; region II mutants are all Nod^d ; and region III mutants include both Nod^d (IIIa) and Nod^- (IIIb) phenotypes ; region IIIa mutants had a longer delay than region II mutants. For both region II and region IIIa Nod^d mutants, the nodules had unambiguous nitrogenase activity (Nod^d Fix^+).

Plasmid pIJ1089, a pRK290 derivative carrying nod genes of R. leguminosarum required for nodulation of Pisum sativum (Downie et al 1983), was introduced into the mutants to test whether heterologous nod genes could complement the mutant lesions. Normal nodulation of lucerne was restored for region I mutants which carry Tn5 inserts in the common nodABC genes already described (Fischer et al 1985 ; Jacobs et al 1985). Because the nodulation delay of region II mutants is very limited results of complementation were difficult to assess : it is not clear whether these mutants are altered in common or non-common genes. In contrast, normal infection and nodulation of lucerne could not be restored for region III mutants which thus carry Tn5 inserts in non-common (= specific) nod genes. Hence in strain RCR2011, as reported by Kondorosi and co-workers for R. meliloti strain 41 (Kondorosi et al 1984 ; Rostas et al 1986), a cluster of host-specific nodulation genes lies between nodABC and the nifHDK operon. Complete nucleotide sequence of region III was determined (see below). All the nod::Tn5 inserts are located within three open reading frames ; we will call nodFE the two genes located in IIIa and nodH the gene located in IIIb (Debellé and Sharma, submitted).

CYTOLOGICAL CHARACTERIZATION OF MUTANTS

To investigate the role of various nod genes in the control of root infection, twenty eight mutants were characterized cytologically. The root system of lucerne seedlings was observed by light microscopy ; methylene blue staining procedure was used to facilitate observation of infection threads (Vasse and Truchet 1984). Results are summarized in Table 1. In all Nod^- mutants of region I (nodABC) root hairs are straight ; they show no marked distortions (Had^-) and no marked curling (Hac$^-$). Nod$^-$ mutants due to a Tn5 insert in <u>nodH</u> exhibit slight root hair distortions (Had$^+$) but never give rise to shepherd's crooks, that is marked hair curling with a bright spot (infection focus) in the center of the curl (Hac$^-$).

Table 1. Symbiotic phenotypes of nodulation mutants

Tn5 insertions in	M. sativa[a]			T. repens[a]	N. oleracea[b]
	Hac[c]	Inf	Nod	Hac	Nod
nodABC (6)[x]	-	-	-	-	-
region II (7)	++	++	d	-	+
nodFE (10)	+	-	d	+	+
nodH (5)	-/s	-	-	s	+
wild-type control	+	+	+	-	+

x in parenthesis the number of mutants tested
a experiments performed at 20°C, 16 hr of lighting per 24 hr
b experiments performed at 27°C, same daylength
c for symbiotic phenotype symbols see text

Tn5 inserts in <u>nodFE</u> result in a Hac$^+$ Inf$^-$ phenotype : genuine shepherd's crooks are formed with bright spots (Hac$^+$), however, in contrast to the wild-type, no infection threads can be detected five days after inoculation (Inf$^-$). With Nod^d Fix$^+$ mutants in region II, shepherd's crook formation is slightly delayed and this delay is followed by an unusually large number of shepherd's crooks (Hac^{++}) and of infection threads (Inf^{++}).

Thus <u>nodABC</u> and <u>nodH</u> appear to be required for root hair curling, <u>nodFE</u> for infection thread formation, and region II to be involved in the control of the intensity and kinetics of root hair curling and infection thread formation. These detailed cytological studies will be exploited in the following paragraphs to assign to common and specific <u>nod</u> genes the control of early infection.

nodH PRODUCT IS REQUIRED FOR ROOT HAIR CURLING OF LUCERNE

Root hair curling is a very specific reaction. For instance R. meliloti strain RCR2011 is Hac^+ on lucerne the homologous host, but Hac^- on heterologous hosts such as pea or clovers. Similarly R. trifolii strain ANU843 is Hac^+ on clover and Hac^- on lucerne. It was thus surprising to observe that R. meliloti nodABC genes which are required for hair curling are common nod genes : they can be complemented or replaced by nod genes from R. leguminosarum and then give rise to a Hac^+ phenotype on lucerne (Truchet et al 1985). This suggested the existence of another gene(s) controlling the specificity of curling.

Tn5 inserts in nodH are Hac^- or Hac^s on lucerne, but they cannot be complemented for hair curling of lucerne by R. leguminosarum nodulation genes. nodH is thus likely to be involved in the specificity of root hair curling of the homologous host. This is supported by introduction of plasmid pGMI515, which carries region II, nodFE and nodH genes from R. meliloti, into a R. trifolii ANU843 recipient. ANU843(pGMI515) transconjugants form typical shepherd's crooks on lucerne. Plasmids pGMI515 carrying Tn5 insertions in nod regions were introduced into ANU843 : only the transconjugants having a Tn5 insert in nodH are Hac^- on lucerne.

NEGATIVE CONTROL OF ROOT HAIR CURLING ON HETEROLOGOUS HOSTS

The infection phenotypes of the mutants were studied on two heterologous hosts, white and red clovers. Mutants in nodABC and in region II behave like wild-type : root hairs are straight or slightly distorted. In contrast, Tn5 inserts in region III result in marked hair curling on the heterologous hosts : inserts in nodFE provoke formation of genuine shepherd's crooks with a bright refractile spot (Hac^+), whereas inserts in nodH provoke abundant marked hair curling with a bright spot only rarely present (Hac^s).

Thus, whereas common nodABC transcription unit is absolutely required for hair curling (Jacobs et al 1985), nodFE and nodH transcription units control its specificity. nodH acts as a positive determinant to control curling on lucerne the homologous host ; nodFE and nodH negatively control hair curling on an heterologous host (Figure 2). A negative control of host range was also reported in R. trifolii (Djordjevic et al 1985).

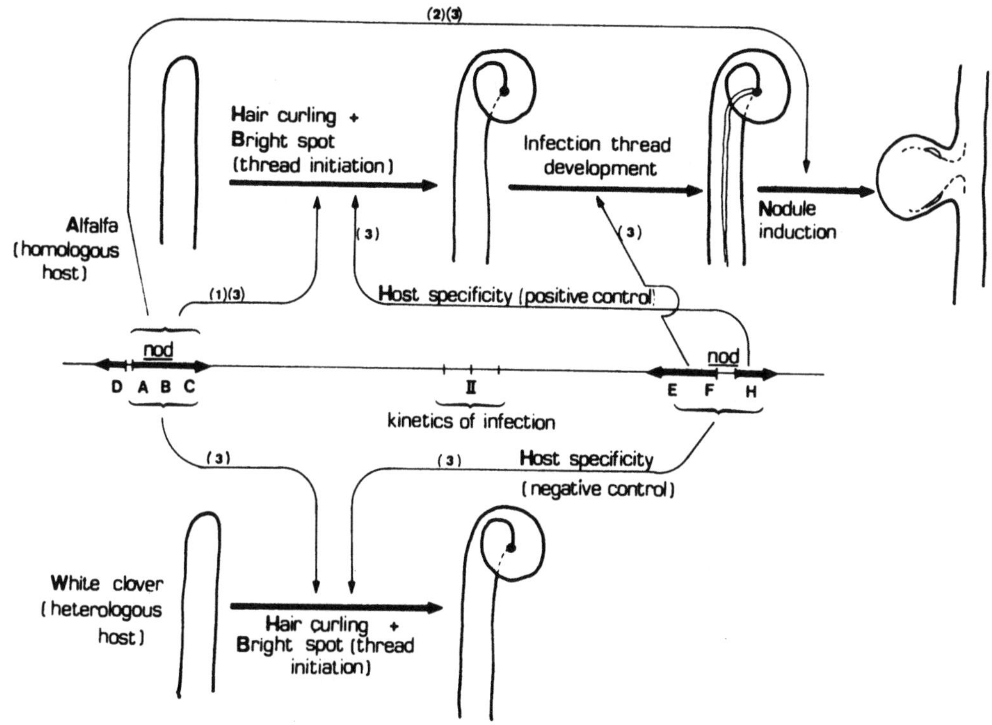

Figure 2. Respective roles of common and specific nod loci of R. meliloti in the control of infection and nodulation. Thick arrows represent steps in the infection process. Thin arrows represent genetic control.
Interpretations from (1) and (2) S.R. Long and co-workers ; (3) This work.

CONTROL OF INFECTION THREAD INITIATION AND GROWTH

The bright refractile spot in the middle of the shepherd's crook is the point at which rhizobia enter the root hair and the site of initiation of the infection thread (Vasse and Truchet 1984 ; Djordjevic et al 1985 ; Truchet et al 1985). The plasmid pRmSL26 contains nodABC genes and region IIa of R. meliloti RCR2011. A. tumefaciens (pRmSL26) transconjugants induce the formation of shepherd's crooks with a bright spot on white and red clover root hairs : bacterial penetration into the root hair and initiation of the infection thread formation may therefore be determined by common nod genes and possibly IIa. On the other hand nodH::Tn5 mutants, which can induce root hair distortions but rarely hair curling on lucerne the homologous host, cannot induce shepherd's crooks with a bright spot. Thus rhizobial entry and initiation of infection thread synthesis in the root hair of the homologous host seem to be determined by both common nod genes and nodH.

Tn5 inserts in transcription unit nodFE result in the formation of shepherd's crooks with a bright spot, but no infection thread can be observed in the root hairs : thus the development of an infection thread from the bright spot, in lucerne root hair, requires nodFE (see Figure 2).

CONTROL OF INDUCTION OF NODULE ORGANOGENESIS

To identify those nod genes cloned in pGMI149 that are directly involved in nodule induction, we looked for a natural experimental system in which the first steps of infection via root hairs could be short-circuited.

We have found that R. meliloti strains can induce nodule formation on an aquatic tropical legume Neptunia oleracea (Vasse J, Dreyfus B, Truchet G, Rosenberg C, Camut S, Dénarié J, in preparation). The infection proceeds not through root hairs but by intercellular penetration. The various mutant constructs were tested for nodulation of this plant.

The only mutants which are Nod^- are those altered in nodABC. Thus the nod loci in region II and the nodFE and nodH genes which are implicated in the infection via root hairs are not required. This suggests that the common nodABC genes are likely to be those directly involved in triggering nodule organogenesis (see Figure 2). If so, this induction would not be specific, since R. meliloti strains deleted for the common nod genes and carrying instead the common nod genes of R. leguminosarum, induce the formation of genuine nodules on lucerne (Truchet et al 1985).

PREDICTED PROTEIN PRODUCTS OF HOST RANGE GENES

The nucleotide sequence of region III was determined by the dideoxy chain termination procedure (Sanger et al 1977). All the Tn5 insertions which result in a detectable nodulation defect fall in the three open reading frames we have called nodE nodF (region IIIa) and nodH (region IIIb) (see Figure 1). nodF and E are transcribed in the same direction, and since they are just separated by a TGA cordon, they probably belong to the same transcription unit. nodH is transcribed in the opposite direction (Debellé and Sharma, submitted). Another open reading frame, transcribed in the same direction as nodFE, starts at around 500 bp downstream from nodE. Tn5 insertions in this region cause delayed nodulation (S.R. Long, personal communication). In agreement with the nomenclature used by Rostas et al (1986) for R. meliloti strain 41 we call this putative gene nodG.

The predicted nodF product is a small protein of 93 residues with a molecular weight of 9760 daltons and a low isoelectric point of 4.15. The amino acid sequence shows some homology with acyl carrier proteins from E. coli and barley. Such an homology has already been reported for the nodF product of R. leguminosarum (Shearman et al 1986). The putative nodE product is a protein of 402 residues with a molecular mass of 41,765. nodFE genes have been sequenced in R. leguminosarum (Shearman et al 1986). The predicted amino acid sequence of R. meliloti nodF gene product has 59% homology with the corresponding product of R. leguminosarum. The homology between the amino acid sequences of the predicted nodE gene products is even higher at 86%. These values are comparable with the homologies between amino acid sequences of common nodABC and nodD gene products of R. meliloti and R. leguminosarum, which are 71.5, 65.3, 70.2 and 81.1% respectively (Jacobs et al 1985 ; Shearman et al 1986). The fact that the nodE gene product, controlling host specificity, is even more conserved than the products of common genes is intriguing. This high conservation suggests that these proteins could have very similar functions in the various Rhizobium species, but perhaps with sufficient differences to result in different modifications of a chemical mediator.

The putative nodG polypeptide has 247 residues and a molecular mass of 26,058. The amino acid sequence shows a clear homology with the ribitol dehydrogenase of Klebsiella aerogenes. nodG product could therefore be involved in sugar metabolism. Computer analysis failed to detect sequences characteristic of membrane bound proteins (signal peptide, highly hydrophobic domain, membrane associated helix) in the predicted nodF, E, and G products.

nodH is required for the specific hair curling and nodulation of lucerne. In contrast to nodFE no gene homologous to nodH has been described so far in either R. leguminosarum or R. trifolii. The codon usage of this gene is very different from those of nodFE and G and the nodH polypeptide is rich in proline (8.5%). It is worth noting that a sequence of four proline residues is present, close to a serine-rich region (Debellé and Sharma, submitted). Similar four proline sequences have been reported in certain glycoproteins of eucaryotes in which poly-L proline residues are hydroxylated into hydroxyproline residues and form part of glycosylation domains. Such four proline sequences have been described in plant hydroxy-proline rich glycoproteins, the synthesis of which is stimulated after wounding and as part of plant defence mechanisms (Chen and Varner 1985 ; Esquerré-Tugayé et al, in this book).

REGULATORY REGIONS OF COMMON AND SPECIFIC nod GENES

Is the host range of Rhizobium meliloti simply determined by gene products of the specific nodulation genes, or is specificity also mediated by differential expression of the specific nodulation genes controlled by specific plant products ? The regulatory regions preceding nodABC, nodFE and nodH were examined to investigate this question.

We have observed a reiterated sequence of 21 nucleotides located at -247, -202 and -210 5' to the nodH, nodF and nodA translational starts respectively (Debellé and Sharma, submitted). A similar sequence (nod box) is present in the upstream promoter sequences of both common and specific nod genes of R. leguminosarum (Shearman et al 1986) and R. trifolii (Schofield and Watson 1986). Rostas et al (1986) have shown that this reiterated sequence is required for nod expression in R. meliloti. This nod box is found upstream of common and specific genes which require the nodD protein and a plant product for their expression (Mulligan and Long 1985 ; Innes et al 1985 ; Shearman et al 1986). It is worth noting that the plant factor(s) required to stimulate expression of nod genes is present in root exudates of the heterologous as well as of the homologous host (Mulligan and Long 1985 ; Roger Innes, personal communication). One class of such factors has been identified as flavonoids. These flavonoids, which induce transcription of the nod genes, do not seem to show a clear-cut species specificity (see the chapters devoted to this topic in this book). All these data suggest that common and specific nod genes are part of the same regulon controlled non-specifically by plant product(s). This hypothesis is in agreement with our observation that R. meliloti nodFE and nodH both exert a positive control on hair curling infection of lucerne and a negative control on clover : these operons have to be expressed, at least at some level, both on heterologous hosts as well as on homologous hosts.

However, another reiterated sequence is found close to the putative ribosome binding sites, at positions -11 to -28 5' upstream of R. meliloti nodH and -4 to -20 upstream of nodF (Debellé and Sharma, submitted). This sequence is neither found upstream of the R. meliloti nodA gene nor upstream of the nod genes of either R. leguminosarum or R. trifolii. Location of this reiteration near the translation start suggests that it could be involved in the regulation of expression of species specific nod genes (translational control ?). Furthermore the mutant M2 of R. trifolii, with a widely extended host range (nodulation of white and subterranean clover as well as peas and beans), has been shown to carry a Tn5 insertion located a few bases 5'upstream of the translational start of nodF (Djordjevic et al 1985 ; Schofield and Watson 1986). It was suggested that the extended host range of this mutant may result from plant-independent, Tn5-promoted transcription of the nodFE genes. These results suggest that the control of the level of gene expression may play a role in modulating host range specificity. More detailed studies concerning nod regulation are clearly required, and the question still remains as to whether the control of host range is determined only by gene product activities or whether a specific regulation system is also involved.

ACKNOWLEDGEMENTS

This work was supported by the European Communities Biomolecular Programm and by the Biological Nitrogen Fixation grant from Société Nationale Elf Aquitaine, Entreprise Minière et Chimique, Rhône Poulenc and Charbonnages de France-Chimie. S.B.S. was supported by a fellowship from S.N. Elf Aquitaine. We thank David Barker for reviewing the manuscript.

REFERENCES

Batut J, Terzaghi B, Ghérardi M, Huguet M, Terzaghi E, Garnerone AM, Boistard P, Huguet T (1985) Localization of a symbiotic fix region on Rhizobium meliloti pSym megaplasmid more than 200 kilobases from the nod-nif region. Mol Gen Genet 199:232-239

Chen J, Varner JE (1985) An extracellular matrix protein in plants : characterization of a genomic clone for carrot extensin. EMBO J 4:2145-2151

Djordjevic MA, Schofield PR, and Rolfe BG (1985) Tn5 mutagenesis of Rhizobium trifolii host-specific nodulation genes results in mutants with altered host-range ability. Mol Gen Genet 200:463-471.

Downie JA, Hombrecher G, Ma QS, Knight CD, Wells B, Johnston AWB (1983a) Cloned nodulation genes of Rhizobium leguminosarum determine host-range specificity. Mol Gen Genet 190:359-365

Egelhoff TT, Fischer RF, Jacobs TW, Mulligan JT, Long SR (1985) Nucleotide sequence of Rhizobium meliloti 1021 nodulation genes : nodD is read divergently from nodABC. DNA 4:241-248

Fischer RF, Tu JK, Long SR (1985) Conserved nodulation genes in Rhizobium meliloti and Rhizobium trifolii. Appl Env Microbiol 49:1432-1435

Innes RW, Kuempel PL, Plazinski J, Canter-Cremers H, Rolfe BG, Djordjevic MA (1985) Plant factors induce expression of nodulation and host-range genes in Rhizobium trifolii. Mol Gen Genet 201:426-432

Jacobs TW, Egelhoff TT, Long SR (1985) Physical and genetic map of a Rhizobium meliloti nodulation gene region and nucleotide sequence of nodC. J Bacteriol 162:469-476

Kondorosi E, Banfalvi Z, Kondorosi A (1984) Physical and genetic analysis of a symbiotic region of Rhizobium meliloti : identification of nodulation genes. Mol Gen Genet 193:445-452

Long SR, Buikema WJ, Ausubel FM (1982) Cloning of Rhizobium meliloti nodulation genes by direct complementation of Nod^- mutants. Nature 298:485-488

Long SR (1984) Genetics of Rhizobium nodulation. In : Kosuge T, Nester EW (eds) Plant-microbe interactions, vol I. Macmillan, New York.

Mulligan JR, Long SR (1985) Induction of Rhizobium meliloti nodC expression by plant exudate requires nodD. Proc Natl Acad Sci USA 82:6609-6613

Rosenberg C, Boistard P, Dénarié J, Casse-Delbart F (1981) Genes controlling early and late functions in symbiosis are located on a megaplasmid in Rhizobium meliloti. Mol Gen Genet 184:326-333

Rostas K, Kondorosi E, Horwath B, Simoncsits A, Kondorosi A (1986) Conservation of extented promoter regions of nodulation genes in Rhizobium. Proc Natl Acad Sci USA 83:1757-1761

Sanger F, Nicklen S, Coulson AR (1977) DNA sequencing with chain-terminating inhibitors. Proc Natl Acad Sci USA 74:5463-5467

Schofield PR and Watson JM (1986) DNA sequence of Rhizobium trifolii nodulation genes reveals a reiterated and potentially regulatory sequence preceding nodABC and nodFE. Nucleic Acids Res 14:2891-2904

Shearman CA, Rossen L, Johnston AWB, Downie JA (1986) The Rhizobium leguminosarum nodulation gene nodF encodes a polypeptide similar to acyl-carrier protein and is regulated by nodD plus a factor in pea root exudate. EMBO J. 5:647-652

Truchet G, Rosenberg C, Vasse J, Julliot JS, Camut S, Dénarié J (1984) Transfer of Rhizobium meliloti Sym genes into Agrobacterium tumefaciens : host specific nodulation by atypical infection. J Bacteriol 157:134-142

Truchet G, Debellé F, Vasse J, Terzaghi B, Garnerone AM, Rosenberg C, Batut J, Maillet F, Dénarié J (1985) Identification of a Rhizobium meliloti pSym 2011 region controlling the host specificity of root hair curling and nodulation. J Bacteriol 164:1200-1210

Vasse JM, Truchet GL (1984) The Rhizobium-legume symbiosis : observation of root infection by bright-field microscopy after staining with methylene blue. Planta 161:487-489

INFECTION MUTANTS OF RHIZOBIUM MELILOTI ARE ALTERED IN ACIDIC EXOPOLYSACCHARIDE PRODUCTION

A. Pühler, M.F. Hynes, D. Kapp, P. Müller and K. Niehaus
Lehrstuhl für Genetik
Fakultät für Biologie
Universität Bielefeld
Postfach 8640
D-4800 Bielefeld 1
Federal Republic of Germany

INTRODUCTION

Rhizobium meliloti induces the formation of root nodules on its symbiotic partner alfalfa (*Medicago sativa*). The morphogenesis of the nodule is a complex interaction between the host plant and the invading bacteria. Genetic analysis of the microsymbiont has revealed interesting details about this symbiotic association. A very large plasmid (pSYM) has bee identified in *R.meliloti* which was found to carry symbiotic genes, such as common nodulation, host specificity and fixation genes (Rosenberg et al 1981; Banfalvi et al 1981; Kondorosi et al 1982). Recently, it was found that *R.meliloti* strain 2011 contains two Megaplasmids (Banfalvi et al 1985; Simon 1984) of 1200 and 1500 Kd (Burkardt et al submitted). The smaller plasmid is identical to the symbiotic plasmid pSYM whereas the larger plasmid is involved in exopolysaccharide biosynthesis (Hynes et al 1986). In this paper we report the isolation and characterization of infection mutants which are altered in acidic exopolysaccharide production. The mapping of the infection genes is described and experiments are presented showing that infection and nodulation mutants can complement each other on a cellular level.

NATO ASI Series, Vol. H 4
Recognition in Microbe-Plant Symbiotic and Pathogenic Interactions. Edited by B. Lugtenberg

ISOLATION AND CHARACTERIZATION OF NODULATION AND INFECTION MUTANTS

Symbiotic mutants of Rhizobium meliloti strain 2011 were obtained by general Tn5 mutagenesis using plasmid pSUP2021 (Simon et al 1983). About 6000 Tn5 mutants were tested for their capacity to form nodules on alfalfa (Medicago sativa). Two classes of symbiotic mutants could be obtained.

1. The R.meliloti mutants 0544, 0701, 1142 and 2526 are unable to form nodules on their host plants and to induce root hair curling. Evidently, these mutants are completely defective in nodulation (Nod^-).

2. The R.meliloti mutants 0540, 2204, 2505, 101.45 and 688 generated ineffective root nodules of characteristically altered morphology. The induced nodules are usually small and of spherical shape and no bacteria are detectable within the nodule cells. Further analysis revealed that at least some of the mutants were still able to curl root hairs but they did not induce infection thread formation. Therefore, mutants inducing such structurally modified nodules are called infection mutants (Inf^-).

Infection mutants were characterized in more detail. It was shown that they are altered in exopolysaccharide biosynthesis. This was first demonstrated by the Cellofluor white (Cfw) staining procedure (Vandenbosch et al 1985; Finan et al 1985). R.meliloti 2011 wild type colonies grown on a medium containing cellofluor white show bright fluorescence under UV light (Cfw^+). In contrast, colonies of the R.meliloti mutants 0540, 2204 and 2505 are Cfw^-. Since Cellofluor white interacts with exopolysaccharides (EPS) it was concluded that the mutants lack EPS. Moreover, attempts to isolate acidic exopolysaccharides from these mutants proved to be unsuccessful (Table 1).

Different results were obtained for the infection mutants 101.45 and 688. These mutants produce EPS and are Cfw^+. The acidic EPS of the R.meliloti mutant 101.45 was isolated (Kamberger 1979). It was found that this mutant produces three times more EPS than the wild type 2011. The structure of the

TABLE 1. Analysis of infection mutants

mutant	characteristics	cellofluor white test	exopoly-saccharide production	complementing cosmids
0540	mutation maps on a 7.8 kb EcoRI fragment of pEPS	Cfw$^-$	EPS$^-$	pRmPM551 pRmPM552 pRmPM5511
2204	mutation maps on pEPS	Cfw$^-$	EPS$^-$	pRmPM551
2505	mutation maps on pEPS	Cfw$^-$	EPS$^-$	pRmPM563
101.45	mutation maps on a 6.4 kb EcoRI fragment of the chromosome	Cfw$^+$	EPS$^+$	pRmPM5814 pRmPM584 pRmPM582
688	mutation maps on a 6.4 kb EcoRI fragment of the chromosome	Cfw$^+$	EPS$^+$	pRmPM5814 pRmPM584 pRmPM582

isolated EPS was analyzed by ^{13}C-NMR spectroscopy. Comparing the NMR spectrum of the wild type EPS to that of the mutant EPS it could be demonstrated that the acidic EPS of strain 101.45 is structurally modified. The mutant EPS is most probably free of pyruvate. In summary, it could be shown that infection mutants either lack EPS or produce a structurally modified EPS.

GENETIC ANALYSIS OF R.MELILOTI NODULATION AND INFECTION MUTANTS

For the *R.meliloti* nodulation mutants 1142 and 2526 it was shown that they could be complemented by plasmid pRmSL26 (Long et al 1982). In addition, the Tn*5* insertions were mapped within an *Eco*RI fragment known to carry the common nodulation genes. Complementation with pRmSL26 was not successful for the remaining nodulation mutants 0544 and 0701. These mutations could be complemented by an *Eco*RI fragment known to carry host specificity genes (Kondorosi et al 1984). This indicates that these nodulation mutants are similar to mutants already described by several other groups. From these analyses it can be deduced that all four nodulation mutants map on the *R.meliloti* pSYM plasmid.

The genetic analysis of infection mutants lacking acidic

EPS was carried out by using a mobilizable cosmid bank of R.meliloti 2011 in E.coli (Hynes et al 1986). This cosmid bank was transferred en masse into the R.meliloti infection mutants 0540, 2204 and 2505. Complementing cosmids could be isolated for all of these mutants (Table 1). Mutants complemented by individual cosmids regained their ability to induce wild type nodules on alfalfa and to produce acidic EPS resulting in a positive Cfw reaction. The mutations induced by Tn5 map on the second Megaplasmid of R.meliloti designated pEPS. The mapping was carried out by constructing Agrobacterium tumefaciens strains carrying pSYM or pEPS from R.meliloti (Hynes et al 1986). Using the complementing fragments of the mutants 0540, 2204 and 2505 in hyridization experiments with DNA of the above mentioned strains the assignment to pEPS could be demonstrated (Hynes et al 1986).

In contrast, the infection mutants with an altered EPS structure do not map on the pEPS plasmid. Again, complementing cosmids were isolated for the mutants 101.45 and 688. An EcoRI subclone of 6.4 kb of these cosmids was able to complement the mutations in both mutants. This time, no hybridization was found to total DNA isolated from A.tumefaciens (pSYM) and A.tumefaciens (pEPS). Therefore, we assume that the mutations in the infection mutants 101.45 and 688 map on the R.meliloti chromosome.

MIXED INOCULATION USING NODULATION AND INFECTION MUTANTS

We asked the question whether nodulation and infection mutants can help each other to form effective nodules on alfalfa. Therefore, we carried out mixed inoculation experiments. The scheme of such an experiment is given in Figure 1. In proper combination, the mixture of nodulation and infection mutants can form fully effective nodules which fix nitrogen. Results are summarized in Table 2. Effective nodules were produced when nodulation mutants were mixed with infection mutants lacking EPS. The resulting nodules were analyzed in more detail and it was found that they contain both partners in a ratio of 1:1.

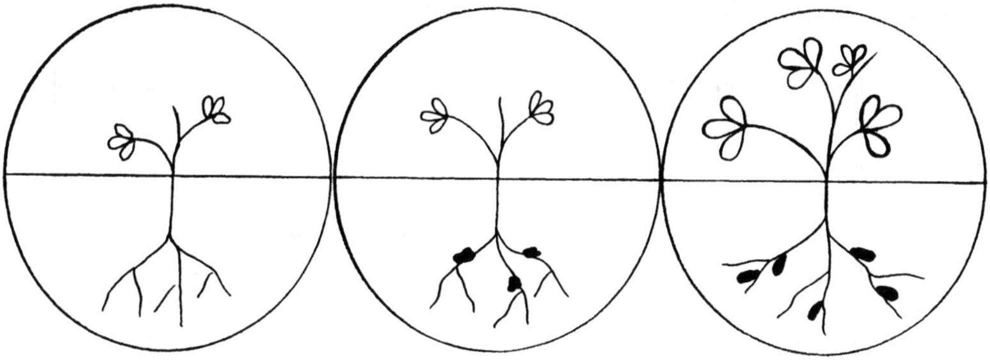

Fig 1. Scheme of a mixed inoculation experiment. Using the plate assay R.meliloti mutants were tested for their capacity to form nodules on alfalfa. Nodulation mutants do not form nodules at all. Infection mutants form nodules of irregular shape which are devoid of bacteria. In proper combination, nodulation and infection mutants are able to form fully effective nodules.

This may indicate that the presence of both mutants is necessary during all steps of nodule development.

Again, the infection mutant 101.45 was found to react differently. It was unable to complement nodulation mutants in a mixed inoculation experiment. In this case, one has to conclude that the mutation of the infection mutant 101.45 is dominant.

DISCUSSION

In this paper special infection mutants of R.meliloti are described which cause ineffective nodules devoid of bacteria. These Inf^- mutants could be subdivided into two groups. Members of the first group, namely the mutants 0540, 2204 and 2505, are deficient in acidic exopolysaccharide production. Mutants with similar properties were already described for the R.meliloti

TABLE 2. Mixed inoculation with nodulation und infection mutants

		nodulation mutants				infection mutants (EPS^-)			infection mutants (EPS*)
		0544	0701	1142	2526	0540	2204	2505	101.45
nodulation mutants	0544	0	0	0	0	++	++	++	+
	0701		0	0	0	++	++	++	+
	1142			0	0	++	++	++	+
	2526				0	++	++	++	+
infection mutants (EPS^-)	0540					+	+	+	+
	2204						+	+	+
	2505							+	+
infection mutants (EPS*)	101.45								+

EPS^-, lacking acidic EPS; EPS*, enhanced EPS showing a modified structure

strain SU47 (Finan et al 1985). In contrast to our findings it was reported that these mutants do not curl root hairs. Further exopolysaccharide deficient mutants of R.meliloti that form ineffective nodules were reported in a recent paper (Leigh et al 1985). Again, the formation of a shepherd's crook was claimed to be absent or delayed. Evidently, it is necessary to compare the different exopolysaccharide deficient R.meliloti mutants and to analyze in detail their hair curling activity. Members of the second group (infection mutants 101.45 and 688) still produce exopolysaccharides. The acidic EPS of mutant 101.45 was further analyzed by ^{13}C-NMR spectroscopy and found to be most probably pyruvate free.

The genetic analysis of the reported infection mutants was started by the isolation of complementing cosmids. Using subfragments of these cosmids it was found that exopolysaccharide deficient mutants map on the second Megaplasmid, the pEPS plasmid of R.meliloti. This result shows that the second Megaplasmid of R.meliloti plays an important role in effective nodulation of alfalfa. In contrast, infection mutants with modified EPS map on the chromosome. Further work is necessary to study in more detail the function of EPS genes located on the R.meliloti chromosome and on the pEPS plasmid.

A very special feature of Inf^- mutants lacking EPS is their ability to produce effective nodules when mixed with nodulation mutants. One has to assume that in such a mixed inoculation the Inf^- mutant carries out the nodulation step and that the Nod^- mutant is then able to perform the infection step. It should be mentioned that we did not succeed in complementing Inf^- mutants with isolated wild type EPS. This means that up to now we do not know whether EPS on its own is an essential substance for infection thread formation. Of special interest is that Inf^- mutants showing a modified EPS structure cannot be complemented by Nod^- mutants in a mixed inoculation experiment. One can speculate that the presence of the modified EPS in the mixed inoculation experiment prevents the formation of the infection thread. It is tempting to speculate about a possible role of acidic EPS of the microsymbiont in infection thread formation. One assumption is that microbial EPS constitutes a structural component of the infection thread. Since alfalfa distinguishes between the wild type and a structurally modified EPS the specificity of the EPS seems to play an important role.

It should be mentioned that exopolysaccharide deficient mutants of other Rhizobium species inducing also nodules devoid of bacteria were isolated for Rhizobium trifolii (Chakravorty et al 1982; Chen et al 1985) and Rhizobium phaseoli (Vandenbosch et al 1985). This indicates that EPS genes in general are involved in effective nodule formation of different legumes.

ACKNOWLEDGEMENTS

We thank Dr. U. Priefer and B. Kosier for critical reading and Ms. S. Malmivaara for typing of the manuscript. This work was financed by the Agrigenetics Research Corporation, Boulder, Colorado.

REFERENCES

Banfalvi Z, Sakanyan V, Koncz C, Kiss A, Dusha I, Kondorosi A (1981) Location of nodulation and nitrogen fixation genes on a high molecular weight plasmid of R.meliloti. Mol Gen Genet 184:318-325

Banfalvi Z, Kondorosi E, Kondorosi A (1985) Rhizobium meliloti carries two megaplasmids. Plasmid 13:129-138

Burkardt B, Schillik D, Pühler A (submitted) Physical characterization of Rhizobium meliloti Megaplasmids. Plasmid

Chakravorty AK, Zurkowski W, Shine J, Rolfe BG (1982) Symbiotic nitrogen fixation: Molecular cloning of Rhizobium genes involved in exopolysaccharide synthesis and effective nodulation. J Mol Appl Genet 1:585-596

Chen H, Batley M, Redmond J, Rolfe BG (1985) Alteration of the effective nodulation properties of a fast-growing broad host range Rhizobium due to changes in exopolysaccharide synthesis. J Plant Physiol 120:331-349

Finan TM, Hirsch AM, Leigh JA, Johansen E, Kuldau GA, Deegan S, Walker GC, Signer ER (1985) Symbiotic mutants of Rhizobium meliloti that uncouple plant from bacterial differentiation. Cell 40:869-877

Hynes MF, Simon R, Müller P, Niehaus K, Labes M, Pühler A (1986) The two megaplasmids of Rhizobium meliloti are involved in the effective nodulation of alfalfa. Mol Gen Genet 202: 356-362

Kamberger W (1979) An ouchterlony double diffusion study on the interaction between legume lectins and rhizobial cell surface antigens. Arch Microbiol 121:83-90

Kondorosi A, Kondorosi E, Pankhurst CE, Broughton WJ, Banfalvi Z (1982) Mobilization of a Rhizobium meliloti megaplasmid carrying nodulation and nitrogen fixation genes into other rhizobia and Agrobacterium. Mol Gen Genet 188:433-439

Kondorosi E, Banfalvi Z, Kondorosi A (1984) Physical and genetic analysis of a symbiotic region of Rhizobium meliloti: identification of nodulation genes. Mol Gen Genet 193:445-452

Leigh JA, Signer ER, Walker GC (1985) Exopolysaccharide-deficient mutants of Rhizobium meliloti that form ineffective nodules. Proc Natl Acad Sci USA 82:6231-6235

Long SR, Buikema WJ, Ausubel FM (1982) Cloning of Rhizobium meliloti nodulation genes by direct complementation of Nod$^-$ mutants. Nature (London) 298:485-488

Rosenberg C, Boistard P, Dénarié J, Casse-Delbart F (1981)

Genes controlling early and late functions in symbiosis are located on a megaplasmid in R.meliloti. Mol Gen Genet 184: 326-333
Simon R (1984) High frequency mobilization of gram-negative bacterial replicons by the in vitro constructed Tn5-Mob transposon. Mol Gen Genet 196:413-420
Simon R, Priefer U, Pühler A (1983) A broad host range mobilization system for in vivo genetic engineering: Transposon mutagenesis in Gram-negative bacteria. Biotechnology 1:784-791
Vandenbosch KA, Noel KD, Kaneko Y, Newcomb EH (1985) Nodulation initiation elicited by noninfective mutants of Rhizobium phaseoli. J Bacteriol 162:950-959

INTERCELLULAR COMMUNICATION AND RECOGNITION IN THE RHIZOBIUM-LEGUME SYMBIOSIS

B. G. Rolfe, J.W. Redmond[1], M. Batley[1], H. Chen, S.P. Djordjevic, R.W. Ridge, B.J. Bassam, C.L. Sargent, F.B. Dazzo[2], and M.A. Djordjevic

Genetics Department
Research School Biological Sciences
Australian National University
Canberra, Australia, 2601

[1] School of Chemistry
Macquarie University
North Ryde, Sydney, Australia, 2113

[2] Department of Microbiology and Public Health
Michigan State University
East Lansing, Michigan, 48824, U.S.A.

Introduction

A complex, multi-step interaction between the soil bacterium Rhizobium and specific leguminous plants results in the induction of nitrogen-fixing root nodules (Rolfe et al 1981). The bacteria respond to plant-secreted compounds which stimulate the expression of the nodulation genes (Innes et al 1985; Mulligan and Long 1985; Rossen et al 1985). The bacteria are thought to respond by releasing compounds which induce marked root hair distortion and by elaborating receptors for the binding of specific plant-derived lectins (Dazzo et al 1985). Only actively growing root hair cells that occur behind the root tip at the time of inoculation, are highly susceptible to infection by compatible Rhizobium strains (Bhuvaneswari et al 1980). The cell wall of the root hair can be penetrated after as little as 24h exposure to the Rhizobium strain (Callaham and Torrey 1981; Ridge and Rolfe 1985). The nucleus of the root hair cell migrates to the site of wall penetration and an infection thread is synthesised by the plant presumably in response to the perturbation of the root hair cell wall (Dart 1974). The bacteria are carried towards the cortex inside the infection thread where they actively divide. Shortly before, or concurrent with

NATO ASI Series, Vol. H 4
Recognition in Microbe-Plant Symbiotic and Pathogenic Interactions. Edited by B. Lugtenberg

the initiation of infection thread synthesis, cortical cells (which are normally non-dividing) are stimulated to divide, probably by diffusable substances released by the bacterium (Bauer et al 1985). The bacteria are packaged into plant-elaborated membranes where, after a period of time, they begin to fix atmospheric nitrogen (Vincent 1980). These steps are summerised in Fig 1.

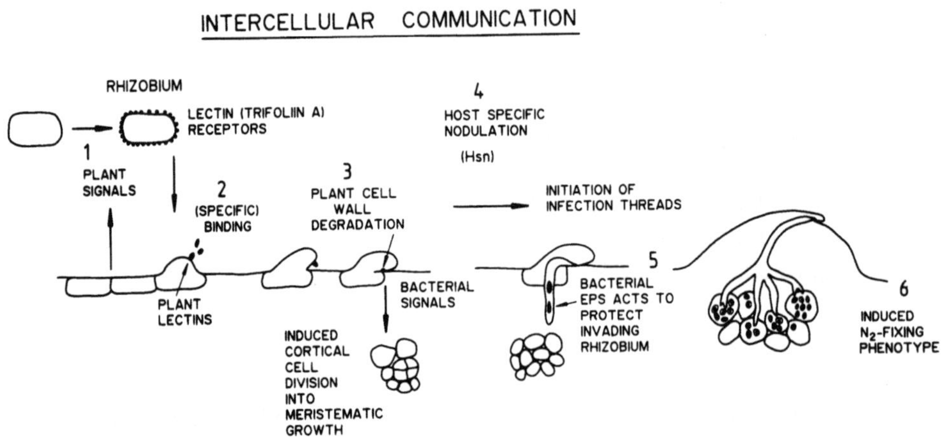

Figure 1. Diagramatic representation of the proposed intercellular communication events which occur between R.trifolii and clover plants

Rhizobia also exhibit host specificity, for example, fast-growing R.trifolii strains, infects only a narrow range of Trifolium species. No detectable signs of infection occur when most R.trifolii strains are inoculated onto pea or alfalfa (Yao and Vincent 1976 ; Djordjevic et al 1986). In contrast, the fast-growing Rhizobium strain NGR234 possesses an unusually extensive host range which includes a variety of tropical and temperate legumes as well as the tropical tree Parasponia andersonii (Trinick and Galbraith 1980). In this volume we will attempt to define some of the intercellular recognition and communication processes which occur during the infection of legumes by these two different Rhizobium strains. Recent genetic analysis of the nodulation and host-range genes of these two Rhizobium species, indicate that crucial exchanges of signals occur between the bacterium and the plant during the first 24h of the infection process which affects the success of bacterial infection.

We have identified at least five important steps which occur in the early stages of the infection process (Fig 1) and several of these will be discussed briefly.

Rhizobium nod genes are induced by plant-secreted factors

In the R.trifolii strain ANU843, at least six and probably more than 10 bacterial nodulation (nod) genes are required for the host-specific nodulation of clovers (Djordjevic et al 1985, 1986; Innes et al 1985; Schofield et al 1984; Schofield and Watson 1986). The nod genes thus far defined by DNA sequence analysis (nodABCDEF) reside on a 14kb fragment derived from the native 180kb Sym plasmid that resides in this strain (Schofield and Watson 1986) (Fig 2). DNA sequence, mutational analysis and nod gene expression data have shown that these defined genes are arranged into three operons (nodABC, nodD and nodFE) (Innes et al 1985; Djordjevic et al 1985; Schofield and Watson 1986). A similar arrangement of nod genes occurs in R.leguminosarum (Downie et al 1985; Rossen et al 1984; Shearman et al 1986) and in R.meliloti (Kondorosi et al 1984; Torok et al 1984;Kondorosi et al 1985; Egelhoff et al 1985; Mulligan and Long 1985). Other nod genes in R.trifolii located in regions II, IV and V have also been shown by mutational analysis and by the analysis of specific subcloned fragments, to be important for normal clover nodulation (region II) and for conferring clover host range ability (regions IV and V) (Djordjevic et al 1985,1986; Innes et al 1985). Only the nod D gene is expressed to appreciable levels in normal culture media (Innes et al 1985; Mulligan and Long 1985 Sherman et al 1985). High levels of ß-galactosidase activity can be detected in R.trifolii strains containing E.coli lac operon fusions to the nodABC or nod FE genes or to nod genes located in region II, IV or V, only when these strains are exposed to compounds secreted by the legume plant (Innes et al 1985; Mulligan and Long 1985; Rossen et al 1985). These plant-secreted compounds have been isolated from white clover plants (Redmond et al 1986) and shown to be hydroxyflavones (7,4'dihydroxyflavone, 7,4'- dihydroxy-3'methoxyflavone and 4'-hydroxy- 7-methoxyflavone) and these are discussed in more detail in the

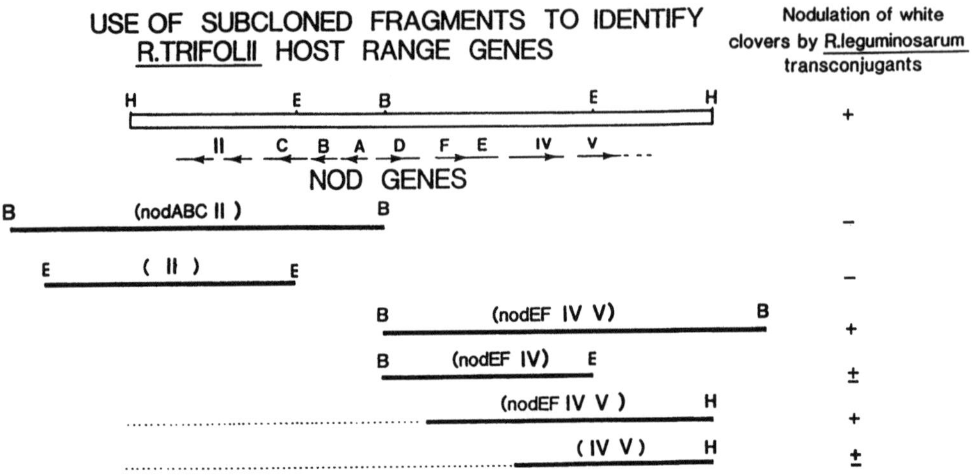

Fig 2. The location and direction of transcription for the defined nod genes (nodABCDFE) and for the nod genes which occur in regions II, IV and V are indicated. Transfer of the 14kb R.trifolii Nod region to R.leguminosarum recipients results in the extention of the host range to include white clovers. Transfer to the same R.leguminosarum recipient of DNA fragments coding the nodFE operon and regions IV and V also resulted in the extention of the host range of the recipient strain to include white clovers (Djordjevic et al 1986). In contrast, transfer of a BamHI-EcoRI sub-clone to R.leguminosarum which codes only the nodFE operon and region IV, conferred the ability to induce infection threads on white clovers but few nodules resulted. The introduction of DNA fragments coding for the common nod genes (nodABCD) did not affect the host range of the R.leguminosarum strain (Djordjevic et al 1986).

accompaning manuscript (see Redmond et al this volume). Both the nodD gene product and the plant secreted hydroxyflavones are required for the expression of inducible nod genes in R.trifolii, in R.leguminosarum (which nodulates peas) (Rossen et al 1985 ; JA Downie personal communication) and in R.meliloti (which nodulates alfalfa) (Mulligan and Long 1985; SR Long, personal communication). R.trifolii strains containing lac fusions to the inducible nod genes, have shown that these strains will respond to exudate from pea, alfalfa and other legumes (Rolfe et al 1985; Redmond et al 1986). The stimulatory hydroxyflavones are released from the zone of emerging root hairs and that the bacterial nod genes are expressed rapidly. These results are consistent with the predictions from studies on the transient infectability of the root hair cells of soybeans (Bhuvaneswari et al 1980). Their study suggested that the R.japonicum cells had to respond to the presence of the plant root environment within 2h exposure to the plant. It is therefore possible that the root hair cells themselves are the site of release of the plant signal compounds.

R.trifolii cells show a transient ability to respond to stimulatory flavones

In vitro assays using the most effective stimulatory molecule secreted by white clovers (7,4'dihydroxyflavone, DHF) shows that nod gene inducibility is growth-phase dependent and occurs only during early log phase (Rolfe et al submitted) (Fig 3). Using cells from the same culture, ß-galactosidase activity of lac operon fusions to either nodA or nod F are induced 12 and 2.5 fold respectively over control when cells at each time point are incubated for 3h in the presence of DHF. However early lag, late log and stationary phase cells do not respond to the presence of DHF. It is interesting to note that control cells in early log phase which are not exposed to DHF also show a small but reproduceable amount of ß-galactosidase activity indicating that some nodA and nodF gene expression is occurring in the absence of DHF. Moreover, the expression of the nodD lac fusion is also growth-stage dependent regardless of the presence or absence of DHF. Again, a distinct peak of ß-galactosidase expression occurs in early log phase thus indicating that nodD expression (previously regarded as being constitutive, see Rossen et al 1985; Innes et al 1985) occurs only for a short period. The decline of nodD expression reflects decline of the expression of the nodA and nodFE operons as would be expected if the expression of these genes was dependent upon the presence of the (regulatory) nodD product. Experiments done in our laboratory indicate that the E.coli ß-galactosidase activity is stable for several hours after it is translated. This means that actual mRNA synthesis of the nodA,nodF and the

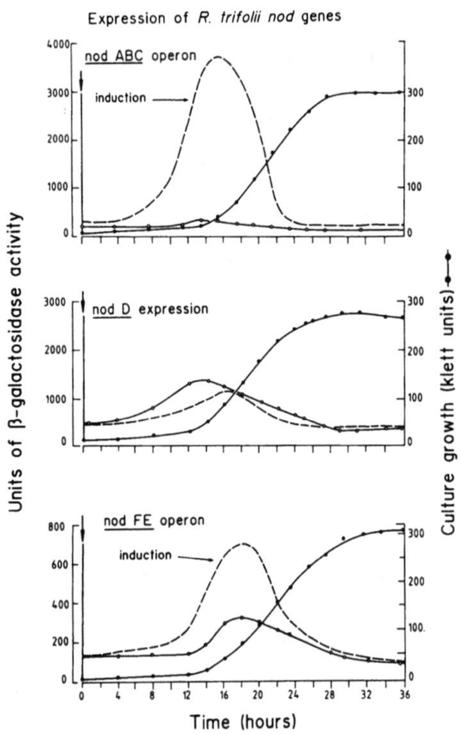

Fig 3. In vitro expression of R.trifolii nod genes. A single culture of cells grown in BMM medium was used to generate the results in each figure. The expression of the ß-galactosidase operon fusions to the nodD, nodA and nodF genes was monitored throughout the growth cycle (closed circles). Aliquots of cells were removed at 4h intervals and were either lysed immediately (open circles) to measure the background ß-galactosidase activity or were exposed to 5 x 10^{-7} M DHF for a further 3h in water (broken lines) before lysis. The method of Mulligan and Long (1985) was used to measure ß-galactosidase activity except that cell lysates were incubated at 42C for 30min to inactivate the native ß-galactosidase before OMPG was added as substrate.

nodD genes may be substantially less than that indicated by the levels of ß-galactosidase activity. In other words, the point where the ß-galactosidase induction curve starts to decline may, in reality, mean a 100% reduction in the levels of mRNA synthesis. The ability to induce the nodA and nodF genes would also be dependent upon the the stability of the nodD gene product and the availabity of the plant stimulator. Data presented here is consistent with the hypothesis that the nodD product is regulatory and requires the presence of the plant stimulatory compounds to convert it to an active form which is involved in the initiation of expression of the nodABC and nodFE operons and probably the expression of the other inducible nod genes in regions II, IV and V.

Involvement of plant lectins in the early infection events

The second recognisable effect of the incubation of R.trifolii cells in the root environment is the production of receptors on the surface of the R.trifolii cells which bind the clover lectin Trifoliin A. Using florescently-labelled antibodies to Trifoliin A, Dazzo and co-workers have shown that strain ANU843 cells incubated in the presence of the roots of white clover seedlings for less than 16h produce a receptor which specifically binds Trifoliin A (Dazzo et al 1985). Recently they have shown that the Sym plasmid-cured derivative of ANU843 (strain ANU845) does not bind trifoliin A under these conditions (Dazzo et al 1985) but this Trifoliin A binding ability can be restored to this strain by the introduction of the 14 kb nodulation fragment of R.trifolii strain ANU843 (located on plasmid pRt032). In addition, specific Tn5-induced nodulation mutants located in either nodA,B,C,D E or F or in region II nod genes (but not region IV) greatly reduces the ability of these cells to bind Trifoliin A in the rhizosphere (Dazzo et al 1985). These results indicate that the nod genes of R.trifolii have a positive role in the elaboration of the Trifoliin A binding receptors and that the ability to bind Trifoliin A is a useful bioassay for nod gene expression. Since non-homologous Rhizobium species do not bind Trifoliin A and these species do not show detectable signs of infection on clovers, the elaboration of these receptors may form the basis of a species-specific signal from the bacterium to the plant which may be crucial for infection to occur.

The effect of Rhizobium host range genes

On white clover plants, non-homologous bacteria such as R.leguminosarum or R.meliloti either do not induce marked root hair curling or induce curling but fail to infect. This occurs despite the finding that expression of the nod genes of non-homologous bacteria are stimulated in the presence of many legume hosts (Rolfe et al 1985; Redmond et al 1986) and that several nod genes are functionally conserved between species of Rhizobium (Kondorosi et al 1984; Fisher et al 1985; Djordjevic et al 1985a). Recent results suggest that several bacterial host range genes control the outcome of the infection events. Transfer of R.trifolii DNA fragments containing the nodFE operon and regions IV and V to R.leguminosarum strains (Fig 2) resulted in the ability of the transcongugant strains to initiate both infection threads and nodules on white clover plants (Djordjevic et al 1986). The transfer of smaller fragments coding, for example, only the nodFE operon and the region IV locus, enhanced the infectability of the R.leguminosarum transconjugant on white clovers but few nodules resulted (Djordjevic et al, 1986). In short, these results indicate that the host range genes act as a single dominant trait when transferred to the R.leguminosarum strain and that all the R.trifolii host range genes (nodFE and region IV and V nod genes) are required to confer efficient extension of host range ability.

Recent work with the broad host range strain NGR234 also indicates that at least some of the host range genes are clustered (as they are in R.trifolii), that they act as a single dominant trait, and that they are linked to a nodD-like gene in this strain (Bassam et al 1986; Bachem et al 1986). Transfer of a 6.7kb fragment from NGR234 to R.trifolii strain ANU843 resulted in the extension of the host range of the recipient to include the sub-tropical legume siratro (Macroptilium atropurpureum) (Bassam et al 1986). The 6.7kb fragment coded for (a) a nodD-like gene which could complement nodD mutations in R.trifolii, and (b) two flanking loci which were both required to confer the extended host range phenotype to R.trifolii. Transfer of this fragment to R.trifolii did not confer extended host range ability to other legumes nodulated by NGR234 and required that the R.trifolii recipient have functional nodABC, region II and nodFE genes (Bassam et al 1986). In this case the extended host range ability was additive as it required functional host range genes in the recipient.

However, host range genes may not always act as a dominant trait and in fact, may interfere with the natural host range of the recipient strain. For instance, (1) transfer of R.trifolii host range genes (on plasmid pRt032) to R.meliloti does not confer clover nodulation ability to this strain (Djordjevic et al 1985a); (2) transfer to R.trifolii of a plasmid pGMI515 (which is an RP4 based plasmid which codes R.meliloti host range genes) confers alfalfa nodulation ability to the R.trifolii recipient ANU843 but interferes markedly with the ability of strain ANU843 to nodulate one of its normal host plants white clover (unpublished; J. Denarie personal communication); (3) similarly, transfer of the R.leguminosarum recombinant plasmid pIJ1089 to strain ANU843 also interferes with the ability of this strain to nodulate white clover plants; (4) transfer of the NGR234 pSym to R.meliloti confers the ability to nodulate several tropical legumes but either interferes or abolishes the ability of the recipient strains to nodulate alfalfa (Djordjevic et al 1985a) and (5) mutations in the nodFE operon of R.trifolii strain affects the ability to nodulate white clovers but permits an extention of the host range of the mutant strain to nodulate pea plants (Djordjevic et al 1985b). This means that the the ability of R.trifolii strain ANU843 to nodulate peas is masked by the presence of the dominant genes in the nodFE operon. Hence we conclude (a), that the appropriate bacterial response which permits the Rhizobium strain to infect one host and not another is controlled by the host range genes and (b), the effect of possessing the incorrect host range genes can either elicit an inappropriate (rejective) response from the plant or, alternatively, the corresponding host range genes are required to elicit a favourable response from the plant host.

Can rhizobia induce the plant hypersensitive reaction?

Once the cell wall of the root hair cell has degraded, and an infection thread has been synthesised, the plant is still able to respond in an unfavourable manner to the presence of the bacterium. Infection studies have been done in our laboratory using a Tn5-induced mutant of NGR234 (ANU2861) which overproduces bacterial EPS (Chen et al 1985). This mutant has altered host range properties on various tropical legumes. Strain ANU2861 is unable to infect siratro or Desmodium species but appears to induce a normal nodulation response on other tropical legume plants (Chen et al 1985). When the infective ability of this mutant is compared to the wild type strain on siratro plants several different responses are noted (Fig 4). Like the wild type strain NGR234 the mutant strain ANU2861 can initiate root hair curling and the first signs of infection thread formation within 20h exposure to the plant, but then the infection fails to procede from this point onwards (Ridge et al submitted). Curiously, even though further infection by this mutant is minimal, this strain

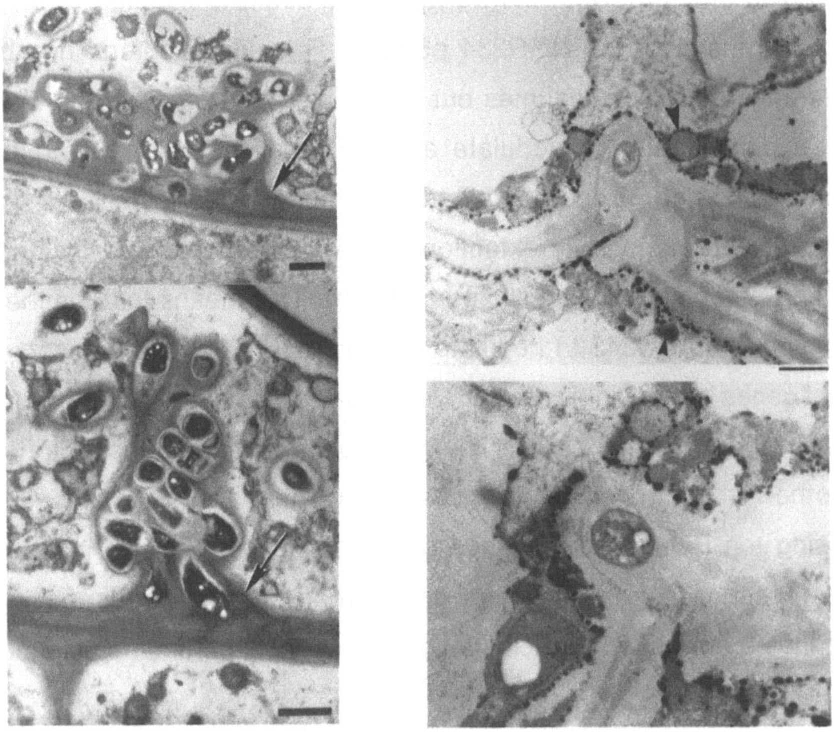

Fig 4. Transmission electron micrographs of the initial penetration sites on curled root hairs of Macroptilium atropurpureum (A) parent strain NGR234, arrow points to the junction of new and old cell wall material; (B) mutant ANU2861 causes curling, initiates the beginning of an infection thread which readily aborts with the accumulation of osmophilic droplets on the outer surface of the plasma membrane at the penetration site.

is still able to initiate some cortical cell division indicating that the signal from the bacterium to initiate cortical cell division is still released. In addition, the presence of the mutant strain on siratro plants induces the plant to produce electron-dense material around the infection site of the mutant strain which are not apparent when the parent strain infects (see Fig 4)(Ridge et al submitted). This response is similar to the response shown by certain plants which are exposed to avirulent bacteria or fungi (Dickerson and Lucas 1982). These avirulent interactions are said to initiate the "hypersensitive response" in the plant which enable the plant host to reject an invading organism.

The role of oligosaccharides in early infection events

Study of the phenotype of non-mucoid (Muc⁻) mutants of NGR234 has shown that they are generally unable to make detectable amounts of acidic exopolysaccharide (EPS) when grown on normal culture media and bacterial colonies are notably non-mucoid in appearence (S.P. Djordjevic et al submitted). These mutants are unable to induce nitrogen-fixing nodules on various tropical legumes although the phenotype varies depending upon the plant tested (Chen et al 1985). Recently, methods have been developed which enables (a) large amounts of purified EPS and the oligosaccharide repeat unit (which composes the backbone of the EPS structure) to be isolated from culture grown wild type cells and (b) for the rapid structural determination of the EPS and oligosaccharide repeat unit (S.P. Djordjevic et al 1986 a b). Purified EPS or oligosaccharide repeat unit was added to plants together with either of several Muc⁻ mutants (ANU2811, ANU2820 and ANU2840) which are normally only able to induce non-nitrogen-fixing nodules on siratro plants and disorganised, callus-like structures on Leucaena plants which contain osmophilic inclusions and few bacteria. The addition of purified EPS or oligosaccharide repeat unit to the plants with any of these Muc⁻ mutants of NGR234 resulted in the correction on the nodulation-defective phenotype of these mutants on siratro and Leucaena plants as nitrogen-fixing nodules occurred on both plant species. The effect of EPS or oligosaccharide repeat unit could also be obtained if the Muc⁺ strain ANU265 (strain NGR234 cured of its native Sym plasmid) was coinoculated with these Muc⁻ strains onto Leucaena plants. No genetic transfer was detected in any experiment when the contents of nodules were analysed and the bacteria in the nodules retained their original phenotype when reintroduced onto plants. These results demonstrate that acidic EPS plays a significant role in the establishment of functional nitrogen-fixing nodules.

Rhizobium infection induces a systemic plant response which affects subsequent infectability of the plant

Finally, after plant infection has been initiated a systemic response can be induced by the bacterium which can affect subsequent infections elsewhere on the plant root system. This has been shown by using split root assays with subterranean clovers. Plants with two equally-robust lateral root systems ("split roots") can be generated easily by damaging the root tip of the main root system (Kosslak and Bohlool 1984; Sargent et al submitted). Split root assays are advantageous in that they partition two root systems of the one plant in space and time and allow the effect of the inoculation of one root system to be compared with the inoculation of the second root system. Any positive or negative systemic effects of inoculating one root system can be assessed by the nodulation response on the second root system. Using this procedure we have shown that the inoculation of strain ANU843 on both root systems at the same time leads to equal numbers of nodules on both root systems. However, a delay in the inoculation of the second root system by time periods greater than 24h leads to an inhibition of the nodulation capacity of the delayed inoculum. The results showed that at least 24h exposure of the bacterium was required to elicit this systemic response. Futher experiments suggested that the bacterium had to possess the Sym plasmid to initiate the inhibitory response as strain ANU845 (Sym plasmid cured) inoculated onto the first root system does not affect the nodulation response given by strain ANU843 on the second root system regardless of the delay period. Moreover, recent experiments showed that the inoculation of the wild type strain on one root system can inhibit the nodulation capacity of *nodFE* or region IV mutants of ANU843 on the second root system when inoculated at the same time (Fig 5).The biological and molecular basis of this "autoregulation" response(s) and how it discriminates between infection by these bacteria has not yet been defined.

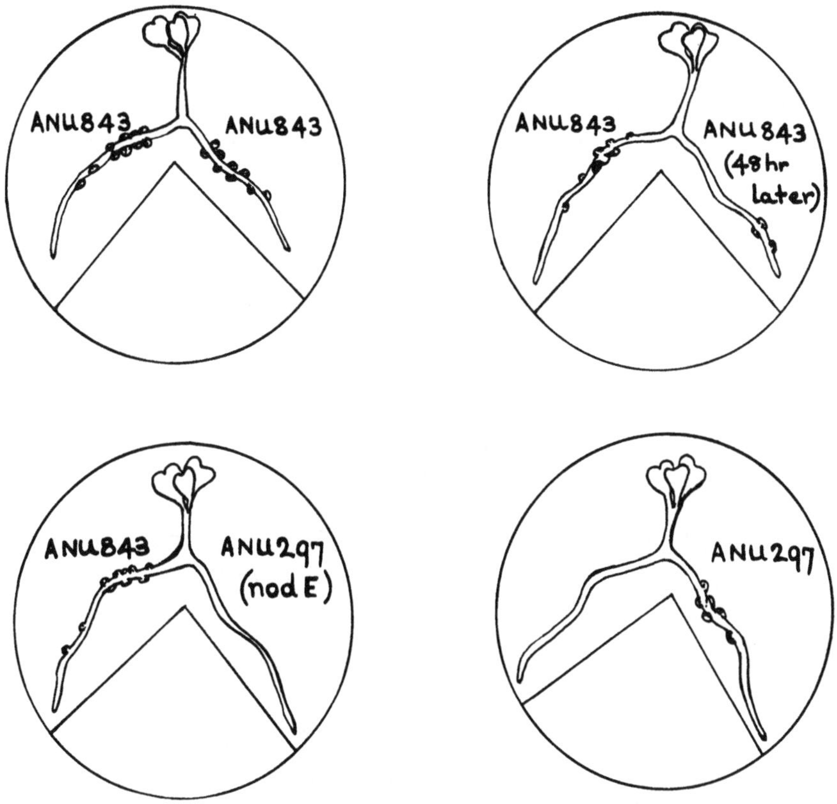

Fig 5.Diagramatic representation of the results observed using split root assays with subterranean clover plants. When inoculated alone onto subterranean clover plants both strain ANU843 and a nodE mutant derivative of this strain (nod297) induce nitrogen fixing nodules. Simultaneous inoculation of ANU843 on both root leads to equal numbers of nodules on both sides (A). Delayed inoculation of ANU843 on the second root (B) leads to a marked inhibition of the ability to nodulate this side (see text). Inoculation of one root with ANU843 and the other with the nodE mutant results in nodules occurring exclusively on the root inoculated with ANU843 (C) while the mutant inoculated alone induces many nodules (D).

Acknowledgements

This work is partially supported by an Agrigenetics-sponsered research grant (to BGR) by the Australian Wool Board (MAD and BGR) and by the Australian Meat and Live-stock Research and Development Corporation. BJB and CLS are recipients of a Commonwealth Post-Graduate Research Award and SPD and HC by Australian National University Post-Graduate Research Awards. The technical support of J. McIver, M. Oakes, A. Motten and E. Gartner is acknowledged. FBD was supported by the following grants: NIH GM 34331-02 and USDA 85-CRCR-1-1627.

References

Bachem CWB, Banfalvi Z, Kondorosi E, Schell J, Kondorosi A (1986) Identification of host range determinants in the Rhziobium species MPIK3030. Mol gen Genet (in press)

Bassam BJ, Rolfe BG, Djordjevic MA (1986) Macroptilium atropurpureum (siratro) host specificity genes are linked to a nodD-like gene in the broad host range Rhizobium strain NGR234 Mol gen Genet (in press)

Bauer WD, Bhuvaneswari TV, Calvert HE, Law IJ, Malik NAS, Vesper SJ (1985) Recognition and infection by slow-growing rhizobia. In:Evans HJ, Bottomly PJ and Newton WE (eds). Nitrogen Fixation Research Progress p247-253 Martinus Nijhoff, Amsterdam.

Bhuvaneswari TV, Turgeon GB, Bauer WD. (1980) Early events in the infection of soybean (Glycine max L. Merr). Plant Physiol. 66: 1027- 1031

Callaham DA, Torrey JG (1981) A structural basis for infection of root hairs of Trifolium repens by Rhizobium Can J Bot 59 :1647-1651

Chen H, Batley M, Redmond JW, Rolfe BG (1985) Alteration of the effective nodulation properties of a fast-growing broad host range Rhizobium due to changes in exopolysaccharide synthesis J. Plant Physiol. 120:331-349

Dart PJ (1974) The infection process. In: Quispal A (ed) The Biology of Nitrogen Fixation p381-429 North Holland Amsterdam

Dazzo FB, Hollingsworth RI, Sherwood JE, Abe M, Hrabak EM, Gardinol AE, Pankratz S, Smith KB ,Yang H (1985) Recognition and infection of clover root hairs by Rhizobium trifolii In: Evans HJ, Bottomley PJ, Newton WE (eds) Nitrogen Fixation Research Progress p239-245 Martinus Nijhoff Amsterdam

Dickinson CH, Lucas JA (1982) Plant Pathology and Plant Pathogens. Blackwell Scientific Publications Oxford

Djordjevic MA, Schofield PR, Ridge RW, Morrison NA, Bassam BJ, Plazinski J Watson JM, Rolfe BG (1985a) Rhizobium nodulation genes involved in root hair curling (Hac) are functionally conserved. Plant Mol Biol 4:147-160

Djordjevic MA, Schofield PR, Rolfe BG (1985b) Tn5 mutagenesis of Rhizobium trifolii host-specific nodulation genes result in mutants with altered host range ability. Molec gen Genet 200: 463-471

Djordjevic MA, Innes RW, Wijffelman CA, Schofield PR, Rolfe BG (1986) Nodulation of specific legumes is controlled by several distinct loci in Rhizobium trifolii Plant Mol Biol (in press)

Djordjevic SP, Batley M, Redmond JW, Rolfe BG (1986)The structure of the exopolysaccharide from Rhizobium SP. strain ANU280 (NGR234) Carbohydr. Res. in press

Djordjevic SP, Batley M, Redmond JW (1986) Preparative gel chromatography of acidic oligosaccharides using a volatile buffer J. Chromatogr. 354: 507-510

Downie JA, Knight CD, Johnston AWB, Rossen L (1985) Identification of genes and gene products involved in the nodulation of peas by Rhizobium leguminosarum Molec gen Genet 198: 255-262

Egelhoff TT, Fisher RF, Jacobs TW, Mulligan JT, Long SR (1985) Nucleotide sequence of the Rhizobium meliloti 1021 nodulation genes: nodD is read divergently to nodABC. DNA 4: 241-248

Fisher RF, Tu JK, Long SR (1985) Conserved nodulation genes in R.meliloti and R.trifolii Appl Environ Microbiol. 39: 1432-1435

Innes RW, Keumpel PL, Plazinski J, Canter-Cremers H, Rolfe BG, Djordjevic MA (1985) Plant factors induce expression of nodulation and host-range genes in Rhizobium trifolii Mol gen Genet 201: 426-432

Kondorosi E, Banfalvi Z, Kondorosi E (1984) Physical and genetic analysis of a symbiotic region of Rhizobium meliloti: identification of nodulation genes Mol gen Genet 193:445-452

Kondorosi A, Horvath B, Gottfert M, Putnoky P, Rostas K, Zoltan G, Kondorosi E, Torok I, Bachem C, John M, Schmidt J, Schell J (1985) Identification and organisation of Rhizobium meliloti genes relevant to the initiation of nodules. In: Evans HJ, Bottomley PJ, Newton WE (eds.) Nitrogen Fixation Research Progress p73-78 Martinus Nijhoff Amsterdam

Kosslak RM, Boohlool BB (1984) Suppression of nodule development on one side of a split root system of soybeans caused by prior inoculation of the other side Plant Physiol 75:125-130

Mulligan JT, Long SR 1985 Induction of Rhizobium meliloti nodC expression by plant exudate requires nodD Proc Natl Acad Sciences USA 82:6609-6613

Ridge RW, Rolfe BG. (1985) Rhizobium sp degradation of legume root hair cell wall at the site of infection thread origin. Appl Environ Microbiol 50:717-720.

Redmond JW, Batley M, Djordjevic MA, Innes RW, Keumpel PL, Rolfe BG (1986.) Flavones induce the expression of the nodulation genes in Rhizobium Nature (in press)

Rolfe BG, Djordjevic M, Scott KF, Hughes JE, Badenoch-Jones J Gresshoff PM, Cen Y,Dudman W, Zurkowski W Shine J (1981) Analysis of the nodule forming ability of fast-growing Rhizobium strains. In: Gibson AH, Newton WE eds Current Prespectives in Nitrogen Fixation, p142 -145 Australian Academy Sciences, Canberra

Rolfe BG, Innes RW, Schofield PR, Watson JM, Sargent CL, Keumpel PL, Plazinski J, Canter-Cremers H, Djordjevic MA.(1985) Plant-secreted factors induce the expression of R. trifolii nodulation and host-range genes In: Evans HJ, Bottomley PJ, Newton WE eds. Nitrogen fixation Research Progress Martinus Nijhoff Amsterdam

Rossen L, Johnston AWB, Downie JA (1984) DNA sequence of the Rhizobium leguminosarum nodulation genes nodAB and C required for root hair curling Nucleic Acids Res. 12 :9497-9508

Rossen L, Shearman CA, Johnston AWB, Downie JA (1985) The nodD gene of Rhizobium leguminosarum is autoregulatory and in the presence of plant exudate induces the nodA,B,C genes

Schofield PR, Ridge RW, Rolfe BG, Shine J, Watson JM (1984) Host-specific nodulation is encoded on a 14kb DNA fragment in Rhizobium trifolii Plant Mol Biol 3:3-11

Schofield PR, Watson JM (1986) DNA sequence of the Rhizobium trifolii nodulation genes reveals a reiterated and potentially regulatory sequence preceding the nodABC and nodFE genes Nucleic Acids Res 14:2891-2905

Torok I, Kondorosi E Stepowski, Posfai J, Kondorosi A (1984) Nucleotide sequencee of Rhizobium meliloti nodulation genes. Nuclic Acids Res 12: 9509-9524

Trinick MJ, Galbraith J (1980) The Rhizobium requirements of the non-legume Parasponia in relation to the cross inoculation concept of legumes. New Phytol 86:17 26

Vincent JM (1980) Factors controlling the legume-Rhizobium symbiosis In Newton WE Orme-Johnston WH (eds) Nitrogen Fixation 2 University Park Press p103-129 Baltimore

Yao PY, Vincent JM (1976) Factors responsible for the curling and branching of clover root hairs by Rhizobium. Plant and Soil 45:1-16

PROMOTERS AND OPERON STRUCTURE OF THE NODULATION REGION OF THE RHIZOBIUM LEGUMINOSARUM SYMBIOSIS PLASMID pRL1JI

Herman P. Spaink, Robert J.H. Okker, Carel A. Wijffelman, Elly Pees and Ben Lugtenberg
University of Leiden, Department of Plant Molecular Biology,
Nonnensteeg 3, 2311 VJ Leiden,
The Netherlands

INTRODUCTION

Bacteria of the genus Rhizobium, which are able to establish a symbiosis with leguminous plants, invade the roots of their hosts where they induce the formation of nodules in which they fix atmospheric nitrogen. Each Rhizobium strain has only a narrow range of host plants on which it is successfully able to establish a symbiosis. In fast-growing Rhizobia, which include R.leguminosarum with peas and vetches as their hosts, R.trifolii with clovers as their hosts, and R.meliloti with alfalfa as a host, many genes involved in nodulation (nod) and nitrogen fixation (nif) are localized on large plasmids designated symbiosis (Sym) plasmids (Johnston et al 1978; Hooykaas et al 1981; Banfalvi et al 1981; Rosenberg et al 1981). The nod genes, which constitute only a relatively small portion of the Sym plasmid (Downie et al 1983; Schofield et al 1984), appear to be of prime importance in the determination of the host-specificity.

R.leguminosarum strain 248 contains the Sym plasmid pRL1JI (Hirsch et al 1980) from which mutants defective in nodulation have been isolated at the John Innes Institute (e.g. see Downie et al 1985) as well as in our laboratory (Wijffelman et al 1985). In the latter case seventeen Tn5 mutants, defective in nodulation on Vicia sativa, have been isolated, all of which were located on a 6.6 kb EcoRI fragment, situated between the nifKDH and nifA genes. Based on the phenotypes of the mutants, this fragment could be functionally divided into two regions, now designated as nodulation subregions I and III (Fig. 1). Mutants in region I

NATO ASI Series, Vol. H 4
Recognition in Microbe-Plant Symbiotic and Pathogenic Interactions. Edited by B. Lugtenberg

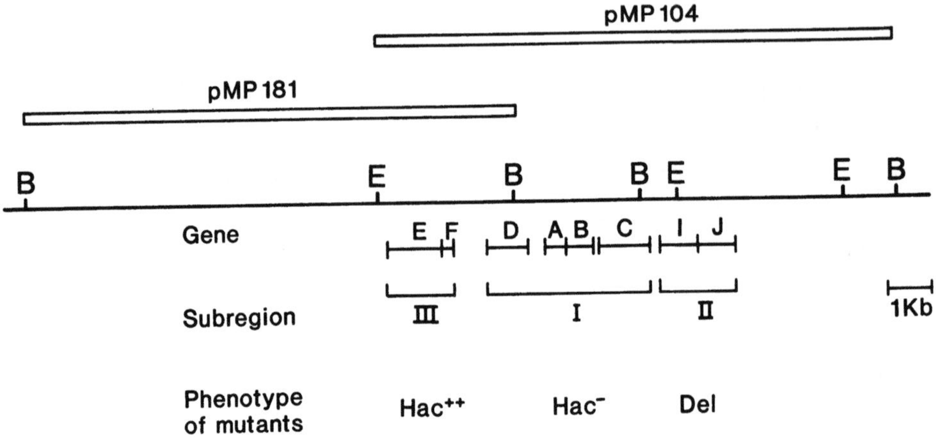

Fig. 1. The nodulation region of pRL1JI. Nomenclature and position of the indicated genes is according to Shearman et al (1986). pMP104 and pMP180 are subclones of pRL1JI in the vector pMP92, a derivative of pTJS75 (Schmidhauser et al 1985). B,BamHI; E,EcoRI.

do not induce root hair curling (Hac^-), one of the initial steps in nodulation, and completely fail to nodulate. In this region four open reading frames have been identified, probably corresponding to the four nod genes designated as nodA,B,C and D (Rossen et al 1984; Downie et al 1985; Shearman et al 1986). At the predicted protein level these genes are highly homologous with the nodA,B,C and D genes of R.meliloti and R.trifolii (Török et al 1984; Jacobs et al 1985; Egelhof et al 1985; Schofield and Watson 1986). These genes are referred to as common nod genes since mutations in these genes can be complemented by Sym plasmids of other fast-growing Rhizobium species (Kondorosi et al 1984; Fisher et al 1985; Djordjevic et al 1985b; Wijffelman et al 1985). Mutants in region III give rise to exaggerated root hair curling (Hac^{++}) and in delayed and infrequent nodulation. In this region two open reading frames have been reported, designated as nodF and nodE, the predicted protein products of which are highly homologous with those of the reported open reading frames of nodF and nodE of R.trifolii (Shearman et al 1986; Schofield and Watson 1986). However, mutations in these genes cannot be complemented by Sym plasmids of other Rhizobium species and therefore they are assumed to play a role in host-specificity. Besides the regions I and III, both located in the 6.6 Kb EcoRI fragment, another region adjacent to region I, (Fig. 1) has been reported. Two genes have been identified in this region II, designated as nodI and nodJ (Shearman et al 1986). Mutants in this region show a delayed nodulation phenotype on pea (Downie et al 1985).

The research presented in this paper is focused on the regulation of the genes situated in the nod region of pRL1JI. Firstly, the operon structure of the nod genes A,B,C,I and J is discussed by a comparison with the corresponding genes of the Sym plasmid of R.trifolii in complementation studies. Secondly, the nod region of pRL1JI was scanned for promoters using small subclones in a promoter probe vector.

THE GENES NODA,B,C,I AND J REPRESENT ONE TRANSCRIPTION UNIT

The nodulation regions of pRL1JI and of pRt843, the Sym plasmid from R.trifolii strain ANU843 (Schofield et al 1983), are compared with each other in Fig. 2. The Tn5 mutants 262 and 261 in region II of pRt843, which show very poor nodulation on Trifolium plants (Djordjevic et al 1985a), could be complemented by the presence of wild type pRL1JI. Conversely, a strain containing the mutant plasmid pRL631, which is pRL1JI::Tn1831 deleted for the nodB and nodC genes as well as for region II (del nod31, Fig. 2), was restored in nodulation ability on V.sativa by the presence of pRt843. However, restoration of nodulation was very poor when the R.trifolii mutant strains 262 or 261 were supplemented with the deletion mutant pRL631. These observations demonstrate the importance of region II for nodulation and show that region II, like region I, has common nodulation properties.

The mutants in regions I and II indicated in Fig. 2 of the Sym plasmids of R.leguminosarum and R.trifolii were used in complementation studies to investigate the transcriptional organization of the genes localized in these regions. Tn5 mutants in one of the genes nodA,B or C of pRL1JI could not be complemented for nodulation on V.sativa by pRt843 mutants with a Tn5 insertion in either of the nod genes A,B,C or in region II (mutants 262 and 261) genes. Mutations induced by Tn5 are usually polar, i.e.

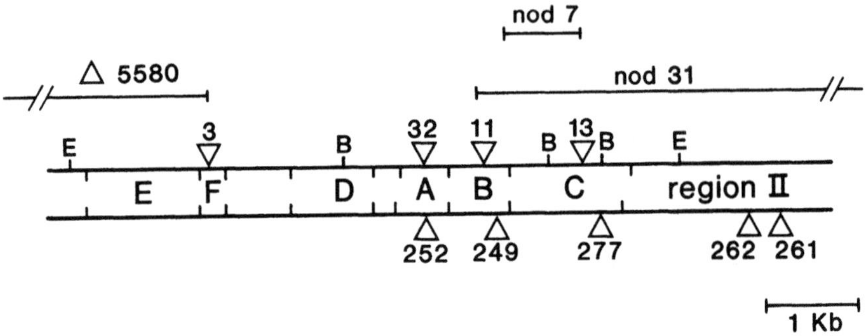

Fig. 2. Comparison of the nodulation regions of the R.leguminosarum Sym plasmid pRL1JI and the R.trifolii Sym plasmid pRt843. The localization of mutations in pRL1JI (Wijffelman et al 1985) and in pRt843 (Djordjevic et al 1985a), used for the complementation studies described in the text, is indicated by triangles. Del5580, is pRL1JI::Tn1831 containing a deletion of approximately 60 kb. The mutations nod3, nod7 and nod31 are present in the plasmids pRL603, pRL607 and pRL631, respectively.

transcription of an operon is greatly reduced downstream from the mutation. Therefore the most likely interpretation of the results is that the nodA,B and C genes constitute one operon with the genes nodI and J of region II. This interpretation is supported by the observation that the region II mutant strains 262 and 261 supplemented with pRL607, which is pRL1JI containing a non-polar deletion (nod7) in nodC (Fig. 2), were able to nodulate V.sativa.

Strain ANU843 containing pRL1JI is able to nodulate Pisum sativum ssp. Afghanistan, which is normally only nodulated by the Rhizobium strain TOM, a field isolate from Turkey (Winarno and Lie 1979), and not by pRL1JI-containing strains. However, broadened host range does not occur when pRL1JI is introduced in the R.trifolii mutant strains 262 or 261. The additional nod locus present in region II of R.trifolii responsible for the broadened host range, designated as nodK, therefore is probably also a part of the nodABCIJ transcription unit. The cosmid clone pRmSL26 of the R.meliloti Sym plasmid (Long et al 1982) enabled R.leguminosarum strain 248 to nodulate Afghan peas as well. This demonstrates that a nodK locus is also functionally present on the Sym plasmid of R.meliloti strain 1021.

DEVELOPMENT OF PROMOTER PROBE VECTOR PMP190

One of the approaches which is very useful for studying gene regulation is the use of expression indication vectors (for a discussion see Okker et al 1986). Such vectors contain an indicator gene, coding for an easily detectable enzyme which is deprived of its own promoter and consequently is not expressed. However, insertion of a promoter-containing DNA fragment in a restriction site located upstream of the indicator gene can restore the expression and the indicator gene is then under control of regulatory properties of the cloned promoter region. In promoter probe vectors the indicator gene is located behind a translational initiation signal, distinguishing it from translational fusion vectors in which expression of the indicator gene is also dependent on a translational fusion to occur. Promoter probe vectors are most suited for our objective, scanning for and subsequently defining promoter regions.

Some of the requirements of promoter probe vectors suited for use in Rhizobium are: 1) The vector must have a wide host range. 2) Background expression of the indicator gene should be as low as possible, whereas cloned promoters should induce high indicator activity, e.g. a ribosome binding site operating efficiently in Rhizobium should be present. 3) The vector should contain many unique restriction sites upstream of the indicator gene, should be as small as possible and should code for suitable antibiotic resistances.

With these requirements in mind we have developed the vector pMP190 which is outlined in Fig. 3. pMP190 is of the incompatibility (Inc) Q class with a copy number of about 15 and contains the lacZ gene of E.coli coding for β-galactosidase as an indicator gene. It was cloned as a 3.4 kb restriction fragment, thereby minimalizing the size of the vector. Background expression of pMP190 is extremely low in E.coli as β-galactosidase activity is less then 0.1 units in a del(lac) strain. However,

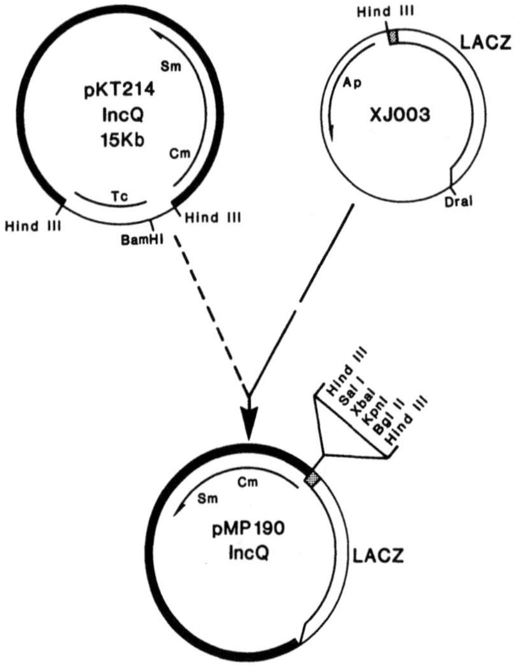

Fig. 3. Construction of the promoter probe vector pMP190. A 3.4 kb HindIII-DraI fragment of pXJ003 (Legocki et al 1984) containing the lacZ gene without its promoter and containing the ribosomal binding site of the chloramphenicol acetyl transferase gene was cloned in the vector pKT214 (Bagdasarian et al 1979). Several restriction sites have been removed and the polylinker of pIC20H (Marsh et al 1984) has been inserted, resulting in the vector pMP190, with 5 unique cloning sites upstream of lacZ.

it is considerably higher in Rhizobium: a R.trifolii strain showed 50 units β-galactosidase activity including the endogenous Rhizobium background of 15 units. The vector pMP190 with an insert of random DNA showed a β-galactosidase activity of at least 1 unit in E.coli and of 150 units in Rhizobium, which we found to be the minimal background activity of the insert-containing vector. The latter feature has been useful in cloning experiments as it enabled us to distinguish insert-containing vectors from the recirculized vector on indicator plates.

PROMOTERS IN THE NOD REGION OF pRL1JI

In order to isolate the wild type clones of pRL1JI, indicated in Fig. 1, a Tn5 mutant of pRL1JI in nodF (pRL603; Fig. 2) was used. pMP104, the 12 kb clone inserted in a vector of the IncP1 class, enabled a strain of R.trifolii cured for its Sym plasmid (LPR5045) to nodulate on both Pisum sativum and Vicia hirsuta. However, no nodulation was observed on V.sativa, indicating that this fragment does not contain all nod genes necessary for nodulating the latter host in this Rhizobium background. pMP180, the 11 kb fragment inserted in an IncPl class vector, enabled strain RBL5580, which is pRL1JI with a deletion of the region leftwards of nodD including nifKDH (Fig. 2), to nodulate V.sativa. Therefore the 19.3 kb fragment indicated in Fig. 1 is considered to contain all genes of pRL1JI necessary for nodulating this host.

A KpnI-BglII fragment covering the 5' ends of the nodA and nodD genes (Rossen et al 1985) was subcloned in the promoter probe vector pMP190. The construct with the presumed nodA promoter directed towards lacZ was mobilized into Rhizobium strain RBL5560 (containing pRL1JI). The resulting derivative did not show any β-galactosidase activity higher than the background activity when grown in the usual bacterial culture media. However, results from previous experiments in our laboratory (Van Brussel et al 1986) indicated that R.leguminosarum nod genes could be induced by root exudates of V.sativa. Therefore β-galactosidase levels of the strain containing the promoter nodA construct were tested after overnight incubation with this root exudate and a 90-fold induction of β-galactosidase activity was observed. This result indicated that the used vector system was suitable for monitoring promoter activity in Rhizobium. Induction of the nodA promoter by exudate was not possible in a Rhizobium strain which did not contain a Sym plasmid. Induction was neither observed in the presense of the nodD mutant pRL602 (pRL1JI nodD ::Tn5; Wijffelman et al 1985).

Subsequently the nod region was scanned systematically for promoters. For this purpose the fragments indicated in Fig. 4 were cloned in pMP190 and tested in Rhizobium for β-galactosidase activity in the presence and absence of V.sativa root exudate. In these experiments the β-galactosidase activities observed in the presence of an intact pRL1JI and the nodD mutant pRL602 were compared with each other. Two other plant-inducible promoter regions were identified by this screening procedure (Fig. 4). One of these is designated as the nodF promoter, according to its presence upstream of the nodF open reading frame (Shearman et al 1986; our unpublished data). We designated the other re-demonstrated by the full inducibility of the constructs pMP158 and 157 (Fig. 4), containing the nodD gene, in a Rhizobium strain without a Sym plasmid (Table 1). In Agrobacterium tumefaciens, a close relative of Rhizobium which is also able to nodulate leguminous plants if provided with a Sym plasmid (Hooykaas et al 1982), we also observed comparable inducibility of the nodA and nodF promoters, provided that the nodD gene is present (Table 1).

TABLE 1. β-Galactosidase activity of cloned promoter regions of pRL1 in pMP190[a]

TESTED PLASMID CONSTRUCT	BACTERIAL HOST STRAIN	$U.10^{-3}$ β-galactosidase - EXUDATE	+ EXUDATE
pMP154 (p.nodA)	RBL5561	0.3	0.3
	RBL5560	0.2	18
pMP168 (p.nodF)	RBL5561	0.4	0.4
	RBL5560	0.4	12
pMP155 (p.nodH)	RBL5561	0.9	0.9
	RBL5560	0.9	8
pMP158 (p.nodA,nodD)	LPR5045	0.3	27
	LBA4301 (A.tum.)	0.3	26
pMP157 (p.nodF,nodD)	LPR5045	0.7	16
	LBA4301 (A.tum.)	0.3	23
pMP86 (p.nodD)	RBL5561	6	6
	RBL5560	2	2

[a]Plasmids with subcloned fragments were mobilized from E.coli to Rhizobium or Agrobacterium tumefaciens using pRK2013 (Ditta et al 1980) as a helper plasmid and with rifampicin ($20mg.1^{-1}$), streptomycin ($1\ g.1^{-1}$) and chloramphenicol ($10mg.1^{-1}$) as selective markers. Rhizobium acceptor strains were the isogenic strains LPR5045 (without Sym plasmid) (Hooykaas et al 1982), RBL5560 (with Tn5 marked R.leguminosarum Sym plasmid pRL1JI) and RBL5561 (with pRL1JI nodD::Tn5) (Wijffelman et al 1985). Strain RBL5560 shows symbiotic properties comparable to wild type R.leguminosarum strains. Agrobacterium receptor strain was LBA4301 (Rec^-, without Ti plasmid) (Hooykaas et al 1982). The constructs were tested for β-galactosidase activity according to methods described previously (Zaat et al, submitted).

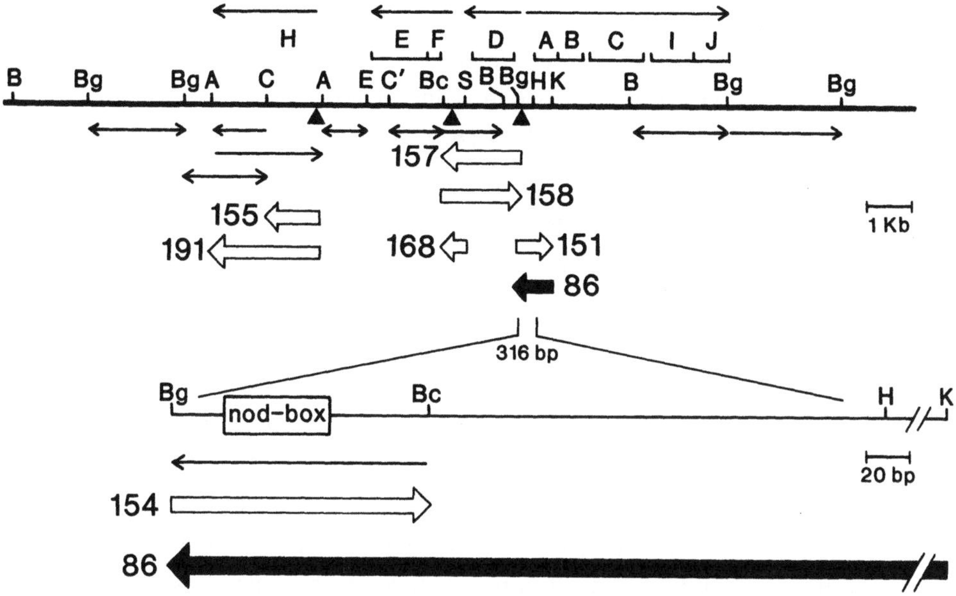

Fig. 4. Promoter activity of subcloned fragments from the nod region of pRL1JI in the promoter probe vector pMP190. The position and size of the nodABCIJ and nodFE genes is according to Shearman et al (1986). The presence of a nodH gene is tentatively indicated. The indicated transcription units are concluded from the work described in this paper. The presence of nod boxes (Fig. 5) is indicated by triangles. White arrows, plant inducible promoter activity; black arrows, constitutive promoter activity; thin arrows, no promoter activity. Restriction sites are only indicated when relevant as borders of subclones. A,AccI; B,BamHI; Bc,BclI; Bg,BglII; C,ClaI; C',ClaI; dam methylated; E,EcoRI; H,HindIII; K,KpnI; S,SmaI.

gion the nodH promoter, although no evidence for the presence of a functional nod locus directed by this promoter can be presented yet. This region might be corresponding to region V of the R.trifolii Sym plasmid (Innes et al 1985). The size of the nodH locus might be large considering the fact that the 2.5 kb AccI fragment present in pMP191 promotes a β-galactosidase activity which is comparable to the smaller indicated nodH fragment. The observed β-galactosidase activities of three constructs containing the inducible promoters of nodA,F and H were compared with each other (Table 1). All three promoters need the presence of a functional nodD gene for induction. No other Sym plasmid functions are required for the nodA and nodF inducible promoter activity, as is

Besides these three inducible promoters also the nodD promoter has been localized (Fig. 4) which has high constitutive activity when pRL1JI, containing a functional nodD gene, is absent but whose activity is reduced 3-fold in the presence of pRL1JI (Table 1). This result confirms the autoregulatory property of nodD of pRL1JI, which has been reported by Rossen et al (1985). The other cloned fragments, indicated in Fig. 4 as thin arrows, all showed β-galactosidase activities between 150 and 700 units, which were neither influenced by the presence of exudate nor by the presence of the Sym plasmid. Therefore we do not consider these activities as significant promoter activities. The transcriptional organization of the nodABCIJ genes as one operon, as concluded from complementation studies, is supported by the observation that we did not locate promoter activity between the nodC and nodIJ genes. The absence of promoter activity between the nodF and nodH inducible promoters indicates that the nodFE genes also represent one transcription unit.

Plant-inducible nod promoters have already been reported for R.leguminosarum (Rossen et al 1985; Shearman et al 1986) and for R.meliloti (Mulligan and Long 1985) using translational fusions and for R.trifolii (Innes et al 1985) using Mud(lac) insertions. The nodD product appears to be the only Sym plasmid-coded product required for induction in R.leguminosarum and R.trifolii (Spaink et al, in preparation) while in R.meliloti also other Sym plasmid genes might be involved. Our results suggest a simple model for the regulation of Sym plasmid-localized nod genes of R.leguminosarum in the initial stages of nodulation. All nod genes identified until now, except the nodD gene, appear to be transcribed actively in the presence of a plant factor provided that the nodD product is present as well.

NOD BOXES ARE PROBABLY PART OF THE PLANT-INDUCIBLE PROMOTERS

The promoter region between nodA and nodD of pRL1JI has been studied in more detail by subcloning small fragments of this region in the promoter probe vector pMP190. A 114 bp BclI-BglII fragment possessed full inducible promoter capacity (Fig. 4).

However, nodD promoter activity was not present in the reverse orientation, indicating that at least part of the nodD constitutive promoter region is situated at the right hand side of the nodA inducible promoter (Fig. 4).

Considering the similarities in regulation of the nodA,F and H promoters, we expected to find homologous sequences in the promoter regions. Comparison of the sequences of the 114 bp fragment with that of the smallest inducible clones of nodF and nodH indeed revealed a region of high homology (Fig. 5). Such homologous sequences have also been reported upstream of nod genes of other Rhizobium species and have been designated as nod boxes (Rostas et al 1986; Schofield and Watson 1986; Scott 1986).

```
P.NODA  GGGTTGAATATCCATTCCATAGATGATTGCCATCCAAACAATCAATTTTACCAATCTTTCGGATCACTTATAGAAAACCCGGAACTTGATC
P.NODF  CGAGCCACAATCCATAGTGTGGATGCTTTTGATCCACACAATCAATTTTACCAATGATGCCATATGATCCATAGCAGGGCAGCCGCGCGGC
P.NODH  GTGGGCGACATCCATATCGTGGATGATAGCTATCCCAACAATCAATTTTACTAATCTGTTTGGATTTATTAGCACGCGCTGGAGGACACGC
CONS.             ATCCAY  UYUGATG Y  Y ATCCAAACAATCUATTTTACCAATCY
```

Fig. 5. Nod boxes detected in the nod region of the R.leguminosarum Sym plasmid pRL1JI. The subclones present in pMP154, 168 and 155, which showed promoter activity in the presence of plant exudate, were further subcloned in two orientations into phage M13 derivatives and both strands were sequenced at least twice with the dideoxy method (Sanger et al 1977). The parts of the sequences which were highly homologous with each other are shown. These three homologous parts were compared with each other and with homologous sequences of other Rhizobium species (Schofield and Watson 1986; Rostas et al 1986; Egelhoff et al 1985; Scott 1986). Consensus (cons.) indicates that at least ten of the twelve sequences contain the indicated base. The nod boxes of p.nodA and p.nodF were compared with R.trifolii p.nodA and p.nodF, respectively, and the homology is indicated by underlining of the identical bases. In order to facilitate comparisons, the nod boxes are orientated with the right end pointing to the coding sequences. Abbreviations: Y, pyrimidine; U, purine.

The consensus sequence, indicated in Fig. 5, covers almost half of the 114 bp inducible promoter fragment and it therefore is very likely to be an important part of the inducible promoter. At present we are testing several Bal31 deleted promoter fragments for promoter activity. Preliminary results indicate that the nod box cannot be deleted without loss of inducible promoter activity (Spaink et al, in preparation).

ACKNOWLEDGEMENTS

We thank Ms I. Mulders, Mr T. Tak and Mr F. Goedegebuur for their skilful technical assistance.

REFERENCES

Bagdasarian M, Bagdasarian MM, Coleman S and Timmis KN (1979) New vectors for cloning in Pseudomonas. In: Timmis KN and Pühler A (eds). Plasmids of medical, environmental and commercial importance. Elsevier/North Holland 1979 Amsterdam: 411-422

Banfalvi Z, Sakanyan V, Koncz C, Kiss A, Dusha I and Kondorosi A (1981) Location of nodulation and nitrogen fixation genes on a high molecular weight plasmid of R.meliloti. Molec Gen Genet 184: 318-325

Ditta G, Stanfield S, Corbin D and Helinski DR (1980) Broad host range DNA cloning system for gram-negative bacteria: construction of a gene bank of Rhizobium meliloti. Proc Natl Acad Sci U.S.A. 77: 7347-7351

Djordjevic MA, Schofield PR and Rolfe BG (1985a) Tn5 mutagenesis of Rhizobium trifolii host-specific nodulation genes result in mutants with altered host-range ability. Mol Gen Genet 200: 463-471

Djordjevic MA, Schofield PR, Ridge RW, Morrison NA, Bassam BJ, Plazinski J, Watson JM and Rolfe BG (1985b) Rhizobium nodulation genes involved in root hair (Hac) are functionally conserved. Plant Mol Biol 4: 147-160

Downie JA, Hombrecher G, Ma Q-S, Knight CD, Wells B and Johnston AWB (1983) Cloned nodulation genes of Rhizobium leguminosarum determine host-range specificity. Molec Gen Genet 190: 359-365

Downie J, Knight CD, Johnston AWB and Rossen L (1985) Identification of genes and gene products involved in the nodulation of peas by Rhizobium leguminosarum. Molec Gen Genet 198: 255-262

Egelhoff TT, Fisher RF, Jacobs TW, Mulligan JT and Long SR (1985) Nucleotide sequence of Rhizobium meliloti 1021 nodulation genes: nodD is read divergently from nodABC. DNA 4: 241-248

Fisher RF, Tu JK and Long SR (1985) Conserved nodulation genes in Rhizobium meliloti and Rhizobium trifolii. Appl Env Microbiol 49: 1432-1435

Hirsch PR, van Montagu M, Johnston AWB, Brewin NJ and Schell J (1980) Physical identification of bacteriocinogenic nodulation and other plasmids in strains of Rhizobium leguminosarum.

J Gen Microbiol 120: 403-412
Hooykaas PJJ, van Brussel AAN, den Dulk-Ras H, van Slogteren GMS and Schilperoort RA (1981) Sym plasmid of Rhizobium trifolii expressed in different Rhizobium species and Agrobacterium tumefaciens. Nature 291: 351-353
Hooykaas PJJ, Snijdewint FGM and Schilperoort RA (1982) Identification of the Sym plasmid of Rhizobium leguminosarum strain 1001 and its transfer to and expression in other Rhizobia and Agrobacterium tumefaciens. Plasmid 8: 73-82
Innes RW, Kuempel PL, Plazinski J, Canter-Cremers H, Rolfe BG and Djordjevic MA (1985) Plant factors induce expression of nodulation and host-range genes in Rhizobium trifolii. Molec Gen Genet 201: 426-432
Jacobs TW, Egelhoff TT and Long SR (1985) Physical and genetic map of a Rhizobium meliloti nodulation gene region and nucleotide sequence of nodC. J Bact 162: 469-476
Johnston AWB, Beynon JL, Buchanan-Wollaston AV, Setchell SM, Hirsch PR and Beringer JE (1978) High frequency of transfer of nodulating ability between strains and species of Rhizobium. Nature 276: 635-636
Kondorosi E, Banfalvi Z and Kondorosi A (1984) Physical and genetic analysis of a symbiotic region of Rhizobium meliloti: identification of nodulation genes. Mol Gen Genet 193: 445-452
Legocki RP, Yun AC and Szalay AA (1984) Expression of β-galactosidase controlled by a nitrogenase promoter in stem nodules of Aeschynomene scabra. Proc Natl Acad Sci U.S.A. 81: 5806-5810
Long SR, Buikema WJ and Ausubel RM (1982) Cloning of Rhizobium meliloti nodulation genes by direct complementation of nod⁻ mutants. Nature 298: 485-488
Marsh JL, Erfle M and Wijkes EJ (1984) The pIC plasmid and phage vectors with versatile cloning sites for recombinant selection by insertional inactivation. Gene 32: 481-485
Mulligan JT and Long SR (1985) Induction of Rhizobium meliloti nodC expression by plant exudates requires nodD. Proc Natl Acad Sci U.S.A. 82: 6609-6613
Okker RJH, Spaink H, Wijffelman CA, van Brussel AAN and Lugtenberg B (1986) Regulation of bacterial genes involved in bacterium-plant interactions by plant signal molecules. In: Bailey JA (ed) Biology and molecular biology of plant pathogen interactions. Springer Verlag, in press
Rosenberg C, Boistard P, Dénarié J and Casse-Delhart F (1981) Genes controling early and late functions in symbiosis are located on a mega-plasmid in Rhizobium meliloti. Molec Gen Genet 184: 326-333
Rossen L, Johnston AWB and Downie JA (1984) DNA sequence of the Rhizobium leguminosarum nodulation genes nodAB and C required for root hair curling. Nucl Acid Res 12: 9497-9508
Rossen L, Shearman CA, Johnston AWB and Downie JA (1985) The nodD gene of Rhizobium leguminosarum is autoregulatory and in the presence of plant exudates induces the nodA,B,C genes. EMBO J 4: 3369-3373
Rostas K, Kondorosi E, Horvath B, Simoncsits A and Kondorosi A (1986) Conservation of extended promoter regions of nodulation genes in Rhizobium. Proc Natl Acad Sci U.S.A. 83: 1757-1761

Sanger F, Nicklen S and Coulson AR (1977) DNA sequencing with chain terminating inhibitors. Proc Natl Acad Sci U.S.A. 74: 5463-5467

Schmidhauser J and Helinsky DR (1985) Regions of broad host--range plasmid RK2 involved in replication and stable maintenance in nine species of gram-negative bacteria. J Bact 164: 446-455

Schofield PR, Djordjevic MA, Rolfe BG, Shine J and Watson JM (1983) A molecular linkage map of nitrogenase and nodulation genes in Rhizobium trifolii. Mol Gen Genet 192: 459-465

Schofield PR, Ridge RW, Rolfe BG, Shine J and Watson JM (1984) Host-specific nodulation is encoded on a 14 kb fragment in Rhizobium trifolii. Plant Mol Biol 3: 3-11

Schofield PR and Watson JM (1986) DNA sequence of Rhizobium trifolii nodulation genes reveals a reiterated and potentially regulatory sequence. Nucl Acids Res 14: 2891-2903

Scott KF (1986) Conserved nodulation genes from the non-legume symbiont Bradyrhizobium sp. (Parasponia). Nucl Acids Res 14: 2905-2919

Shearman CA, Rossen L, Johnston AWB and Downie JA (1986) The Rhizobium leguminosarum nodulation gene nodF encodes a polypeptide similar to acyl-carrier protein and is regulated by nodD plus a factor in pea root exudate. EMBO J 5: 647-652

Török I, Kondorosi E, Stepkowski T, Pósfai J and Kondorosi A (1984) Nucleotide sequence of Rhizobium meliloti nodulation genes. Nucl Acids Res 12: 9509-9523

Van Brussel AAN, Zaat SAJ, Canter-Cremers HCJ, Wijffelman CA, Pees E, Tak T and Lugtenberg BJJ (1986) Role of plant root exudates and Sym plasmid localized nodulation genes in the synthesis by Rhizobium leguminosarum of Tsr factor, which causes thick and short roots on common vetch. J Bact 165: 517-522

Wijffelman CA, Pees E, van Brussel AAN, Okker RJH and Lugtenberg BJJ (1985) Genetic and functional analysis of the nodulation region of the Rhizobium leguminosarum Sym plasmid pRL1JI. Arch Microbiol 143: 225-232

Winarno R and Lie TA (1979) Competition between Rhizobium strains in nodule formation: interactions between nodulating and non-nodulating strains. Plant and Soil 51: 135-142

MANIPULATION OF NODULATION SPECIFICITY IN THE PEA-*RHIZOBIUM LEGUMINOSARUM* SYMBIOSIS

S.L.A. Hobbs[1], C.M.O. DeLong[1], S. Denes[2] and V.N. Iyer[2]

[1]Plant Biotechnology Institute, National Research Council of Canada, Saskatoon, Sask., Canada

[2]Dept. Biology, Carleton University, Ottawa, Ontario, Canada

NRCC No. 25114

INTRODUCTION

The improvement of nitrogen fixation through the genetic engineering or selection of the bacterial component of the legume-*Rhizobium* symbiosis is possible (Hardy 1985; Stacey and Upchurch 1984). However, the use of such improved bacteria in the field may not result in improved fixation if competition from indigenous, unimproved bacteria reduces the infection of the host by the inoculum. Studies on field-grown soybeans, for example, indicated that the inoculum accounted for only 0-20% of the nodules formed and that the most competitive inoculum strains in one soil may be the least effective in another (Ham et al 1976).

Devine and Weber (1977) proposed that a genetic system could be devised using plant-microbe interactions in soybean to increase the percentage of nodules formed by an inoculum. The system required: i) plant varieties that would be resistant to infection by indigenous soil rhizobia; ii) strains of rhizobia that could overcome this resistance and infect these varieties; and iii) the ability to manipulate the genome of these bacteria so that such nodulation specificity would be associated with high fixation. In soybean, although genotypes homozygous for the allele *rj1* are generally non-nodulating, there has apparently been little success in finding a bacterial strain that will nodulate these to an economic level (Devine and Kuykendall 1982). Therefore, the potential is limited for using this system to

NATO ASI Series, Vol. H 4
Recognition in Microbe-Plant Symbiotic and Pathogenic Interactions. Edited by B. Lugtenberg

produce specific nodulation in the field. Such a system can, however, also be envisioned in any legume and this paper looks at the use of naturally occurring nodulation specificity in pea.

AFGHANISTAN-TOM INTERACTION

Pea cv. Afghanistan will not nodulate with most strains of Rhizobium leguminosarum (Lie 1971) apparently due to the presence of homozygous recessive alleles designated sym2 (Holl 1975). However, one strain, designated TOM, isolated from Turkish soil, will nodulate this genotype (Lie 1978). This ability to nodulate Afghanistan has been shown to be conferred by a 2.0 kb region on plasmid pRL5JI (Brewin et al 1980; Götz et al 1985). Hence the potential exists in pea to use this nodulation specificity system by manipulating the genome of the host plant and that of the Rhizobium. Afghanistan is an unimproved genotype by breeders' standards while TOM may not be the best strain for nitrogen fixation. However, the gene associated with non-nodulation could be moved into a variety that has good agronomic characteristics, and the pRL5JI plasmid (or part thereof) could be moved into superior Rhizobium strains. In this manner a pea variety might be produced that is nodulated only by improved inocula and not by indigenous bacteria in the field.

To determine whether such a scheme is feasible, several areas of investigation are presented here:

1. The infection behaviour of TOM in the presence of other strains and in soil;
2. The extent of nodulation of Afghanistan in normal soils;
3. The transfer of non-nodulation to a North American variety;
4. The transfer of pRL5JI (or part thereof) to other strains;
5. The retention of the nodulation specificity by the new strain-genotype combinations in the presence of other Rhizobium strains.

NODULATION OF AFGHANISTAN IN SOIL

It has been reported that Afghanistan fails to nodulate in certain Dutch soils even when inoculated with TOM and that this failure may be due to competition from certain rhizobial strains, even though those strains cannot themselves nodulate Afghanistan (Winaro and Lie 1979). To discover whether this phenomenon is widespread in North American soil, Afghanistan was grown in soil at Saskatoon, Saskatchewan and inoculated with TOM. Nodulation occurred readily and nodules effectively reduced acetylene, averaging 23.4 µmol g^{-1} h^{-1} in 1984 and 30.9 µmol g^{-1} h^{-1} in 1985 (where g = root+nodule dry weight). Jensen et al (1986) reported that Afghanistan nodulates readily with strains present in some Danish soils. However, in our tests, Afghanistan that had not been inoculated with TOM or that had been inoculated with 'Nitragin' (a commercial inoculum which is a mixture of many strains) showed no significant nodulation. Attempts to nodulate Afghanistan indoors with North American strains isolated from the soil also failed.

Soil was sampled from six locations across the pea growing belt in Saskatchewan. Plants indoors were inoculated with TOM, and/or with 5ml of soil extracts obtained by shaking 1g of soil in 10ml of water. The results are summarized in Table 1. Trapper (a North American variety) was effectively nodulated by all soil extracts indicating the presence of R. leguminosarum and acetylene reduction was not significantly increased by adding the TOM inoculum. Afghanistan was not effectively nodulated by soil extracts from any of the locations whereas when TOM was added acetylene reduction rates were comparable to those in Trapper.

Acetylene reduction in Afghanistan inoculated with mixtures of non-infective and infective strains was not significantly different from that obtained when TOM was added alone (Hobbs 1985) and plant growth was the same on the TOM or TOM + 128C52 (a normally non-nodulating strain) inocula. Hence, little or no competition was indicated by the North American non-infective strains tested.

It is apparent that Afghanistan is able to nodulate with TOM in the presence of North American strains. However, another potential problem for the use of the Afghanistan-TOM symbiosis as a delivery system for inoculum was noted in the rj1 system in soybeans. The presence of the bacteria capable of infecting non-

TABLE 1. EFFECT OF TOM AND SOIL EXTRACT INOCULA ON ACETYLENE REDUCTION OF TRAPPER AND AFGHANISTAN.

INOCULUM	TRAPPER	AFGHANISTAN
	----- $\mu mol\ g^{-1}\ h^{-1}$ -----	
TOM	30.2	29.0
TOM + SOIL EXTRACT	29.6	33.1
SOIL EXTRACT	24.2	0

nodulating soybeans also allowed infection and nodulation by those bacteria that were normally non-infective (Devine et al 1980), the latter producing 32% of the nodules under these conditions. Obviously such a specific inoculum in the field would not preclude infection by indigenous bacteria. To determine whether this effect existed in the Afghanistan-TOM symbiosis, Afghanistan plants were grown in the presence of TOM and 128C52. Isolates were made from a random sample of the nodules formed. None resembled 128C52 according to intrinsic antibiotic resistance and all were capable of re-infecting Afghanistan, indicating that the presence of TOM does not allow infection by normally non-nodulating strains.

TRANSFER OF NON-NODULATION TO TRAPPER

By backcrossing and single-seed descent, the non-nodulation character can be tranferred to Trapper. One selection of non-nodulating Trapper (Trapper-nn), tested in the absence of nitrate, formed no nodules with 128C52 and plant growth was reduced, as was the case with Afghanistan. With TOM, Trapper-nn formed effective nodules and plant growth and habit were similar to Trapper. While there is some indication that the <u>sym2</u> gene is not the only genetic difference between Trapper and Afghanistan that can affect nodulation (Hobbs, unpublished data), these results indicate that the backcrossing of the non-nodulation character into improved varieties is possible.

TRANSFER OF pRL5JI TO 128C52

Several constructs that contained all or part of the pRL5JI plasmid were supplied by the John Innes Institute. Relevant plasmids are listed in Table 2. Selection for spontaneous antibiotic resistant mutants of 128C52 produced 128C52 Rif-r, which had an acetylene reduction rate similar to that of the original strain. The four plasmids containing all or part of pRL5JI along with an antibiotic marker were introduced into this strain through conjugative bacterial mating.

Although Afghanistan was nodulated by 128C52 Rif-r derivatives containing pIJ1016, pIJ1095, pIJ1111, or pIJ1251, only the 128C52 Rif-r+pIJ1016 (D5) produced nodules in numbers similar to those found when TOM was used as inoculum. Three isolates were examined further.

PERFORMANCE OF TRAPPER-NN AND DERIVED STRAINS

Isolates D5-C1, D5-C2 and D5-C4 were examined in nitrate-free conditions in Turface and sand mixture that had had a back-

TABLE 2. BACTERIAL PLASMIDS.

PLASMID	RELEVANT CHARACTERISTICS	REFERENCE
pLAFR1	Cloning vector. Tet-r	Friedman et al (1982)
pRL5JI	Confers ability to nodulate Afghanistan	Brewin et al (1980)
pIJ1016	pRL5JI Kan-r	Brewin et al (1983)
pIJ1111	pLAFR1 + less than 24 kb DNA from pRL5JI	Hombrecher et al (1984)
pIJ1095	pLAFR1 + 30 kb DNA from pRL5JI	Hombrecher et al (1984)
pIJ1251	pIJ1095 Kan-r	

ground population of 128C52 added. Trapper, Trapper-nn, and Afghanistan were planted in this growing medium and inoculated with 128C52, TOM, D5-C1, D5-C2, or D5-C4. Acetylene reduction was monitored for four weeks, starting three weeks after planting, using a modification of the direct soil injection technique described by Mahon and Salminen (1981). This method had been found to correlate well with acetylene reduction rates in previous experiments. Data were transformed where Bartlett's test indicated that variances were not homogeneous and analysed by analysis of variance and Duncan's multiple range test.

The results are given in Table 3. Trapper formed effective nodules with 128C52 and acetylene reduction with 128C52 inoculum was the same as that of the control. Trapper-nn and Afghanistan did not form effective nodules with the 128C52 inoculum nor with the background population. TOM nodulated with all three genotypes and acetylene reduction was very similar in Trapper and Trapper-nn. The derived strains nodulated and reduced acetylene in conjunction with Trapper-nn and Afghanistan, although acetylene reduction was lower than with TOM. Acetylene reduction in Trapper-nn inoculated with the D5 strains was lower than that in Trapper + 128C52. There are several possible reasons for this. First, pIJ1016 has genes coding for fixation as well as nodulation and it may be that these fixation genes, rather than those of 128C52, were active in the D5 isolates. Thus, if these did not act as efficiently as the 128C52 genes, reduced acetylene reduction could have resulted. Second, not all the host genes associated with nodulation and fixation may have been passed to Trapper-nn, as Trapper and Afghanistan apparently differ in more than one gene affecting these characters. Third, genes for nodulation specificity on pIJ1016 may not act as efficiently as those controlling normal nodulation in 128C52 or there may be an interaction between the two sym plasmids.

In Trapper, although only D5-C2 was significantly lower than control, acetylene reduction was generally lower with inocula (except for 128C52) (Table 3). This may indicate that the inoculation technique (liquid inoculum placed directly onto the seed) may have been a factor leading to decreased acetylene reduction,

TABLE 3. ACETYLENE REDUCTION OF GENOTYPE-INOCULUM COMBINATIONS IN NITRATE-FREE CONDITIONS WITH BACKGROUND RHIZOBIAL POPULATION.

	TRAPPER	TRAPPER-NN	AFGHANISTAN
	----- ethylene/acetylene x 1000 ----		
CONTROL	9.1 ab[1]	0.1 c	0.1 c
128C52	11.8 a	0.3 c	0.1 c
TOM	5.9 bc	5.5 a	2.7 a
D5-C1	5.1 bc	2.1 b	0.3 c
D5-C2	2.9 c	2.9 b	0.9 b
D5-C4	7.5 ab	2.6 b	0.7 b

[1] Means in the same column not followed by the same letter are significantly different at the 5% probability level according to Duncan's multiple range test

although so too might competition between the inoculated and background strains for nodulation of Trapper.

Nodules isolates were made from some treatments and screened for antibiotic resistance. Of the nodules from the D5-C1 treatment, 100% from Trapper-nn or Afghanistan showed the resistance known to be in that strain whereas almost 100% from Trapper did not. This indicates that the D5-C1 inoculum produced most nodules with Trapper-nn or Afghanistan whereas the background population formed most of the nodules on Trapper.

There was no significant inoculum effect on shoot weight at maturity in Trapper whereas for Trapper-nn or Afghanistan there were significant inoculum differences. Trapper acetylene reduction with different inocula was not correlated to final shoot weight nor to inocula effects in the other genotypes (Table 4). However, the non-nodulating types not only had a high correlation between genotypes for acetylene reduction with the different inocula, but also between acetylene reduction and shoot dry weight within a genotype (Table 4). In addition, the dry weights were correlated between genotypes. This indicates that nodulat-

TABLE 4. CORRELATIONS OF ACETYLENE REDUCTION AND SHOOT WEIGHT IN GENOTYPES GROWN UNDER NITRATE-FREE CONDITIONS WITH DIFFERENT INOCULA (n=6) WITH A BACKGROUND RHIZOBIAL POPULATION.

	2	3	4	5	6
TRAPPER ACET. REDN. (1)	-0.633	-0.415	0.400	-0.687	-0.608
TRAPPER-NN ACET. REDN. (2)		0.939**	-0.547	0.902*	0.877*
AFGHANISTAN ACET. REDN. (3)			-0.347	0.812*	0.840*
TRAPPER DRY WEIGHT (4)				-0.622	-0.680
TRAPPER-NN DRY WEIGHT (5)					0.875*
AFGHANISTAN DRY WEIGHT (6)					1.000

*,** Significant at 5% and 1% probability levels, respectively capabilities, even in in the presence of a background population of Rhizobia.

ion specificity could be used to obtain predictable results under low nitrate conditions from strains varying in their nitrogen fixing capabilities, even in the presence of a background population of rhizobia.

To examine the ability of derived strains to nodulate in the presence of a wider background population of rhizobial strains, Trapper and Afghanistan were grown in soil in which peas had been previously grown and were inoculated with 128C52, TOM, or the derived strains. Uninoculated controls of Trapper nodulated prolifically with the indigenous bacteria, Afghanistan did not nodulate. Nodule isolates from the plants inoculated with derived strains were screened for antibiotic resistance. On Afghanistan, 91-100% of the nodules showed the resistance known to be in the derived strains, on Trapper only 32-46% did. This shows that nodule composition can be controlled even in the presence of a natural indigenous population of soil bacteria.

THE FUTURE

While this approach of using plant and bacterial genes to develop nodulation specificity is clearly feasible, no plant-bacterium combinations have yet been produced that give higher fixation potential than is normally found. However, such combinations may be developed by: further genetic engineering or selection of rhizobial strains; investigating and manipulating plant nodulation genes at the plant breeding or molecular level; or developing inoculation methods that allow rapid establishment of inocula. The use of genetic engineering in bacteria for improved fixation, combined with nodulation specificity to ensure the establishment of that strain in the nodules of the host plant, may be a short-term application for biotechnology. However, at the time of writing, there remain many questions surrounding the desirability of the release of any genetically engineered microorganisms into the environment. Satisfactory answers must be quickly found to these questions so that potentially useful science does not become socially unacceptable.

REFERENCES

Brewin NJ, Beringer JE, Johnston AWB (1980) Plasmid-mediated transfer of host-range specificity between two strains of *Rhizobium leguminosarum*. J Gen Microbiol 120:413-420

Brewin NJ, Wood EA, Young JPW (1983) Contribution of the symbiotic plasmid to the competitiveness of *Rhizobium leguminosarum*. J Gen Microbiol 129:2973-2977

Devine TE, Kuykendall LD (1982) Genetic control of specificity of nodulation. In: Clark KW, Stephens JHC (eds) Proc 8th North American Rhizobium conference, p 7

Devine TE, Kuykendall LD, Breithaupt BH (1980) Nodulation of soybeans carrying the nodulation-restrictive gene, *rj1*, by an incompatible *Rhizobium japonicum* strain upon mixed inoculation with a compatible strain. Can J Microbiol 26:179-182

Devine TE, Weber DF (1977) Genetic specificity of nodulation. Euphytica 26:527-535

Friedman AM, Long SR, Brown SE, Buikema WJ, Ausubel FM (1982) Construction of a broad host range cosmid cloning vector and its use in the genetic analysis of Rhizobium mutants. Gene 18:289-296

Götz R, Evans IJ, Downie JA, Johnston AWB (1985) Identification of the host-range DNA which allows Rhizobium leguminosarum strain TOM to nodulate cv. Afghanistan peas. Mol Gen Genet 201:296-300

Ham GE, Lawn RJ, Brun WA (1976) Influence of inoculation, nitrogen fertilizers and photosynthetic source-sink manipulations on field-grown soybeans. In: Nutman PS (ed) Symbiotic nitrogen fixation in plants. Cambridge University Press, Cambridge, p 239

Hardy RWF (1985) Applications of nitrogen fixation in agriculture and forestry. In: Evans HJ, Bottomley PJ, Newton WE (eds) Nitrogen fixation research progress. Martinus Nijhoff, Dordrecht, p 683

Hobbs SLA (1985) Nodulation specificity in Pisum sativum L. In: Evans HJ, Bottomley PJ, Newton WE (eds) Nitrogen fixation research progress. Martinus Nijhoff, Dordrecht p 33

Holl FB (1975) Host plant control of the inheritance of dinitrogen fixation in the Pisum-Rhizobium symbiosis. Euphytica 24:767-770

Hombrecher G, Götz R, Dibb NJ, Downie JA, Johnston AWB, Brewin NJ (1984) Cloning and mutagenesis of nodulation genes from Rhizobium leguminosarum TOM, a strain with extended host range. Mol Gen Genet 194:293-298

Jensen ES, Sorensen LH, Engvild, KC (to be published) Danish Rhizobium leguminosarum strains nodulating 'Afghanistan' pea (Pisum sativum). Physiol Plant

Lie TA (1971) Symbiotic nitrogen fixation under stress conditions. In: Lie TA, Mulder EG (eds) Biological nitrogen fixation in natural and agricultural environments. Plant and Soil special volume, p 127

Lie TA (1978) Symbiotic specialisation in pea plants: The requirement of specific Rhizobium strains for peas from Afghanistan. Ann Appl Biol 88:462-465

Mahon JD, Salminen SO (1980) Non-destructive semi-quantitative determination of acetylene reduction in the field. Plant and Soil 56:335-340

Stacey G, Upchurch RG (1984) Rhizobium inoculation of legumes. Trends in biotechnology 2:65-70

Winaro R, Lie TA (1979) Competition between Rhizobium strains in nodule formation: interaction between nodulating and non-nodulating strains. Plant and Soil 51:135-142

BRADYRHIZOBIUM JAPONICUM GENES INVOLVED IN SOYBEAN ROOT-NODULE DEVELOPMENT

Joseph W. Lamb, Brigitte Regensburger, Hans-Martin Fischer, Michael Göttfert, Linda Meyer, Sabine Ebeling, Daniel Studer, Matthias Hahn and Hauke Hennecke

Mikrobiologisches Institut
Eidgenössische Technische Hochschule, ETH-Zentrum
Universitätstrasse 2
CH-8092 Zürich, Switzerland

INTRODUCTION

Bradyrhizobium japonicum is a member of the slow-growing genus of the rhizobia. It is able to form determinate, N_2 fixing root nodules on soybean (Glycine max L. Merr.). Research in our laboratory is concerned with the structural and functional analysis of bacterial genes involved in nodulation and symbiotic nitrogen fixation. We have chosen B.japonicum strain 110 for this study, as this strain is able to derepress nitrogenase activity under free-living, microaerobic growth conditions (ex planta). In this report we describe some promising mutants of B.japonicum which are defective in different steps during bacteroid and root nodule development. Furthermore, we present a progress report on the isolation and characterization of B.japonicum nod genes.

B.JAPONICUM MUTANTS DEFECTIVE IN BACTEROID AND ROOT-NODULE DEVELOPMENT

Two clusters of symbiotic nitrogen fixation genes have been identified in B.japonicum strain 110 (Hennecke et al 1985). Cluster I harbors nine genes within a region of 27 kb. They are arranged in the following order: nifD, nifK, nifE, nifN, nifS, nifB, nifH, fixB, fixC. Cluster II carries two adjacent genes, nifA and fixA, which are located 10 kb downstream of the "common" nodulation genes, nodD, nodA, nodB and nodC (see below). Mutations were constructed in all the nif and fix genes, and the phenotypes of the corresponding mutants were characterized extensively (Hahn et al 1984; Gubler and Hennecke 1986; Fischer et al 1986; S. Ebeling et al manuscript in preparation). The interesting results were that mutants with de-

NATO ASI Series, Vol. H 4
Recognition in Microbe-Plant Symbiotic and Pathogenic Interactions. Edited by B. Lugtenberg

fects in nifD,K,E,N,S,B,H and fixB,C and A behaved similarly: (i) all were Nif^- in free-living, microaerobic culture and Fix^- in symbiosis; (ii) all led to the development of determinate soybean nodules in which the bacteroids proliferated and differentiated almost as normally as the wild-type, with the only exception being that the mutant bacteroids accumulated a higher amount of poly-β-hydroxybutyric acids as compared to the wild-type bacteroids (Table 1). Also, a deletion mutant lacking all the cluster I genes had a similar phenotype (Hahn et al 1984). This shows that none of these nif and fix genes is involved in essential steps during bacteroid development. Hence, the host-plant does not suppress the development of these nitrogen non-fixing mutant symbionts, even though this situation is not beneficial for the host plant. Other genes in the bacterium must, therefore, be responsible for programming determinate root nodule formation.

One of these genes could be nifA. NifA is a nitrogen fixation regulatory gene which positively controls the expression of nifDK and nifH in B. japonicum (Fischer et al 1986). In contrast to the Nif^- and Fix^- mutants described before, $NifA^-$ mutants do not lead to the formation of fully developed soybean nodules (Table 1). The nodulation defective phenotype of $NifA^-$ mutants could be seen at least at two levels.
(i) The frequency of nodulation was increased. Under our nodulation assay conditions the wild-type elicited the formation of about 30 nodules per soybean plant located mainly along the tap root, whereas $NifA^-$ mutants produced twice as many nodules plus countless bump-like structures both of which were widely distributed all over the root system. Thus, the corresponding mutants seem to circumvent a plant-internal regulatory phenomenon that has been termed "autoregulation", in which effective nodulation events inhibit further nodulation on other parts of the roots (Pierce and Bauer 1983). Autoregulation has been shown to be prevalent even before the onset of symbiotic nitrogen fixation (Pierce and Bauer 1983; Kosslak and Bohlool 1984). Thus, a nifA-controlled early bacterial function may exist which signals other root tissues to suppress further nodule development. It should be pointed out that this particular NifA phenotype cannot be an indirect consequence of nitrogen starvation in the soybean plant (due to lack of symbiotic N_2 fixation), since mutations in other B.japonicum nif and fix genes (see above) do not lead to such a phenotype. Recently, Carroll et al (1985)

Table 1. Nodulation-defective phenotypes of selected B.japonicum insertion mutants.

B.japonicum mutant	Phenotype (N_2 fixation)	Nodules	Bacteroids	Other comments
wild-type	Nif^+ Fix^+	normal	normal	
nifD,K,E,N,S,B,H, and fixB,C,A	Nif^- Fix^-	~ normal	~normal, more PHB	
nifA	Nif^- Fix^-	small, necrotic	heavily degraded	pleiotropic regulatory mutant
3652	Nif^+ Fix^-	large, white	few, degraded	
2960	Nif^- Fix^-	large, green	few, many vesicles	
259 3160	Nif^- Fix^-	bumps, white	none	pleiotropic regulatory mutants

provided evidence for a soybean (Glycine max) gene that could also be involved in autoregulation of nodule formation.

(ii) Mutations affecting nifA also lead to dramatic disturbances in the development of infected nodule plant cells. From light and electron microscopic investigations it appears as if $NifA^-$ mutants, once released from the infection thread, are treated like plant pathogens. We could observe symptoms of nodule tissue degradation and the typical signs of severe, premature bacteroid degradation, such as disintegration of peribacteroid membranes, formation of larger vesicles, and presence of lysed bacteroid remnants (Figure 1). Thus, nifA may also control functions that help the bacterial symbiont to overcome plant defense reactions and to persist as nitrogen fixing bacteroids. Again, it is important to note that nitrogen non-fixing soybean nodules infected by other B.japonicum nif or fix mutants do not show such peculiar phenotypes.

Other genes involved in soybean root-nodule and bacteroid development were identified as a result of a random Tn5 mutagenesis of the B.japonicum genome (Regensburger et al 1986). In the following we discuss three different mutant classes (Table 1).

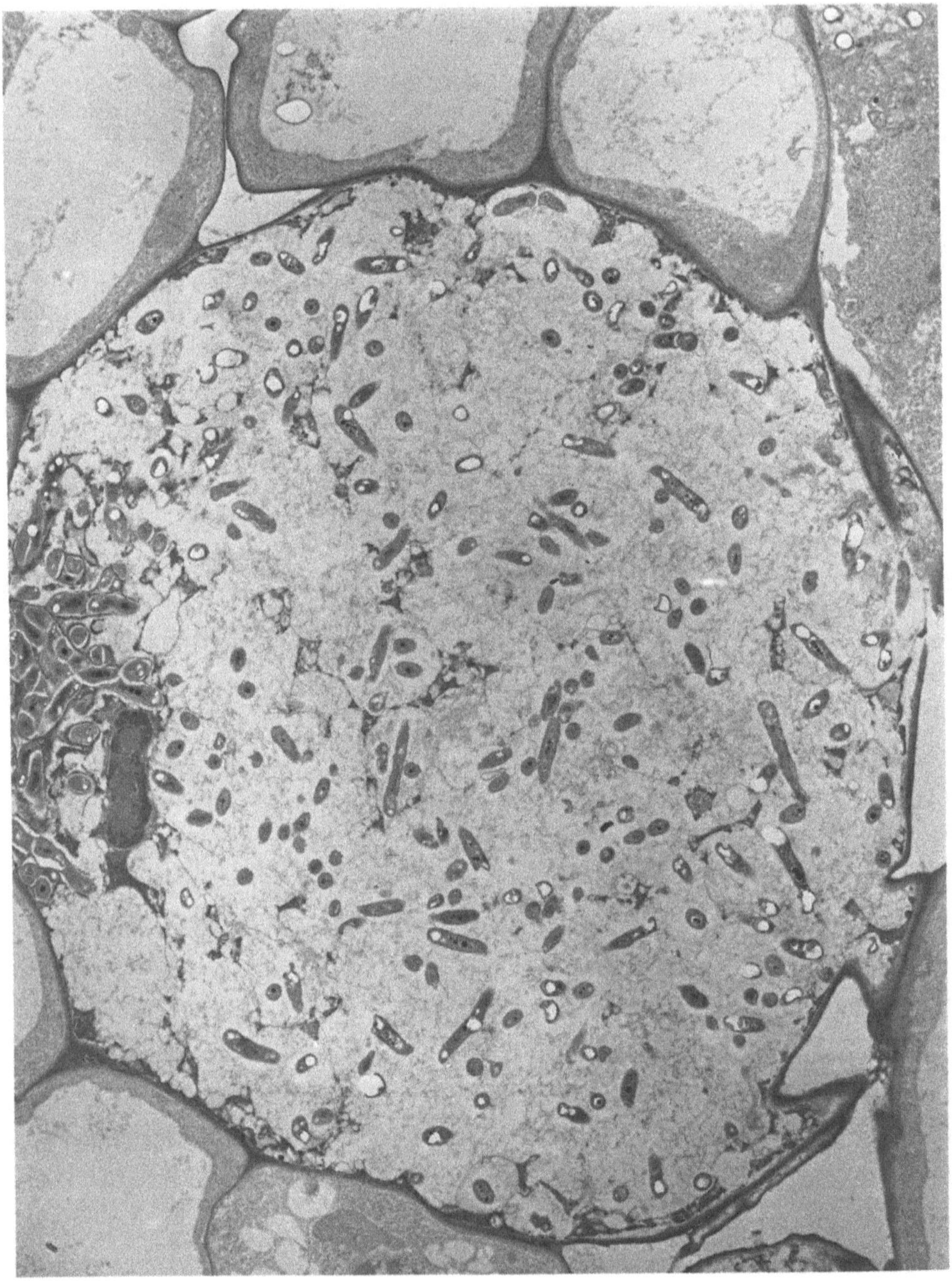

Fig. 1. Degenerating soybean root-nodule cell infected by a <u>nifA</u> mutant (deletion mutant A9; Fischer et al 1986). This is at day 12 after infection, at which time wild-type-infected nodule cells would be entirely filled with completely differentiated bacteroids, surrounded by peribacteroid membranes (not shown).

Mutant 3652. This mutant is of interest because it has wild-type acetylene reduction activity in free-living culture but only marginal activity in symbiosis (phenotype $Nif^{+}Fix^{-}$). Electron microscopic observation showed that the mutant is deficient in bacteroid development and/or persistence. Hence, the affected gene is not a direct determinant for nitrogen fixation. Rather, it is involved in some step during bacteroid establishment in symbiosis. It will be of interest to see whether or not this gene is controlled by nifA.

Mutant 2960. This mutant is $Nif^{-}Fix^{-}$. It also forms few bacteroids (which do not appear to become degraded), yet the infected plant cells keep on producing empty vesicles, i.e. membrane systems that would normally surround the bacteroids. This shows that probably very early after release of bacteria from the infection thread the soybean host plant is programmed to synthesize "peribacteroid" membranes independent of concomitant bacteroid proliferation.

Mutants 259 and 3160. These mutants (phenotype $Nif^{-}Fix^{-}$) elicit the most aberrant soybean root-nodules. The nodules are bump-like structures without clear differentiation into nodule cortex cells and infected core cells. The majority (if not all) of the nodule cells are completely uninfected. The protein patterns of the mutants were inspected by two-dimensional gel electrophoresis after labelling free-living, microaerobic cultures with L-(^{35}S) methionine. The mutants lacked a group of proteins the synthesis of which was known to be under oxygen control. Such strains can be regarded as potential regulatory mutants, in which the Tn5 insertions are located in regulatory genes other than nifA.

In conclusion, the small selection of mutants presented here is highly promising with regard to the identification and characterization of new, essential functions in bacteroid and root-nodule development.

ISOLATION AND CHARACTERIZATION OF B.JAPONICUM NOD GENES

Using cloned nodulation genes of R.phaseoli as hybridisation probes we have isolated DNA regions from the genome of Bradyrhizobium japonicum which share homology with nodulation genes (Lamb and Hennecke 1986). One region has been analysed in detail. This region was located 10 kb distant from the previously isolated nifA and fixA genes (Figure 2). This cluster of symbiotic genes has so far not been linked to cluster I containing the nitrogenase structural genes or to any other genes essential for nitrogen fixation.

The region containing nod-homologous sequences has areas homologous to the R.leguminosarum genes nodD, nodA, nodB, nodC, nodI, and nodJ. These sequences are arranged in the order shown in Figure 2. Insertions have been made in the nodA and nodD homologous sequences. The strains carrying these insertions were nodulation deficient. The nodA and nodD genes are approximately 900 bp apart, and are transcribed from opposite DNA strands.

Using a synthesized 22-mer oligonucleotide probe, based on the consensus sequence of "nod-boxes" found in fast-growing Rhizobium species (Shearman et al 1986; Rostas et al 1986; Schofield and Watson 1986), we detected 2 strongly and 4 weakly homologous DNA fragments in B.japonicum genomic DNA. One of the strongly hybridizing sequences is located between the nodD and nodA genes and has the following sequence: (see next page)

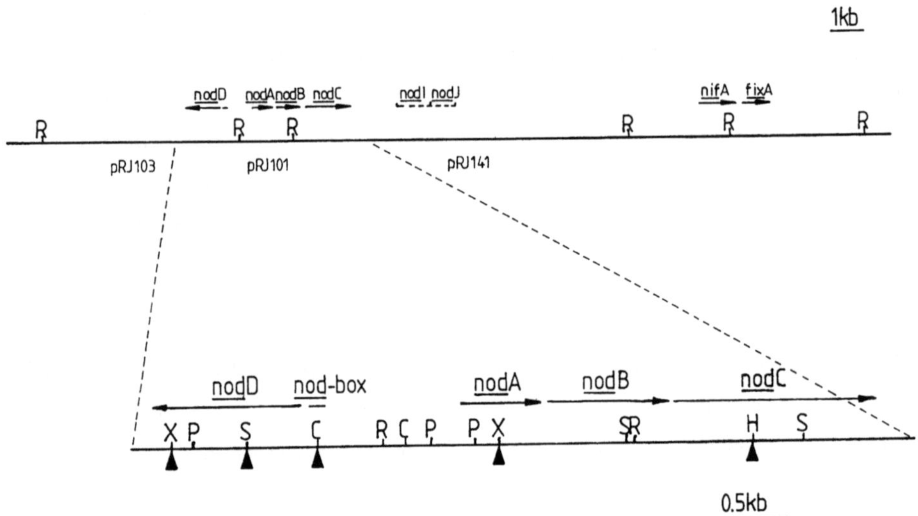

Fig. 2. The B.japonicum (strain 110) nodulation region, showing linkage between nodulation genes and the nifA, fixA genes (Lamb and Hennecke, 1986). The dark triangles denote sites of insertions of Tn5 restriction fragments specifying kanamycin resistance. See text for further details. Restriction sites are: C = ClaI, H = HindIII, P = PstI, R = EcoRI, S = SalI, X = XhoI.

5'-CATGTGATGAGTCTATCCATCGTGTGGATGTGTTCTATCGAAACAATCGATTTTACCAAACTGGGGG-3'
⟵GTA

nodD start

The underlined sequences share homology to the nod-box sequences of the fast-growing rhizobia. The gap between nodD and nodA (∼ 900 bp), and the position of the nod-box suggests that a further gene is located upstream of nodA. If this is the case, then the arrangement in B.japonicum will be similar to that of Bradyrhizobium sp. parasponia (Scott 1986).

Of the remaining "nod-box" sequences two have been located near a copy of a repeated sequence RSα12 (Kaluza et al 1985). This region is more than 80 kb away from the nif cluster harboring at least nine nif/fix genes (cluster I). A deletion covering this region was found to produce a delayed Nod phenotype.

Of the nodulation genes known from fast-growing rhizobia we have not yet been able to identify the genes nodF,E,G, and H. As some of these genes may be concerned with host-specific functions we may not find the obvious analogous genes that we have found for the "common" nodulation genes (nodD,A,B,C,I.J). Future work will be directed towards an analysis of potential nod genes located in the vicinity of the newly identified nod-box regions. It will be of interest to see whether or not any of these nod-box-associated regions carry determinants for host-specific nodulation on soybeans.

ACKNOWLEDGEMENT

We thank our colleague B. Thöny for making available a DNA sequence that led to the identification of nifN. We are also grateful to A. Downie, A. Johnston, L. Rossen and I. Evans for supplying us with specific R.leguminosarum nod probes. H. Paul is thanked for typing the manuscript. This work was supported by grants from the ETH Zürich and the Agrigenetics Research Corporation.

REFERENCES

Carroll BJ, McNeil DL, Gresshoff PM (1985) Isolation and properties of soybean (Glycine max) mutants that nodulate in the presence of high nitrate concentrations. Proc Natl Acad Sci USA 82:4162-4166

Fischer H-M, Alvarez-Morales A, Hennecke H (1986) The pleiotropic nature of symbiotic regulatory mutants: Bradyrhizobium japonicum nifA gene is involved in control of nif gene expression and formation of determinate symbiosis. EMBO J 5: in press

Gubler M, Hennecke H (1986) FixA,B and C genes are essential for symbiotic and free-living, microaerobic nitrogen fixation. FEBS Lett 200:186-192

Hahn M, Meyer L, Studer D, Regensburger B, Hennecke H (1984) Insertion and deletion mutations within the nif region of Rhizobium japonicum. Plant Mol Biol 3:159-168

Hennecke H, Alvarez-Morales A, Betancourt-Alvarez M, Ebeling S, Filser M, Fischer H-M, Gubler M, Hahn M, Kaluza K, Lamb JW, Meyer L, Regensburger B, Studer D, Weber J (1985) Organization and regulation of symbiotic nitrogen fixation genes from Bradyrhizobium japonicum. In: Evans HJ, Bottomley PJ, Newton WE (eds) Nitrogen fixation research progress. Martinus Nijhoff Publishers, Dordrecht, p 157-163

Kaluza K, Hahn M, Hennecke H (1985) Repeated sequences similar to insertion elements clustered around the nif region of the Rhizobium japonicum genome. J Bacteriol 162:535-542

Kosslak RM, Bohlool BB (1984) Suppression of nodule development on one side of a split root system of soybeans caused by prior inoculation of the other side. Plant Physiol 75:125-130

Lamb JW, Hennecke H (1986) In Bradyrhizobium japonicum the common nodulation genes, nodABC, are linked to nifA and fixA. Mol Gen Genet 202:512-517

Pierce M, Bauer WD (1983) A rapid regulatory response governing nodulation in soybean. Plant Physiol 73:286-290

Regensburger B, Meyer L, Filser M, Weber J, Studer D, Lamb JW, Fischer H-M, Hahn M, Hennecke H (1986) Bradyrhizobium japonicum mutants defective in root-nodule bacteroid development and nitrogen fixation. Arch Microbiol 144:355-366

Rostas K, Kondorosi E, Horvath B, Simoncsits A, Kondorosi A (1986) Conservation of extended promoter regions of nodulation genes in Rhizobium. Proc Natl Acad Sci USA 83:1757-1761

Schofield PR, Watson JM (1986) DNA sequence of Rhizobium trifolii nodulation genes reveals a reiterated and potentially regulatory sequence preceding nodABC and nodFE. Nucl Acids Res 14:2891-2903

Scott KF (1986) Conserved nodulation genes from the non-legume symbiont Bradyrhizobium sp. (Parasponia). Nucl Acids Res 14:2905-2919

Shearman CA, Rossen L, Johnston AWB, Downie JA (1986) The Rhizobium leguminosarum nodulation gene nodF encodes a polypeptide similar to acyl-carrier protein and is regulated by nodD plus a factor in pea root exudate. EMBO J 5:647-652

NODULATION OF SOYBEAN: *BRADYRHIZOBIUM JAPONICUM* PHYSIOLOGY AND GENETICS

Gary Stacey, Larry J. Halverson, Tony Nieuwkoop, Zsofia Banfalvi, Maria G. Schell, David Gerhold, Nirupama Deshmane, Jae Seong So and Karl M. Sirotkin
Department of Microbiology and Graduate Program of Ecology
The University of Tennessee
Knoxville, Tennessee 37996-0845, USA

INTRODUCTION

The establishment of a successful symbiotic association between rhizobia and legumes is a complex process requiring functions of both symbiont and host. The symbiotic relationship resulting from rhizobia-plant interactions is an intimate one requiring close coordination. In addition, rhizobia-host combinations exhibit specificity; particular rhizobia species-host species combinations are favored, others are excluded. The intimacy and specificity of rhizobia-plant interaction suggest the necessity for the exchange of regulatory signal molecules between host and symbiont.

The evidence for signal exchange in plant-symbiont and plant-pathogen interactions has been recently reviewed (Halverson and Stacey 1986b). There are numerous examples of plant and/or bacterial signals which play important regulatory roles in the establishment of plant-microbe interactions. For example, a myriad of pathogen produced chemicals, as well as plant derived chemicals produced by pathogen action, are recognized by plant hosts and serve to induce disease resistance responses (Halverson and Stacey 1986b). Other plant produced compounds, such as α-tocopherol (Castle and Day 1984; Day and Castle 1982; Day et al 1981), are recognized by the pathogen and serve to induce parasitism. Recently, acetosyringione and α-hydroxyacetosyringione were shown to induce genes necessary for tumor formation in *Agrobacterium tumefaciens* (Stachel et al 1985). These plant produced chemicals are apparently derived from natural precursors of lignin biosynthesis.

NATO ASI Series, Vol. H 4
Recognition in Microbe-Plant Symbiotic and Pathogenic Interactions. Edited by B. Lugtenberg

Agrobacterium and species of rhizobia are grouped together in the family Rhizobiaceae. Recent evidence indicates that low molecular weight compounds produced by host roots induce gene function in rhizobia essential for nodulation (Mulligan and Long 1985). A functional nod D gene product must be present within the cell for the root exudate factor to induce gene expression (Mulligan and Long 1985). The factors that mediate nod gene induction, apparently via nod D, are present in root exudates of all legumes (Mulligan and Long 1985). Therefore, it is difficult to see how this root-exudate factor could be involved in mediating the specificity seen in most rhizobia-legume interactions.

THE ROLE OF LECTIN IN BRADYRHIZOBIUM JAPONICUM-SOYBEAN INTERACTION

The role of lectins in the nodulation of legumes has been and still is a controversial area. Most previous work has focused on a postulated role of lectins in recognizing and attaching rhizobia to the plant root surface (reviewed in Bauer, 1981; Dazzo and Gardiol 1984; Graham 1981; Stacey and Brill 1980). Partial evidence in support of this hypothesis is that lectins are found in or on legume roots (for example, Dazzo et al 1978; Stacey et al 1980) and, at least in some cases, lectins of a particular legume will only interact with the surface of the respective symbiont (for example, Bohlool and Schmidt 1974; Dazzo and Hubbell 1975). The lectin recognition hypothesis, therefore, serves to explain attachment of rhizobia to the root and, at least in part, the specificity of legume-rhizobia interaction. This hypothesis has been attacked, however, on many occasions. Evidence has been presented showing that rhizobia that do not infect a legume species still bind lectin from that species (Chen and Phillips 1976; Law and Strijdom 1977; Pueppke 1984). In addition, lectins do not bind to all strains within a particular species of rhizobia (Bohlool and Schmidt 1974). Resulting arguments about the lectin recognition hypothesis rage pro and con with no apparent solution.

It is for this reason that we sought to take a different approach to the problem of determining the early steps of host recognition between B. japonicum and soybean. Instead of assuming a mechanism and testing it experimentally, we sought to analyze the phenotype of mutants defective in nodulation with the hope that they would suggest a mechanism. Our efforts have largely, although not exclusively, focused on characterizing a single mutant of B. japonicum strain USDA 110, mutant HS111 (Halverson and Stacey 1984, 1985, 1986a). When compared to the wild type, nodulation induced by strain HS111 is delayed (Halverson and Stacey 1984).

In order to characterize the phenotype of strain HS111, we made use of the techniques developed by Bhuvaneswari and colleagues (1980). Soybean plants are grown in clear plastic growth pouches to enable the root tip (RT) and location of the smallest emergent root hairs (SERH) to be marked. The region between the RT and SERH marks made at the time of inoculation (2-3 days post-germination) is the area in which the majority of nodules develop (Bhuvaneswari et al 1980; Halverson and Stacey 1984). Bhuvaneswari et al (1980) showed, by repeated inoculation, that the RT-SERH region is only infectible for approximately 4-6 h due to the acropetal development of the root. Therefore, nodules that appear within the RT-SERH zone must have been initiated within 6 h following inoculation. This fact gives a method by which the delayed nodulation phenotype of mutant HS111 can be classified as either a defect in initiation of nodule formation or a defect in nodule growth following initiation. If the former were correct,nodules would appear only below the RT mark made at the time of inoculation. If the latter were correct, nodules would appear in the RT-SERH zone.

When the phenotype of HS111 was examined by this method, nodules were formed only below the RT mark, indicating an inability to rapidly induce successful nodulation (Halverson and Stacey 1984). In contrast, other mutants such as SM1 (Stacey et al 1982) show a phenotype consistent with a defect in nodule growth. Bhagwat and Thomas (1981) have observed a similar defect to that of strain HS111 with certain wild type strains of rhizobia when inoculated onto cowpea. They observed that the phenotype of these wild type strains was changed to allow nodulation in the RT-SERH zone if they were pretreated in cowpea root exudate before inoculation onto cowpea. Pretreatment of mutant HS111 with soybean root exudate causes a phenotypic reversal of the defect in nodulation to a near wild type phenotype (Halverson and Stacey 1984). These results suggest that mutant HS111 and certain wild type strains may be defective in nodulation unless activated by a plant produced factor. This activation apparently requires >6 h in the presence of the root.

The question is what is the factor present in soybean root exudate that is required to induce efficient nodulation by mutant HS111. Experiments indicate that the factor is protease and heat labile (Halverson and Stacey 1985). Further work showed that soybean seed lectin (specific for D-galactose/N-acetyl-D-galactosamine) would mimic the lectin enhancing effects of soybean root exudate (Halverson and Stacey 1985) and that a root derived protein with binding specificity for D-galactose/N-acetyl-D- galactosamine could be detected in soybean root exudates (Halverson and Stacey 1985). When this presumptive root lectin was removed from root exudate by affinity chromatography, the root exudate had no effect on the nodulation ability of mutant HS111 (Halverson and Stacey 1985). Treatment of mutant HS111 with soybean seed lectin in the presence of the hapten, D-galactose, shows no effect on the nodulation characteristics of the mutant (Halverson and Stacey 1985). Lectin-cell contact, therefore, appears to be necessary for enhancement of nodulation. Clearly, soybean root lectin appears to be necessary for efficient nodulation by mutant HS111. By taking a different approach to the study of B. japonicum-soybean interaction, we came full circle back to the study of the role of lectins in nodulation.

One of the attractive features in favor of lectin involvement in rhizobia-legume interaction is the fact that lectin binding is inherently specific; natural haptenic sugars must be present at the surface of the rhizobium. We have tested root exudates from several soybean cultivars and other plant species and found that only soybean root exudate will enhance the nodulation characteristics of mutant HS111 (Halverson and Stacey 1986a). Moreover, seed lectins from Arachis hypogaea, Dolichos biflorus, and Ricinus communis, all of which (like soybean lectin) have specificity for D-galactose or N-acetylgalactosamine, have no effect on the nodulation characteristics of mutant HS111 (Halverson and Stacey 1986a). This result suggests, not surprisingly, that the B. japonicum cell surface receptor for soybean lectin is more complex than a single sugar. Alternatively, soybean lectin may possess other attributes, besides specific cell attachment, which are important for the effects seen with mutant HS111. Soybean lectin is a potent effector of nodulation as tested by bioassay with strain HS111. Dilution of soybean seed lectin to 3.3×10^{-12} M (10 molecules per B. japonicum cell) still affects the nodulation ability of mutant HS111 after pretreatment.

One criticism of the work described above concerning lectin effects on nodulation is that only a single mutant was utilized. Are other mutants affected by treatment with soybean lectin or, more importantly, is wild type B. japonicum affected? Following our normal assay condition as applied to mutant HS111, pretreatment of wild type B. japonicum USDA 110 with soybean root exudate or seed lectin does not affect nodulation. The wild type normally produces the majority of the nodules at or above the RT mark made at the time of inoculation. Another effect, however, of lectin on mutant HS111, besides the shift in the location of nodule formation, is that 2-3x more nodules are formed by lectin-treated cells. Pretreatment of wild type B. japonicum USDA 110 with root exudate or soybean seed lectin also increases the number of nodules formed on 6 week old plants approximately two-fold (Table 1).

Table 1. Effect of Soybean Root exudate and seed lectin[a] pretreatment on the nodulation characteristics of wild type B. japonicum USDA 110.

Pretreatment	Sample Size[b]	Avg. no. of nodules $\pm$ SEM[c]
No addition	37	33.8 $\pm$ 2.2
Root Exudate	40	60.3 $\pm$ 2.9
Lectin	40	62.0 $\pm$ 2.3

[a]Lectin added at a concentration of 10 ug/ml

[b]Number of plants grown in vermiculite for 6 weeks

[c]Standard Error

During the course of this work, we developed a working hypothesis to explain how lectin may influence the nodulation characteristics of B. japonicum. We hypothesize that, in the presence of the plant host (lectin), the proportion of the population of B. japonicum cells capable of efficient nodulation is increased. If this hypothesis were correct, one might suspect that the effect of lectin would be more pronounced on B. japonicum inoculum of low cell number. As shown in Figure 1, dilution of B. japonicum USDA 110 to a level of $\leq 10^5$ cells/plant results in a nodulation profile resembling that of mutant HS111; that is, nodulation occurs primarily below the RT mark made at the time of inoculation. At cell concentrations of 10^3-10^4 cells/plant, pretreatment with soybean root exudate or seed lectin results in a statistically significant change in nodulation to resemble that seen at cell concentrations $>10^5$ cells/plant (i.e., nodulation near the RT mark, Figure 1). Thus, using wild type inoculum at low cell density, an effect of lectin can be demonstrated. The results with the wild type resemble those seen with mutant HS111 at higher cell concentrations.

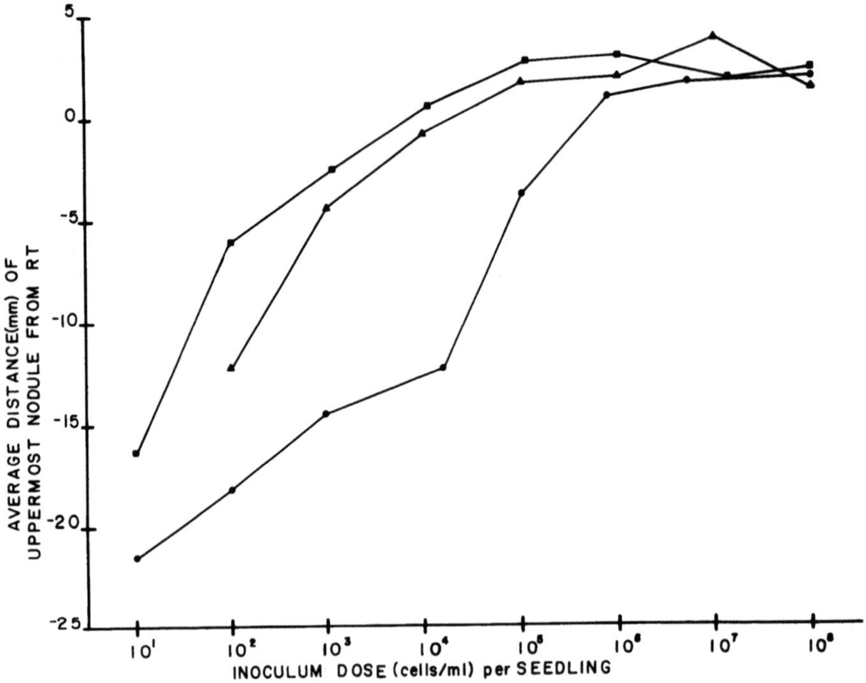

Figure 1. Effect of inoculum concentration and root exudate or lectin pretreatment on nodulation of soybean by B. japonicum. Symbols: ● , control; ▲ , root exudate; ■, lectin.

Whereas the above experiments clearly show an effect of root exudate and lectin on nodulation, they give little insight into the physiological basis for the change in the nodulation phenotype of wild type or mutant HS111. Analysis of the phenotype of mutant HS111 indicates that this mutant binds lectin to the same extent as wild type. Additionally, cells of mutant HS111 attach to soybean roots and curl root hairs to an equal extent as the wild type. This latter result suggests that the defect in nodulation exhibited by HS111 is found at an infection stage after root hair curling.

Since nodulation by mutant HS111 appears delayed based on the placement of nodules below the RT-SERH region, and since this region is infectible for 6h (Bhuvaneswari et al 1980), it appears that longer than 6h preincubation in root exudate is required to induce efficient nodulation in mutant HS111. Therefore, the enhancement of nodulation by *B. japonicum* produced by lectin pretreatment is unlikely to be solely due to adsorption of lectin to the cell surface. Instead, it is likely that lectin induces a physiological change in the cells. To determine whether lectin induces *de novo* RNA or protein synthesis, cells of mutant HS111 were pretreated with soybean seed lectin in the presence or absence of rifampin or chloramphenicol. After pretreatment, cells were inoculated onto soybean seedlings and the location of nodules scored. Table 2 presents the data from this experiment. Both rifampin and chloramphenicol prevent lectin treatment from enhancing the nodulating ability of mutant HS111. The data suggest that lectin pretreatment induces gene expression which is necessary for *efficient* nodulation. As yet, we cannot conclude that the presence of lectin is *essential* for nodule formation.

Table 2. Rifampin and chloramphenicol inhibition of the nodulation-enhancing characteristics of SBL on mutant strain HS111.

Condition[a]	Average distance (mm) of the uppermost nodule from RT mark ± SEM[b]	% Nodulation only below RT
MM	-19.5 ± 4.3	78
MM + SBL	- 1.8 ± 2.7	39
MM + rifampin	-19.8 ± 5.4	63
MM + SBL + rifampin	-16.7 ± 4.8	64
MM + chloramphenicol	-24.3 ± 3.5	85
MM + SBL + chloramphenicol	-18.9 ± 4.2	69

[a]MM = minimal medium, SBL = 10 ug/ml, rifampin = 25 ug/ml, chloramphenicol = 50 ug/ml.

[b]Minus signs indicate nodulation below the RT mark made at the time of inoculation. SEM = standard error of the mean.

Analysis of the proteins produced by wild type and mutant HS111 cells before and after lectin pretreatment indicates several differences (Figure 2). By 2-dimensional protein electrophoresis, at least 4 new proteins appear after lectin pretreatment, while at least 3 proteins disappear. The significance of these results and the possible functions of these proteins is currently under investigation.

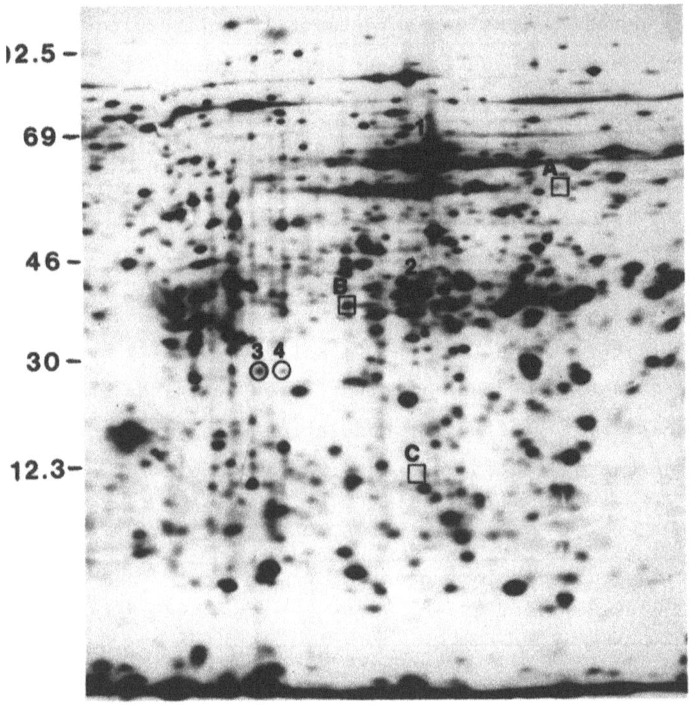

Figure 2. Two-dimensional electrophoretic profile of proteins from B. japonicum pretreated with soybean lectin. Letters (ABC) refer to those proteins whose level is reduced upon lectin treatment. Numbers (1, 2, 3, 4) refer to those proteins whose level increases upon lectin treatment.

MOLECULAR GENETICS OF NODULATION BY B. JAPONICUM

In published work (Russell et al 1985), we described the isolation from B. japonicum of DNA contained in a number of overlapping cosmid clones which show homology to the common nod genes of R. meliloti and R. leguminosarum. Additionally, we showed that a 1.8 Kb Eco RI-Hind III fragment contained within a larger 3.9 Kb Hind III fragment was essential for restoration of Hac (hair curling) function to a Hac$^-$ mutant of Rhizobium fredii (Russell et al 1985). Figure 3 represents a partial restriction enzyme map of the cloned region. This region has been subjected to site-directed mutagenesis with Tn5 (Ruvkin and Ausubel 1981). For clarity, most of the Tn5 mutations are excluded from Figure 3. Thus far, we have identified 3 essential nodulation regions within the cloned region.

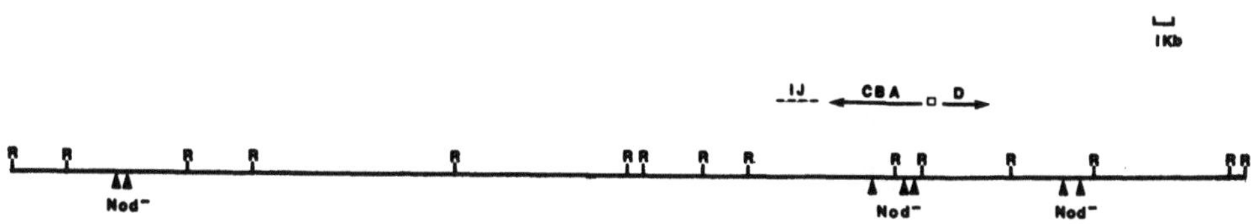

FIGURE 3

Figure 3. EcoRl restriction enzyme map of B. japonicum DNA containing essential nodulations regions. The nod gene regions identified by Tn5 insertion are shown. The position of those genes identified by sequencing (nod ABCD) or by hybridization (nod IJ) are shown. The open square indicates the position of the Nod Box sequence.

One of the identified regions lies within the previously identified 3.9 Kb Hind III fragment (Figure 3). The exact location of Tn5 insertions in this region have not been checked by sequencing. Hybridization of internal gene probes of R. meliloti nod C and nod D to this 3.9 Kb fragment localize these genes as shown (Fig. 3). A R. meliloti gene probe containing both nod A and B shows homology to the region indicated (Fig. 3). Furthermore, sequencing of the DNA within this region indicates extensive homology to the published nod A B C and nod D sequences (Egelhoff et al 1985; Jacobs et al 1985). These data allow the approximate positioning of the nod A B C and nod D genes as shown in Fig. 3 and the assignment of the direction of transcription. The nod IJ genes of R. leguminosarium were cloned into bacteriophage M13 vector, mp 18, and labeled with ^{32}P by primer extension. Homology to these genes was found within the 5.6 K Hind III fragment adjacent to nod C. This is the same relative location as in R. leguminosarum (Downie et al 1985).

Recently, Rostas et al (1985) analyzed the 5' upstream sequences of the nod A B C, hsn A B C, and hsn D transcripts of R. meliloti. A 47 bp conserved region was localized some 200-250 bp 5'of the translational start codons of these three nod regions. The 47 bp conserved region contains three subsequences of 7 bp, 5 bp, and 25 bp. The 25 bp region was synthesized and found to hybridize to genomic DNA of several species of rhizobia. Deletion of the 25 bp region in R. meliloti resulted in an inability to express the adjacent nod genes. This 25 bp region was given the trivial name Nod Box and postulated to be a sequence essential for the induction of nod genes in R. meliloti and other Rhizobium species. The nod D gene of R. meliloti which is known to be expressed constitutively lacks an associated Nod Box sequence.

We labeled the R. meliloti Nod Box oligonucleotide (kindly provided by A. Kondorosi) and hybridized it to the cloned regions shown in Figure 3. Only one copy of this sequence was found by hybridization. Further hybridization to subclones and to double and single restriction digests localized the Nod Box sequence between the nod A B C and nod D gene transcripts. We have now confirmed this location by DNA sequencing. This location is similar to that found in R. meliloti, and is consistent with the postulated role of this sequence in gene regulation. In addition to the common Nod region, we have also identified 2 additional linked Nod regions (Figure 3). Tn5 insertions into these regions result in an inability to nodulate soybean. We have hybridized the nod A B C D and hsn A B C D genes of R. meliloti, as well as the nod I J genes of R. leguminosarum, to these regions and have found no homology. We have used these B. japonicum DNA regions as probes to R. meliloti and R. leguminosarum genomic DNA and found no homology. We suspect that these regions encode essential nod genes specific for B. japonicum. Preliminary experiments indicate that these regions are not expressed in free-living B. japonicum. These regions also lack homology to the presumptive regulatory Nod Box sequence. These data suggest that these regions may be under a different mode of regulation than that seen for the nod A B C genes. Work to examine this possibility is ongoing.

SUMMARY

Examination of the nodulation phenotype of mutant strain HS111 has led us to formulate a new hypothesis of lectin involvement in nodulation.

Our working hypothesis is that in the presence of host produced root lectin de novo RNA and protein synthesis is induced that enhances the ability of B.japonicum to initiate a successful infection. The data, at present, preclude determining whether lectin is absolutely essential for nodulation; as opposed to increasing the efficiency of nodulation. This hypothesis for lectin involvement in nodulation does not necessarily preclude other roles for lectin.

In related work, we have cloned a DNA region of B.japonicum that appears to contain multiple, linked regions which are essential to nodulation. The nod A B C and nod D gene transcripts have been identified and have the same orientation as found in other rhizobia. A Nod Box conserved regulatory sequence has been localized 5' of the start of the nod A B C transcript. The location of this sequence is consistent with the postulated role in regulating nod gene expression. Additional nod genes linked to the nod A B C D region appear to be unique to B.japonicum. Interestingly, these essential Nod regions show no homology to the Nod Box sequence. It is possible that these genes are regulated in a different manner than the nod A B C transcript.

REFERENCES

Bauer WD (1981) Infection of legumes by rhizobia. Ann Rev Plant Physiol 32: 407

Bagwatt AA, Thomas J (1982) Legume-Rhizobium interactions: Cowpea root exudate elicits faster nodulation response by Rhizobium species. Appl Environ Microbiol 43: 800

Bhuvaneswari TV, Turgeon BG, Bauer WD (1980) Early events in the infection of soybean (Glycine max L. Merr) by Rhizobium japonicum. I. Localization of infectious root cells. Plant Physiol 66: 1027

Bohlool BB, Schmidt EL (1974) Lectins: A possible basis for specificity in the Rhizobium-legume root nodule symbiosis. Science 185: 269

Castle AJ, Day AW (1984) Isolation and identification of alpha-tocopherol as an inducer of the parasitic phase of Ustilago violaceae. Phytopathol 74: 1184

Chen AP, Philips DA (1976) Attachment of Rhizobium to legume roots as the basis for specific associations. Physiol Plant 38: 83

Day AW, Castle AJ (1982) The effect of host extracts on differentiation in the genus Ustilago. Bot Gaz 143: 188

Day AW, Castle AJ, Cummins JE (1981) Regulation of parasitic development of the smut fungus Ustilago violacea, by extracts from host plants. Bot Gaz 142: 135

Dazzo FB, Gardiol AE (1984) Host specificity in Rhizobium-legume interactions. In: Verma DPA, Hohn T (eds) Genes Involved in Microbe-Plant Interactions. Springer-Verlag, Wien, New York, p 1

Dazzo FB, Hubbell DH (1975) Cross-reactive antigens and lectin as determinants of symbiotic specificity in the Rhizobium-clover association. Appl Microbiol 30: 1017

Dazzo FB, Yanke WE, Brill WJ (1978) Trifoliin: A Rhizobium recognition protein from white clover. Biochem Biophys Acta 539: 276

Downie AJ, Rossen L, Evans IJ, Shearman C, Johnston AWB (1985) Nodulation genes of Rhizobium leguminosarum. In: Galau GA (ed) Abstracts First International Congress of Plant Molecular Biology. Univ. of Georgia Press, Athens, p 50

Egelhoff TT, Fisher RF, Jacobs TW, Milligan JT, Long SR (1985) Nucleotide sequence of Rhizobium meliloti nodulation genes: nod D is read divergently from nod A B C. DNA 4: 241

Graham TL (1981) Recognition in Rhizobium-legume symbiosis. In: Giles KL, Atherly AG (eds) Biology of the Rhizobiaceae. Academic Press, New York, p 127

Halverson LJ, Stacey G (1984) Host recognition in the Rhizobium soybean symbiosis: detection of a protein factor in soybean root exudate which is involved in the nodulation process. Plant Physiol 74: 84

Halverson LJ, Stacey G (1985) Host recognition in the Rhizobium soybean symbiosis: evidence for the involvement of lectin in nodulation. Plant Physiol 77: 621

Halverson LJ, Stacey G (1986a) Effect of lectin on nodulation by wild type Bradyrhizobium japonicum and a nodulation-defective mutant. Appl Environ Microbiol 51: 753

Halverson LJ, Stacey G (1986b) Signal exchange in plant-microbial interactions. Microbiol Rev 50: 193

Jacobs TW, Egelhoff TT, Long SR (1985) Physical and genetic map of Rhizobium meliloti nodulation gene region and nucleotide sequence of nod C. J Bacteriol 162: 469

Law IJ, Strijdom BW (1977) Some observations on plant lectins and *Rhizobium* specificity. Soil Biol Biochem 9: 79

Mulligan JT, Long SR (1985) Induction of *Rhizobium meliloti nod C* expression by plant exudate requires *nod D*. Proc Natl Acad Sci 82: 6609

Pueppke SF (1984) Adsorption of slow- and fast-growing rhizobia to soybean and cowpea roots. Plant Physiol 75: 924

Rostas K, Kondorosi E, Horvath B, Simoncsits A, Kondorosi A (1985) Conservation of extended promoter regions of nodulation genes in rhizobia. Proc Natl Acad Sci 83: 1757

Russell P, Schell MG, Nelson KK, Halverson LJ, Sirotkin KM, Stacey G (1985) Isolation and characterization of the DNA-region encoding nodulation functions in *Bradyrhizobium japonicum*. J Bacteriol 164: 1301

Ruvkin GB, Ausubel FM (1981) A general method for site-directed mutagenesis in prokaryotes. Nature 289: 85

Stacey G, Brill WJ (1982) Nitrogen-fixing bacteria: colonization of the rhizosphere and roots. In: Mount MS, Lacey GH (eds) Phytopathogenic prokaryotes. Academic Press, New York, p 225

Stacey G, Paau A, Brill WJ (1980) Host recognition in the *Rhizobium*-soybean symbiosis. Plant Physiol 66: 609

Stacey G, Paau AS, Noel KD, Maier RJ, Silver LE, Brill WJ (1982) Mutants of *Rhizobium japonicum* defective in nodulation. Arch Microbiol 132: 219

Stachel SE, Messens E, Van Montagu M, Zambryski P (1985) Identification of the plant signal molecules produced by wounded plant cells that activate T-DNA transfer in *Agrobacterium tumefaciens*. Nature 318: 624

ATTACHMENT OF *RHIZOBIUM LEGUMINOSARUM* TO PEA ROOT HAIR TIPS

J.W. Kijne, G. Smit, C.L. Díaz and B.J.J. Lugtenberg
Department of Plant Molecular Biology,
Botanical Laboratory, Nonnensteeg 3, 2311 VJ Leiden,
The Netherlands

INTRODUCTION

Nodulation in the *Rhizobium*-legume symbiosis is a host plant-specific process. The speciation of the genus *Rhizobium* is primarily based on the existence of different so-called cross-inoculation groups, e.g. *R.meliloti* nodulates alfalfa but not pea, and *R.leguminosarum* nodulates pea, vetch, lentil and sweet pea but not alfalfa.

Attachment of *Rhizobium* to legume root hairs is thought to be an early and necessary step in the root infection process, followed by marked root hair curling and infection thread formation (see also Rolfe et al, this volume). Host-specificity is expressed at some event preceding the outgrowth of the infection thread in the root hair (Li and Hubbell 1969). In the lectin-recognition hypothesis, as developed by Bohlool and Schmidt (1974) and Dazzo and Hubbell (1975), lectin-mediated attachment is considered to be the host-specificity determining step during the establishment of the association. This recognition model implies that within one cross-inoculation group all potential host plants produce similar lectins with unique sugar-binding properties to be complemented by specific sugar sequences at the surface of the homologous rhizobial partner. The available data indicate that the different lectins in leguminous plants indeed are distributed following the cross-inoculation group classification. Most data supporting the lectin-recognition hypothesis have been obtained by study of the clover-*R.trifolii*-association (e.g. Dazzo et al 1985, see also Rolfe et al, this volume). On the other hand, attachment experiments in several laboratories showed that rhizobia can attach non-specifically to various

NATO ASI Series, Vol. H 4
Recognition in Microbe-Plant Symbiotic and
Pathogenic Interactions. Edited by B. Lugtenberg

surfaces, including heterologous root hairs (e.g. Menzel et al 1972). Moreover, recent data obtained with genetically altered rhizobia and agrobacteria point to infection thread formation as the probable specificity barrier, e.g. Agrobacterium tumefaciens carrying the common nod genes of R.meliloti induces marked root hair curling in white and red clover, but no infection threads (see Debellé et al, this volume).

This paper accentuates some aspects of attachment research in the pea-R.leguminosarum symbiosis. It is concluded that the infection pattern of R.leguminosarum in pea roots is positively correlated with the root surface distribution of lectin. The characteristics of pea root hair tip attachment by R.leguminosarum are determined by the bacterial culture conditions. Growth limitation induces attachment ability, and the nature of the limiting factor determines if attachment is lectin-hapten inhibitable or not.

PEA LECTIN LOCALIZATION

Lectins have been localized at the tip of growing root hairs of Trifolium repens (Dazzo et al 1978) and Lotononis bainesii (Law and Strijdom 1984). Pea lectin presence has been observed on pea root hairs (Gatehouse and Boulter 1980; Kato et al 1981), but the part of the root used was not specified.

We investigated pea lectin presence and accessibility at the rhizobial infection site, i.e. growing root hair tips. Indirect immunofluorescence microscopy was used to determine the location of lectin on the surface of 4 and 5 days old axenically grown pea seedlings. Earlier experiments indicated that the major root lectin of peas is similar to seed isolectin 2 (Kijne et al 1980, 1983). A specific seed isolectin 2 rabbit antibody, which cross--reacts with root lectin, was used as a primary antibody (Díaz et al 1984) and next visualized with fluoresceine or tetramethyl rhodamine isothiocyanate-labeled goat antirabbit IgG (Díaz et al 1986).

Lectin was observed on the tips of young developing root hairs and on epidermal cells located just below the young hairs. On both types of cells, lectin was concentrated in dense small patches rather than uniformly distributed. Lectin-positive young hairs were grouped opposite the (proto)xylematic poles of the root stele. Older but still elongating root hairs showed only traces or no lectin at all.

We tested whether the root sites where lectin was observed were susceptible to infection by R.leguminosarum. When a low inoculum dose (less than 10^6 bacteria ml^{-1}, inoculum drop 2 μl) was placed next to lectin-positive epidermal cells or on newly formed root hairs, nodules were formed on the primary roots of 73% and 90% of the plants, respectively. Only a few plants showed nodulation on the primary root when the inoculum was placed on the root zone where lectin was scarcely or not detected (Díaz et al 1986).

Although we have not demonstrated that lectin-positive root hairs are infectable cells, we found a positive correlation between the root area where lectin is localized and susceptibility to infection by Rhizobium. Ninety percent or more of the root nodules are initiated opposite the (proto)xylematic ridges in pea roots (Oinuma 1948, cited by Libbenga and Bogers 1974). A similar xylem-determined pattern was found by Libbenga et al (1973) for the initiation of cell divisions in pea root cortical explants by exogenously applied plant hormones.

These results indicate that in pea roots both nodule initiation and root surface localization of lectin are xylem-determined. Libbenga's hypothetical gradient of plant growth factors in the root cortex might be one of the regulating factors of lectin presence and accessibility at the root surface.

NON-SPECIFIC ATTACHMENT

End log/early stationary phase R.leguminosarum cells, batch-cultivated in TY medium (Beringer 1974), attach abundantly to pea root hair tips after an incubation period of one hour (young lateral roots, immersed in a bacterial suspension of 1.5 x 10^8 cells ml^{-1}; 25 mM phosphate buffer, pH 7.5) (Kijne et al 1985). Under similar conditions, R.trifolii, R.phaseoli and A.tumefaciens also attach abundantly to pea root hair tips. Moreover, the attachment behaviour of Sym plasmid cured strains of R.leguminosarum and R.trifolii is indistinguishable from that of their respective parent strains, which altogether shows that Sym plasmid borne nod-genes (see Rolfe et al, this volume) are not involved in this attachment process.

One of the characteristics of TY-cultivated early stationary phase R.leguminosarum bacteria is the high percentage of fibrillated cells as visualized electronmicroscopically by negative staining. The fibrils have a diameter of 5-6 nm, are up to 10 μm long and extend from the bacterial surface in a relatively low number (up to 10 $cell^{-1}$ for R.leguminosarum 248). After specific purification, cellulase treatment and IR-spectroscopy, the nature of the fibrils was established as being cellulose. We found a strong correlation between fibrillation and the ability to attach abundantly to pea root hair tips, also with Sym plasmid-cured strains. Moreover, fibril-lacking mutants were found to have a strongly decreased attachment ability. As pea lectin-haptenic monosaccharides do not inhibit the attachment process, it is unlikely that pea lectins are involved in cellulose-enhanced attachment (Smit et al, submitted).

Our attachment experiments demonstrate the ability of heterologous rhizobia to attach to pea root hair tips. Attachment of heterologous rhizobia to non-host plants is commonly referred to as non-specific attachment. The involvement of cellulose indicates that non-specific attachment might include specific molecules and molecular interactions, distinguishing it from simple sticking onto a surface. Additional to cellulose, other adhesins might be involved (as is the case in the attachment process of Agrobacterium to its target plant cells, Matthysse et al, this volume), since fibril-lacking mutants of R.leguminosarum are still able to nodulate.

GROWTH LIMITATIONS

The stationary growth phase of R.leguminosarum in TY medium is a consequence of carbon-limitation. Addition of glucose causes an upward shift of both the growth stop and the optimal attachment ability of the cells. We studied if other growth limitations also induce attachment ability and invariably found that growth-limited cells attach in much larger numbers to pea root hair tips than actively dividing (log phase) cells.

For example, a growth limiting factor in yeast-mannitol (YM) medium (Zurkowski and Lorkiewicz 1978) is Mn^{2+}. During batch culture in YM medium, R.leguminosarum shows several transient growth stops which do not appear after additional supply of Mn^{2+}. These growth stops can be visualized most clearly by plotting the number of bacteria ml^{-1} in culture against the corresponding optical density (a so-called n/OD-graph). Each growth stop corresponds to an increase in attachment ability (Fig. 1). Mn^{2+}--limited rhizobia attach much quicker to pea root hair tips than carbon-limited cells; after an incubation time of 30 min already 25% of the developing root hairs shows abundant tip attachment. Attachment is delayed in presence of the pea lectin hapten 3-O-methyl-D-glucose (50 mM) but not in presence of the non--haptens xylose and galactose. Also a Sym plasmid-cured R.leguminosarum strain shows Mn^{2+}-limitation induced attachment. As the lectin hapten delays but not prevents attachment, these results point to the possible operation of both lectin-mediated (thus host plant specific) and non-specific attachment mechanisms during the same attachment assay.

Dazzo et al (1976) and Zurkowski (1980) described attachment experiments in which the clover lectin hapten 2-deoxy-D-glucose largely prevented attachment of R.trifolii to clover root hairs. We found comparable results in our pea root hair tip attachment assay with use of oxygen- or iron-limited R.leguminosarum cells. Using a synthetic medium containing glucose as a carbon-source and glutamate as a nitrogen-source (B^{-}glu medium) we observed several transient growth stops with R.leguminosarum 248, each of which again was correlated with an increase in tip attachment ability (Fig. 2). Bacteria harvested during the growth stop at an OD of 0.4 appeared to be of particular interest, as tip attachment by these cells could be specifically inhibited by 3-O-methyl-D-glucose (inhibition percentage: 75). Chemostat studies with R.leguminosarum (Ratcliffe et al 1983) showed that

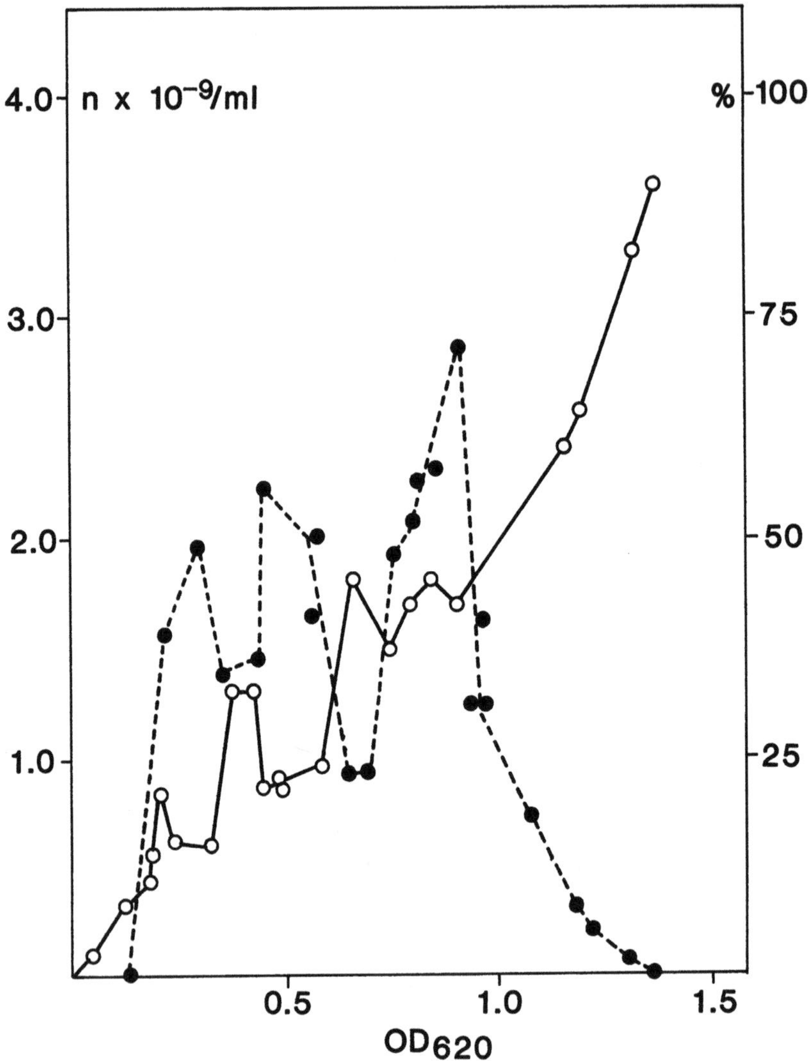

Fig. 1. Correlation between transient growth stops of R.leguminosarum 248 in YM medium and its ability to attach abundantly to pea root hair tips. Flocculation of bacteria during the growth stops causes an apparent decrease in their number/ml. The number of bacteria (n x 10^{-9}/ml) (-o-) and the percentage of developing root hairs with abundant tip attachment (-●-) are both plotted against the optical density (620 nm) of the culture during growth.

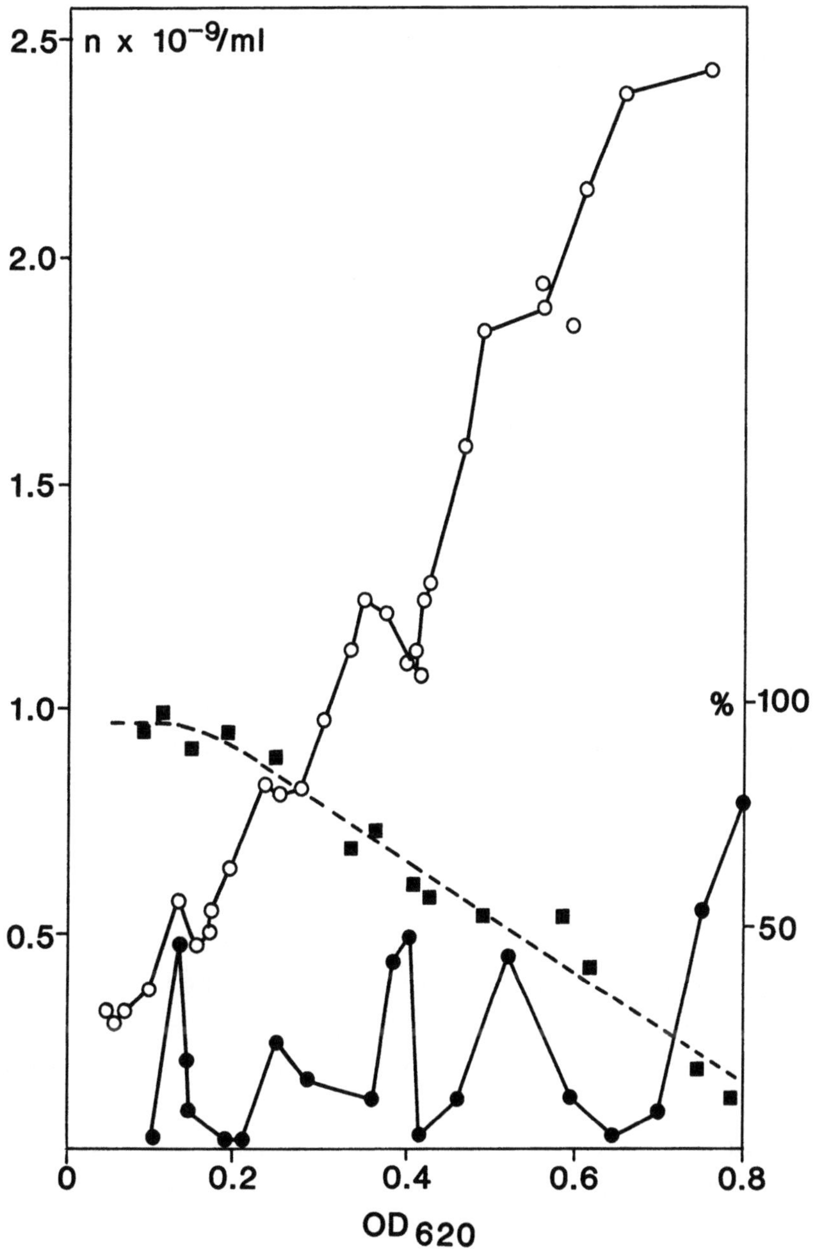

Fig. 2. Abundant root hair tip attachment of R.leguminosarum 248 as related to growth and oxygen concentration in batch culture (B^-glu medium). The number of bacteria (n x 10^{-9}/ml) (-o-) and the percentage of developing root hairs with abundant tip attachment (-●-) are both plotted against the optical density (620 nm) of the culture during growth. The oxygen concentration in the growth medium is given as the percentage of the initial concentration (-■-).

decrease of the oxygen concentration in the culture fluid to about half of the initial value (205 μM = 100%) resulted in a cytochrome shift, apparently representing an adaptation to oxygen limitation. As the pO_2 in B^-glu medium at an OD_{620} of 0.4 is close to half of the initial value, we tested the hypothesis that the attachment induction at this growth stop is a result of oxygen limitation. Indeed, R.leguminosarum 248 cells harvested from a B^-glu continuous culture equilibrated at a limiting pO_2 of 55% showed identical attachment characteristics, including specific inhibition by 3-0-methyl-D-glucose. Interestingly, similar results were obtained with Fe^{3+}-limited batch and continuous cultures of the same strain (Kijne et al, in preparation).

Our results show that growth limitations induce attachment ability in R.leguminosarum. We propose to designate the transient growth stops in batch culture which result in attachment peaks as "attachment windows". The nature of the growth limiting factor determines if attachment is lectin-hapten inhibitable (e.g. oxygen limitation) or not (e.g. carbon limitation).

GENERAL CONCLUSIONS

The observed correlation between lectin surface localization and susceptibility to Rhizobium-infection in pea roots, together with the demonstration of lectin-hapten inhibitable attachment under special cultural conditions support the lectin-recognition hypothesis.

On the other hand, the existence of a non-specific attachment mechanism, the inability of lectin haptens to inhibit attachment completely (see also Dazzo et al 1976) and the genetical data pointing to infection thread formation as the host specific infection step question the lectin-recognition hypothesis in its original form.

However, these results are not conflicting if one assumes that lectins are not necessarily involved in attachment but in infection thread initiation. Oxygen limitation is likely to occur after enclosure of Rhizobium in a root hair curl with the possible expression of lectin receptors at the rhizobial surface as a result. The attachment and lectin receptor window in the experiments of Dazzo's group (R.trifolii 0403, batch-cultured in modified B III - medium, Dazzo et al 1976) might represent an oxygen limitation attachment window (in view of the OD) or an iron limitation attachment window (in view of the limiting Fe^{3+}-concentration in this medium). When these bacteria specifically interact with the root hair cell surface, they will also give positive results in an attachment assay. The stimulation of infection thread formation in clover root hairs by lectin-binding oligosaccharides from capsular and extracellular components of R.trifolii (Abe et al 1984) strongly points to a role of lectins in this process. Physical contact of Rhizobium with the root hair surface is necessary in preceding marked root hair curling, both on experimental (Yao and Vincent 1969) and theoretical (Van Batenburg et al 1986) grounds. This contact might be a non-specific attachment process, as induced by various growth limitations.

Our results underline the desirability of a careful determination of the rhizobial growth conditions near and at the root hair surface of its homologous and heterologous host plants. Mutants in the different attachment mechanisms as well as in the physiological response to nutritional stress will be needed to evaluate the possibilities as outlined above.

REFERENCES

Abe M, Sherwood JE, Hollingsworth RI, Dazzo FB (1984) Stimulation of clover root hair infection by lectin-binding oligosaccharides from the capsular and extracellular polysaccharides of Rhizobium trifolii. J Bacteriol 160: 517-520

Beringer J (1974) R factor transfer in Rhizobium leguminosarum. J Gen Microbiol 84: 188-198

Bohlool BB, Schmidt EL (1974) Lectins: a possible basis for specificity in the Rhizobium-legume root nodule symbiosis. Science 185: 269-271

Dazzo FB, Hubbell DH (1975) Cross-reactive antigens and lectin as determinants of symbiotic specificity in the Rhizobium--clover association. Appl Microbiol 30: 1017-1033

Dazzo FB, Napoli CA, Hubbell DH (1976) Adsorption of bacteria to roots as related to host-specificity in the Rhizobium--clover symbiosis. Appl Env Microbiol 32: 166-177

Dazzo FB, Yanke WE, Brill WJ (1978) Trifoliin: a Rhizobium recognition protein from white clover. Biochim Biophys Acta 539: 276-286

Dazzo FB, Hollingsworth RI, Sherwood JE, Abe M, Hrabak EM, Gardiol AE, Pankratz HS, Smith KB, Yang H (1985) Recognition and infection of clover root hairs by Rhizobium trifolii. In: Nitrogen Fixation Research Progress (eds Evans H J, Bottomley P J, Newton W E), Martinus Nijhoff Publ., pp. 239-245

Díaz CL, Lems-Van Kan P, Van der Schaal IAM, Kijne JW (1984) Determination of pea (Pisum sativum L.) root lectin using an enzyme-linked immunoassay. Planta 161: 302-307

Díaz CL, Van Spronsen PC, Bakhuizen R, Logman GJJ, Lugtenberg EJJ, Kijne JW (1986) Correlation between infection by Rhizobium leguminosarum and lectin on the surface of Pisum sativum L. roots. Planta 168: in press

Gatehouse JA, Boulter D (1980) Isolation and properties of a lectin from the roots of Pisum sativum (garden pea). Physiol Plant 49: 437-442

Kato G, Maruyama Y, Nakamura M (1981) Involvement of lectins in Rhizobium-pea recognition. Plant Cell Physiol 22, 759-771

Kijne JW, Van der Schaal IAM, De Vries GE (1980) Pea lectins and the recognition of Rhizobium leguminosarum. Plant Sci Lett 18: 65-74

Kijne JW, Van der Schaal IAM, Díaz CL, Van Iren F (1983) Mannose specific lectins and the recognition of pea roots by Rhizobium leguminosarum. In: Lectins, Vol. 3 (eds Bøg-Hansen TC, Spengler GA), W. de Gruyter, pp. 521-529

Kijne JW, Smit G, Díaz CL, Lugtenberg EJJ (1985) Attachment of Rhizobium leguminosarum 248 to pea root hairs. In: Nitrogen Fixation Research Progress (eds Evans H J, Bottomley P J, Newton W E), Martinus Nijhoff Publ., p. 260

Law IJ, Strijdom BW (1984) Role of lectins in the specific recognition of Rhizobium by Lotononis bainesii. Plant Physiol 74: 779-785

Li D, Hubbell DH (1969) Infection thread formation as a basis of nodulation specificity in Rhizobium-strawberry clover associations. Can J Microbiol 15: 1133-1136

Libbenga KR, Van Iren F, Bogers RJ, Schraag-Lamers MF (1973) The role of hormones and gradients in the initiation of cortex proliferation and nodule formation in Pisum sativum L. Planta 114: 29-39

Libbenga KR, Bogers RJ (1974) Root nodule morphogenesis. In: The biology of nitrogen fixation (ed Quispel A), North Holland Publ. Cy., pp. 430-472

Menzel G, Uhlig H, Weichsel G (1972) Uber die Besiedlung der Wurzeln einiger Leguminosen und Nichtleguminosen mit Rhizobien und anderen Bodenbakterien, Zbl f Bakt Abt II, 127: 348-358

Ratcliffe HD, Drozd JW, Bull AT (1983) Growth energetics of Rhizobium leguminosarum in chemostat culture. J Gen Microbiol 129: 1697-1706

Van Batenburg FHD, Jonker R, Kijne JW (1986) Rhizobium induces marked root hair curling by redirection of tip growth: a computer simulation. Physiol Plant 66: 476-480

Yao PY, Vincent JM (1969) Host-specificity in the root hair "curling factor" of Rhizobium spp. Aust J Biol Sci 22: 413-423

Zurkowski W (1980) Specific adsorption of bacteria to clover root hairs, related to the presence of plasmid pWZ2 in cells of Rhizobium trifolii. Microbios 7: 27-32

Zurkowski W, Lorkiewicz Z (1978) Effective method for the isolation of non-nodulating mutants of Rhizobium trifolii. Gen Res 32: 311-314

SPECIFICITY OF RHIZOBIUM(GALEGA) - GALEGA -INTERACTION

Päivi Lipsanen and Kristina Lindström
Department of Microbiology
University of Helsinki
SF-00710 Helsinki
Finland

Galega orientalis is a potential forage legume suitable for low temperatures and acid soils in Finland. It is able to fix atmospheric nitrogen when living in symbiosis with its *Rhizobium* partner. There are no rhizobia nodulating *G. orientalis* in Finnish soils, so inoculation is essential. *Rhizobium(Galega)* (i.e. rhizobia nodulating *G. orientalis* and *G. officinalis*) differ from other rhizobia with respect to DNA-homology, phage-typing, and cross-inoculation. *R.(Galega)* is very host-specific: *R.(G.orientalis)* only nodulates *G. orientalis* effectively and *G. officinalis* ineffectively; whereas *R.(G.officinalis)* only nodulates *G. officinalis* effectively and *G. orientalis* ineffectively (Lindström et al 1983).

We studied interactions between *R.(Galega)* and the legumes *G. orientalis*, *G. officinalis*, *Medicago sativa*, *Trifolium pratense*; and also between *R. meliloti*, *R. trifolii*, some tropical fast-growing rhizobia and *G. orientalis*, *G. officinalis* with light microscopy (Methods: Truchet et al (1984), Vasse and Truchet (1984)).

R.(Galega) deforms root hairs of both *Galega*-species to cauliflower- like structures. Infection threads are seen in very short root hairs or starting from cauliflower structures of long root hairs. Both effective and ineffective *Galega*-root nodules have a meristem, an infection zone with infection threads, a central zone with bacteroids, and also a degeneration zone in old nodules. *R.(Galega)* did not cause any reaction on *M. sativa*, but deformed *T. pratense*.

NATO ASI Series, Vol. H 4
Recognition in Microbe-Plant Symbiotic and Pathogenic Interactions. Edited by B. Lugtenberg

R. meliloti and R. trifolii did not cause any changes on Galega plants. Fast-growing tropical strains deformed root hairs and some formed tumors on the roots, but there were no infection threads.

TABLE 1. Cross-inoculation

Bacteria	Plant											
	G.orientalis			G.officinalis			M. sativa			T. pratense		
	D	IT	N	D	IT	N	D	IT	N	D	IT	N
R.(G.or.)												
1261R	+	+	+	+	+	(+)	-	-	-	+	-	-
R.(G.off.)												
B7i	+	+	(+)	+	+	+	-	-	-	+	-	-
R. meliloti												
1021	-	-	-	-	-	-	+	+	+	+	-	-
R. trifolii												
ANU 843	-	-	-	-	-	-	-	-	-	+	+	+
ANU 845	-	-	-	-	-	-	-	-	-	-	-	-
Rhizobium sp												
3021	+	-	o	+	-	o						
3022	+	-	-	+	-	o						
3024	+	-	-	+	-	o						
3025	+	-	-	-	-	-						
3033	+	-	-	+	-	o						
3034	+	-	-	+	-	o						
3039	-	-	-	+	-	o						

D = deformation of root hairs
IT = formation of infection threads
N = nodulation
(+) = ineffective nodules
o = nodule-like structures

REFERENCES

Lindström K, Jarvis DBW, Lindström PE and Patel JJ (1983) DNA-homology, phage-typing, and cross-nodulation studies of rhizobia infecting Galega species. Can J Microbiol 29:781-789

Truchet G, Rosenberg C, Vasse J, Julliot J-S, Camut S, and Denarie J (1984) Transfer of Rhizobium meliloti Sym genes into Agrobacterium tumefaciens: host-specific nodulation by atypical infection. J Bacteriol 157:134-142

Vasse JM and Truchet GL (1984) The Rhizobium-legume symbiosis: observation of root infection by bright-field microscopy after staining with methylene blue. Planta 161:487-489

FLAVONES INDUCE EXPRESSION OF THE NODULATION GENES IN RHIZOBIUM

John W. Redmond[1], Michael Batley[1], Roger W. Innes[2], Peter L. Kuempel[2], Michael A. Djordjevic[3] and Barry G. Rolfe[3]
[1]School of Chemistry, Macquarie University, N.S.W. 2113, Australia
[2]Dept. MCD-Biology, Univ. of Colorado, CO 80309, U.S.A.
[3]Research School of Biological Sciences, A.N.U., Canberra, A.C.T. 2601, Australia

INTRODUCTION

To successfully establish nitrogen-fixing root nodules in a legume plant, a Rhizobium requires at least six plasmid-borne genes (designated nodABCDEF) (Djordjevic et al 1985a; Djordjevic et al 1985b), which are arranged in three separate operons (nodABC, nodD and nodFE) (Torok et al 1984; Rossen et al 1985b). Only one nodulation gene, nodD, is expressed by Rhizobium in culture and its gene product, together with substances secreted by the host plant, induces expression of genes in nodABC and nodFE (Innes et al 1985; Rossen et al 1985b). Lac fusions to presently undefined nod genes, located in regions II and IV of the 14 kb nodulation fragment of R.trifolii, also respond to substances in plant exudate (Innes et al 1985). It is therefore likely that these regions, too, are under control of the nodD product. We report here that the inducing activity of washings of undamaged clover roots is due to flavones, and that several structurally related compounds have similar activities.

THE SITE OF PRODUCTION OF INDUCING SUBSTANCES

Infection of legumes by Rhizobium occurs mainly in the emerging root hair zone (Bauer 1981). That this is also the site of release of the substances which induce nod gene expression

NATO ASI Series, Vol. H 4
Recognition in Microbe-Plant Symbiotic and Pathogenic Interactions. Edited by B. Lugtenberg

was shown by placing seedlings onto a lawn of a R.trifolii strain containing a lacZ gene fusion to the nodA gene. Intense blue areas, resulting from cleavage of 5-bromo-4-chloro-3-indolyl-β-D-galactoside (X-gal), extended up to 20 mm from root tips, indicating localized secretion of nodA gene inducer. Colour was not produced in the absence of either the plants or the bacteria, or when bacteria containing an anti-sense nod::lac fusion were used.

ISOLATION AND CHARACTERIZATION OF INDUCING SUBSTANCES

The stimulatory substances present in aqueous washings of seedlings of white clover were adsorbed to macroreticular polystyrene resin and subsequently eluted with methanol. The activity could be resolved into three fractions by either paper chromatography (1-butanol:acetic acid:water, 5:1:2) or preparative silicic acid thin-layer chromatography (chloroform:methanol, 9:1).

Preparative isolation was carried out by column chromatography on silicic acid (chloroform:methanol, 96:4). Eluates were combined on the basis of thin-layer chromatography to give three active fractions, labelled A,B,C in order of increasing mobility. The fractions appeared homogeneous by thin-layer chromatography, but significant levels of impurities were revealed by NMR spectroscopy. In each case, however, the major (ca. 80%) component was a hydroxyflavone.

The flavone present in fraction A was identical to 7,4'-dihydroxyflavone, as judged by spectroscopic and chromatographic methods. The main components in fractions B and C were identified by proton NMR spectroscopy as geraldone and 4'-hydroxy-7-methoxyflavone, respectively. Sample B had the same chromatographic properties as previously reported for geraldone (Wong and Francis 1968), but 4'-hydroxy-7-methoxyflavone has not previously been isolated. The presence in its mass spectrum of peaks of relative molecular mass (M_r) 268 and 240 and its higher chromatographic mobility, relative to 7,4'-dihydroxyflavone, are consistent with the presence of only a single free hydroxyl group.

INDUCING ACTIVITIES OF SYNTHETIC FLAVONES

Exposure of *R.trifolii* *nod218* cells to 5 x 10^{-7} M 7,4'-dihydroxyflavone for 30 min leads to significant production of β-galactosidase, indicating rapid induction of *nod* gene transcription. The level of β-galactosidase increased progressively for a further 2 to 3 h and, for subsequent experiments, an induction period of 3 h was used.

The structural requirements for inducing activity of flavonoids were assessed by screening several related compounds representing the flavone (I), flavanone (II) and chalcone (III) structural types (Fig. 1). The results in Table I, obtained with readily available materials, are typical maximum induction levels achieved with different flavonoids. Some variations are obtained from run to run, depending on the precise growth phase of the

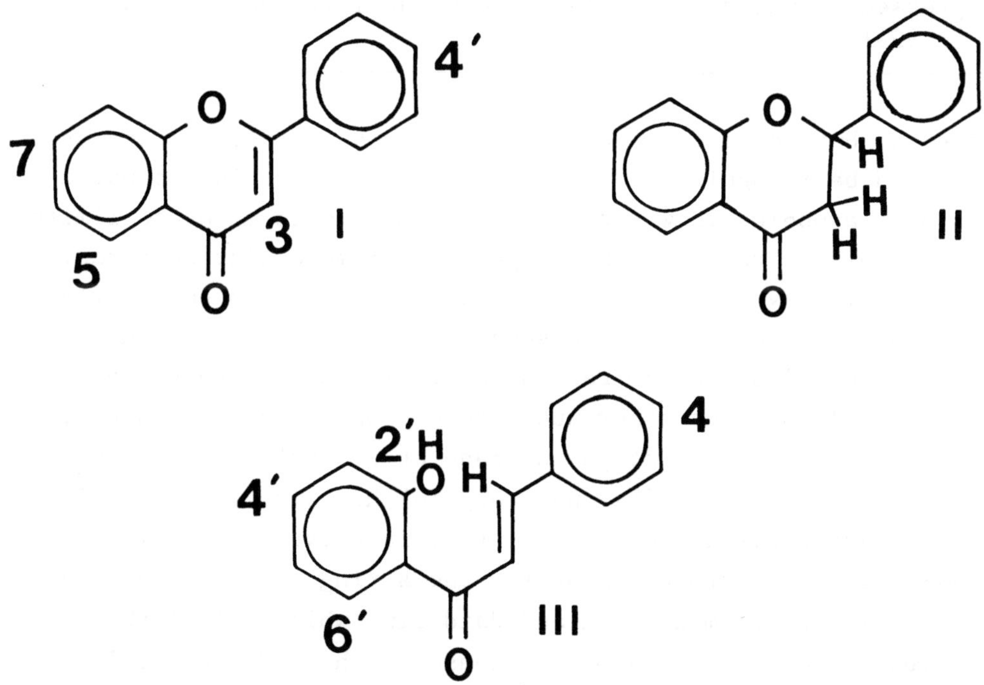

Fig. 1. Flavone (I), flavanone (II), and chalcone (III) types of flavonoids.

TABLE I. MAXIMUM INDUCING ACTIVITIES OF REPRESENTATIVE FLAVONOIDS[a]

HYDROXYL SUBSTITS.	FLAVONE (I)	FLAVANONE (II)	FLAVONOL (3-OH-I)	CHALCONE (III) etc.
7	7000	-	-	-
7,4'	7000	2000	-	500 (chalcone)
7,4'(3-OMe)	5800 (geraldone)	-	-	370 (chalcone)
5,7	1600 (chrysin)	-	-	-
5,7,4'	7000 (apigenin)	5200 (naringenin)	1300 (kaempferol)	-
7,3',4'	-	-	350 (fisetin)	-
5,7,3',4'	4000 (luteolin)	-	500 (quercetin)	390 (taxifolin)

[a]Expressed as units of β-galactosidase activity obtained with R.trifolii nod218

bacteria, but the differences shown are consistent. Furthermore, it should be recognized that, because of the very low solubilities of flavonoids in water, the maximum effects might be limited by insolubility rather than any particular molecular features.

A 7-hydroxy group confers inducing activity on both a flavone and a flavone. An oxygen substituent is present in this position in all but a few rare natural flavonoids (Harborne 1967; Eade et al 1966), but other substituents on the A ring, such as the 8-C-glucosyl group of bayin (Eade et al 1966), can block activity. 5,7-dihydroxyflavone (chrysin) has significant activity, but 6-hydroxyflavone is completely inactive, confirming the primary structural requirement for the A ring. On the other hand, the isoflavones formononetin and daidzein, which fulfil the A ring requirement, are inactive because of their different location of the B ring.

When a 3-hydroxy group is also present, as in the flavonols, activity is greatly depressed, but changes in the substitution pattern of the B ring have much less pronounced effects. In the flavone series, apigenin is consistently more effective than luteolin and this difference is paralleled in the corresponding flavonols (kaempferol and quercetin). Furthermore, other B ring variants (morin and myricetin) have no detectable activity in R.trifolii, indicating that some selectivity of inducing activity can result from variations in the B ring. It is possible that this order of selectivity differs in other Rhizobium backgrounds.

The chalcones tested have low inducing activities which increase when the stock solutions are stored over a period of days. This is attributed to the production of increasing amounts of the isomeric flavanones, as the result of cyclization.

Preliminary experiments indicate that all flavonoid inducers are active in the same concentration range (half maximal effect between 7×10^{-7} to 3×10^{-6} M). This suggests that, regardless of maximal activity, the critical association constants, for uptake or activity, are similar. The level of nod gene induction in the rhizosphere might therefore be determined by competitive binding by the different flavonoids present.

The other genes in the nodulation cluster were also induced by these flavonoids. For example, a concentration of 5×10^{-6} M 7,4'-dihydroxyflavone induced transcription of genes of the nodFE operon, as well as those in regions II and IV of R.trifolii.

THE SIGNIFICANCE OF FLAVONES IN THE RHIZOSPHERE

The stimulatory factors produced by the other legume seedlings have not yet been characterized, but 7,4'-dihydroxyflavone is known to be present in other legumes (Wong and Francis 1968; Markham and Mabry 1981). Such 5-deoxyflavones appear to be confined to legumes (Harborne 1967; Markham and Mabry 1981) and this is consistent with our inability to detect inducing substances from non-legume plants. On the other hand, active 5-hydroxyflavones, such as luteolin and naringenin, are more widely distributed (Harborne 1967; Markham and Mabry 1981). The flavonols, which have only low activity, are most abundant and widespread (Harborne 1967; Markham and Mabry 1981) and it appears that Rhizobium has adapted to the flavonoids with the most restricted distribution. If the profile of flavonoids in the rhizosphere mirrors that in the leaves, the rhizosphere of a legume might provide precisely the right environment for optimal nod gene induction.

ACKNOWLEDGEMENT

We thank Anne Moten, Jan McIver and Brigid Roberts for excellent technical assistance.

REFERENCES

Bauer WD (1981) Infection of legumes by rhizobia. Ann Rev Plant Physiol 32: 407-449

Djordjevic MA, Schofield PR, Rolfe BG (1985a) Tn5 mutagenesis of Rhizobium trifolii host-specific nodulation genes result in mutants with altered host-range ability. Molec Gen Genet 200: 463-471

Djordjevis MA, Schofield PR, Ridge RW, Morrison NA, Bassam BJ, Plazinski J, Watson JM, Rolfe BG (1985b) Rhizobium nodulation genes involved in root hair curling (Hac) are functionally conserved. Plant Molec Biol 4: 147-160

Downie JA, Knight CD, Johnston AWB, Rossen L (1985) Identification of genes and gene products involved in the nodulation of peas by Rhizobium leguminosarum. Molec Gen Genet 198: 255-262

Eade RA, Salasoo I, Simes JJH (1966) Extractives of Australian timbers VII. Bayin. Aust J Chem 19: 1717-1727

Harborne JB (1967) Comparative biochemistry of the flavonoids, Academic Press, London/New York

Innes RW, Kuempel PL, Plazinski J, Canter-Cremers H, Rolfe BG, Djordjevic MA (1985) Plant factors induce expression of nodulation and host-range genes in Rhizobium trifolii. Molec Gen Genet 201: 426-432

Kondorosi E, Banfalvi Z, Kondorosi A (1984) Physical and genetic analysis of a symbiotic region of Rhizobium meliloti: Identification of nodulation genes. Molec Gen Genet 193: 445-452

Markham KR, Mabry TJ (1981) The flavones. In: Harborne JB, Mabry TJ, Mabry H (eds) The flavonoids. Chapman & Hall, London, p 267

Mulligan JT, Long SR (1985) Induction of Rhizobium meliloti nodC expression by plant exudate requires nodD. Proc Nat Acad Sci USA 82: 6609-6613

Rossen L, Johnston AWB, Downie JA (1985a) DNA sequence of the

Rhizobium leguminosarum nodulation genes nodAB and C required for root hair curling. Nucleic Acids Res 12: 9497-9508

Rossen L, Shearman CA, Johnston AWB, Downie JA (1985b) The nodD gene of Rhizobium leguminosarum is autoregulatory and in the presence of plant exudate induces the nodA,B,C genes. EMBO J 4: 3369-3373

Torok I, Kondorosi E, Stepkowski T, Posfai J, Kondorosi A (1984) Nucleotide sequence of Rhizobium meliloti nodulation genes. Nucleic Acids Res 12: 9509-9524 4.

Wong E, Francis CM (1968) Flavonoids in genotypes of Trifolium subterraneum II. Mutants of the Geraldton variety. Phytochem 7: 2131-2137

INDUCTION OF RHIZOBIUM NOD GENES BY FLAVONOIDS: DIFFERENTIAL ADAPTATION OF PROMOTER, nodD GENE AND INDUCERS FOR VARIOUS CROSS-INOCULATION GROUPS

Carel Wijffelman, Bas Zaat, Herman Spaink, Ine Mulders, Ton van Brussel, Rob Okker, Elly Pees, Ruud de Maagd and Ben Lugtenberg
University of Leiden,
Department of Plant Molecular Biology,
Nonnensteeg 3, 2311 VJ Leiden, The Netherlands

INTRODUCTION

The nodulation of leguminous plants by the soil bacterium Rhizobium is a multi-step process in which both plant genes and bacterial genes are involved (Vincent 1980). Many bacterial genes involved in the nodulation process have been shown to reside on plasmids, which are generally called Sym plasmids (Hooykaas et al 1981). These nod genes can be divided in common and host-specific genes, depending on whether they can be complemented by the corresponding genes of other fast-growing Rhizobia (Djordjevic et al 1985a; Fisher et al 1985; Wijffelman et al 1985). The genetic organization and the function of the common nod genes A,B,C,D,I,J of R.leguminosarum, R.trifolii and R.meliloti is very homologous, whereas the organization of the host-specific nodulation functions is more different (Djordjevic et al 1985b; Egelhoff et al 1985; Rossen et al 1984; Schofield et al 1986; Spaink et al, this volume; Török et al 1984). The nodD gene of all three species is the only nod gene which is transcribed constitutively, whereas the other nod operons are not transcribed when cells are grown in the standard culture media (Innes et al 1985; Mulligan et al 1985; Rossen et al 1985; Spaink et al, this volume). Recent data in several laboratories using the nod promoters fused to the E.coli lacZ gene show that the induction of these nod operons requires both a functional nodD gene product and a substance present in root exudate (Innes et al 1985; Mulligan et al 1985; Rossen et al 1985; Shearman

NATO ASI Series, Vol. H 4
Recognition in Microbe-Plant Symbiotic and Pathogenic Interactions. Edited by B. Lugtenberg

et al 1986; Spaink et al, this volume). These inducible common or host-specific nod operons are preceeded by a very strongly conserved sequence of DNA, the nod box, which is located about 200 bp upstream the start codon of the first gene (Rostas et al 1986; Schofield et al 1986; Scott et al 1986; Shearman et al 1985; Spaink et al, this volume). The complete inducibility of a 114 bp clone of the nodA promoter of R.leguminosarum, which contains an intact nod box strongly suggest that these nod boxes are important elements of nod promoters (Spaink et al, this volume).

The nodA,B and C genes of R.leguminosarum are involved in both root hair curling (Hac) and the formation of thick and short roots on Vicia sativa (Tsr) (Downie et al 1985; Van Brussel et al 1986; Wijffelman et al 1985). The formation of thick and short roots is induced by one or more factors secreted by R.leguminosarum. Production or secretion of this Tsr factor requires the presence of root exudate as well as an additional gene, nodD (Van Brussel et al 1986). In this paper we describe some properties of the inducing substance in root exudate of V.sativa. Subsequent experiments show that the inducing substance can efficiently be replaced by several flavonoids, and in all cases only nanomolar concentrations are required for the induction of all three inducible nod promoters (nodA, nodF and nodH) as measured using transcriptional lacZ fusions. Finally and most interestingly, our results show that all components required for the induction of the nod operons, the promoter, the nodD gene product and the inducing substance in exudate, are substantially different for members of various cross-inoculation groups, suggesting that adaptation during evolution has resulted in fine tuning of the molecules participating in the induction of the nodulation process.

PROPERTIES OF THE INDUCING FACTOR IN VICIA SATIVA ROOT EXUDATE

The presence of an inducing factor in two days old V.sativa root exudate was demonstrated by using RBL5560.pMP154. This strain contains the R.leguminosarum Sym plasmid pJB5 as well as pMP154, an expression plasmid in which the R.leguminosarum nodA promoter is cloned in front of the E.coli β-galactosidase structural gene, lacZ (Spaink et al, this volume). Incubation of this strain in the absence and presence of V.sativa root exudate resulted in 400 and 18000 units β-galactosidase, respectively. Ultrafiltration of V.sativa root exudate indicated a molecular weight of the inducing factor of less than 1000 daltons. The inducing activity is heat-stable. These properties suggested the possibility that the inducing substance was of flavonoid nature. Root exudate was concentrated 100-fold by vacuum evaporation, freeze dried and the lyophilisate was extracted with ethanol. The components of this exudate were separated using the two--dimensional cellulose thin layer chromatography system described for the separation of flavonoids (Mabry et al 1970). After separation, the chromatogram was cut and the fragments were tested for inducing activity. More than 80% of the inducing activity of the root exudate was present in one spot on the chromatogram. Flavanones, isoflavones and flavones with one hydroxyl group are substances with approximately the same chromatographic mobility in this system (Mabry et al 1970). A number of commercially available flavonoids were tested for inducing ability. None of the isoflavones daidzein, genistein, genistin, prunetin and 6,7,4' trihydroxyisoflavone had inducing ability, but the flavanones naringenin and eriodictyol, and 7-hydroxyflavone had. The latter substances were able to induce the nodA promoter efficiently at very low, i.e. nanomolar, concentrations (Fig. 1). The most active inducer was naringenin. The inducing activity of this flavanone was compared with that of V.sativa and V.hirsuta root exudate. With both naringenin (100nM) and two days old exudates the first induction was observed after 6 minutes and the induction level increased linearly in time for 10 hrs, after which a plateau was reached which was the same with naringenin and the exudates. The induction by naringenin, like that by exudate, appeared to be nodD dependent.

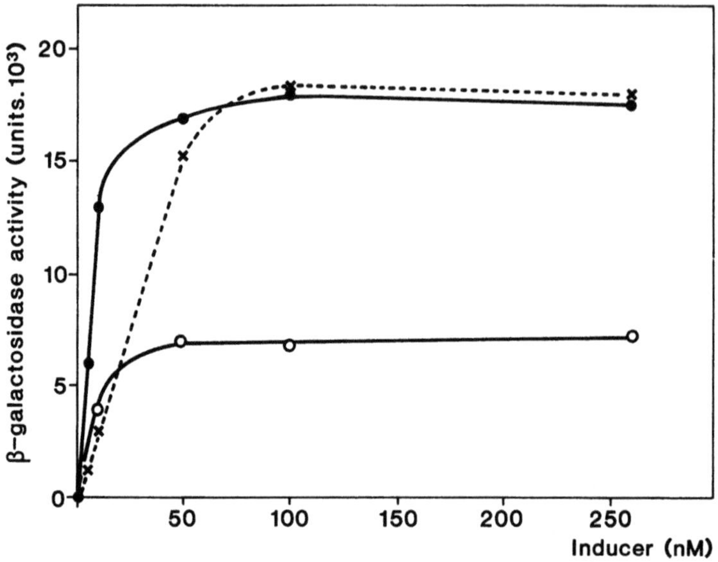

Fig. 1. Level of induction of the R.leguminosarum nodA promoter in strain RBL5560.pMP154 as a function of inducer concentration. Induction levels were measured after incubation for 18 hrs. Symbols: •, Naringenin; x, Eriodictyol and o, 7-OH flavone.

THE ROLE OF EXUDATE IN TSR FACTOR PRODUCTION CAN BE SUBSTITUTED BY NARINGENIN

Gene products of the nodABC operon are required for root hair curling and for the formation of thick and short roots on V.sativa. For the induction of root hair curling the presence of bacteria in the test system is necessary. This is a disadvantage in attempts to purify the factor(s) responsible for these phenomena. During the characterization of the Tsr-factor(s), which does not have this disadvantage, it was demonstrated that the Tsr factor is only produced when the bacteria are cultivated in the presence of plant root exudate (Van Brussel et al 1986). Our present results show that 200 nM naringenin can replace exudate in this respect. Naringenin-induced Tsr factor production mimicks exudate-induced Tsr factor production also in that it requires the presence of the Sym plasmid.

STRUCTURAL REQUIREMENTS OF THE INDUCING FACTOR

At least three inducible nod promoters (p.nodA, p.nodF and p.nodH) are present on the R.leguminosarum Sym plasmid pRL1JI. All three promoters can be induced by both Vicia exudates and naringenin. Concentrations of 400 nM of various flavonoids were tested for inducing ability of these promoters (Table 1) to determine the structural requirements of the inducing substance in R.leguminosarum. The 5,7,4' trihydroxyderivative of flavanone

TABLE 1. STRUCTURAL REQUIREMENTS OF THE INDUCING SUBSTANCE[a]

Putative inducer	Hydroxyl groups 3	5	7	3'	4'	Inducing ability
EXUDATE						
V.sativa						+
V.hirsuta						+
FLAVANONES						
Naringenin		OH	OH		OH	+
Eriodictyol		OH	OH	OH	OH	+
FLAVONES						
Flavone						-
5-hydroxyflavone		OH				-
7-hydroxyflavone			OH			+ (50%)
Chrysin		OH	OH			+ (15%)
Apigenin		OH	OH		OH	+
Luteolin		OH	OH	OH	OH	+
FLAVONOLS						
Kaempferol	OH	OH	OH		OH	+ (3%)
Quercetin	OH	OH	OH	OH	OH	-

[a]Strain RBL5560 with expression plasmids pMP154(p.nodA), pMP157 (p.nodF) or pMP163(p.nodH) were induced with 400 nM of the various flavonoids. β-galactosidase activity was measured after 18 hrs. Maximal induction is indicated as +. The maximum level of induction was 18000, 17000 and 5000 units of β-galactosidase with pMP154, pMP157 and pMP163, respectively. In case no maximal expression level was reached, the level is indicated as the percentage of the maximum level. -, no significant induction.

(naringenin) and of flavone (apigenin) and the 5,7,3',4' tetrahydroxyderivative of flavanone (eriodictyol) and of flavone (luteolin) appeared to be efficient inducers. So the difference between a flavanone and a flavone does not influence the inducing ability of the flavonoid. The presence of a hydroxyl group at the 3 position, as is the case in the flavonols kaempferol and quercetin, reduces the inducing ability strongly. The hydroxylation of the 7 position is important. Flavone and 5-hydroxyflavone have no inducing ability, whereas 7-hydroxyflavone induces the nod promoters to a level of 50% of the level reached with exudate or naringenin. The influence of the 4' hydroxyl group is still not clear. Apigenin (5,7,4' trihydroxy) is as active as exudate whereas chrysin (5,7 dihydroxy) induces only to 15% of the maximum level. However, chrysin is very insoluble and it is possible that the low value found for chrysin is due to insolubility rather than the intrinsic absence of maximal inducing ability.

DIFFERENCES BETWEEN nodA PROMOTERS OF FAST-GROWING RHIZOBIA

Three components are required for the induction of the nodA BCIJ operon of Rhizobium, namely the promoter itself (p.nodA), the nodD gene product, and the inducing substance(s) present in exudate. The three nodA promoters of R.leguminosarum, R.trifolii and R.meliloti have a nod box with a consensus sequence of at least 35 bp in common, and in all three promoters there is a distance of approximately 200 bp between the nod box and the translational start of nodA (Egelhoff et al 1985; Rostas et al 1986; Schofield et al 1986; Shearman et al 1986; Spaink et al, this volume). To compare the induction behaviour of the nodA promoters of these Rhizobia we have also cloned the nodA promoters of R.trifolii ANU843 and R.meliloti 2011 in the expression vector pMP150. For R.trifolii a 700 bp KpnI-BamHI fragment from pRt032 (Schofield et al 1984) was used and for R.meliloti a 500 bp SacI-BamHI fragment from pRmSL26 (Long et al 1982). The resulting expression vectors were pMP193 and pMP194, respectively. These plasmids were introduced in RBL5560 (the same strain in which the previous results were obtained with pMP154) and the induction by V.hirsuta exudate was measured. All three nodA promoters appeared to be inducible, but the level of induction was different. The highest level was reached with pMP154, (18000 units), whereas the induction with the p.nodA of R.trifolii and R.meliloti was considerably lower, 5000 units with pMP193 and 7000 units with pMP194. The flavonoids naringenin, eriodictyol, apigenin and luteolin induced the three nodA promoters to the same level as V.hirsuta exudate did.

THE nodD GENE PRODUCTS

A second component required for the induction of the nod genes is the nodD gene product. To find out whether there are differences between the nodD gene products of R.leguminosarum, R.trifolii and R.meliloti, we transferred the nodD gene, including some other nod genes, into the Sym plasmid-cured R.trifolii strain LPR5045 (Hooykaas et al 1981). For R.leguminosarum we used pMP100, an IncP plasmid in which the 6.6 EcoRI nod fragment was cloned (nod genes ABCDE and F). R.trifolii pRt032 contains all known nodulation genes of R.trifolii whereas pRmSL26 contains nodABC and D of R.meliloti. Finally we transfered the expression plasmid pMP154 to these strains. The resulting strains were tested for induction by several flavonoids (Table 2). Surprisingly, the uninduced levels of the three strains were not identical, although the same expression plasmid was present in all three strains. Moreover, remarkable differences in the induction of these strains by the flavonoids were observed in that (i) the induction pattern by the various compounds is not the same, and (ii) the maximum level of induction differs. When the nodD gene products of R.leguminosarum and R.trifolii are compared, differences in induction levels were observed with eriodictyol and kaempferol. The induction of pMP100 by eriodictyol is comparable to that by V.hirsuta exudate while the induction of pRt032 by eriodictyol is six times reduced compared to the induction by V.hirsuta exudate. In contrast, kaempferol is inactive with pMP100 whereas it is active, although at a lower level, with

TABLE 2. INFLUENCE OF THE SOURCE OF THE nodD GENE ON THE LEVEL OF INDUCTION OF THE nodA PROMOTER OF R.LEGUMINOSARUM[a]

INDUCER	Units of β-galactosidase.10^{-3} with nodD from		
	pMP100 (R.leg)	pRt032 (R.tri)	pRmSL26 (R.mel)
None	0.3	1.1	1.4
Naringenin	18	31	1.8
Eriodictyol	19	5	1.9
Apigenin	19	30	4
Luteolin	18	26	8
Kaempferol	0.4	9	1.4
EXUDATES			
V.sativa	16	15	2
V.hirsuta	20	35	7
T.repens	13	35	9
T.subterraneum	3	18	2

[a]Strains containing the p.nodA-lacZ fusion plasmid pMP154 plus one of the plasmids containing nodD of various cross-inoculation groups were grown overnight in induction medium and induced with 100 nM of one of the flavonoids or with a 2-fold diluted, two days old, root exudate. The β-galactosidase activity was measured after 18 hrs of incubation.

pRt032. The differences are more pronounced with pRmSL26. The flavanones naringenin and eriodictyol, and the flavonol kaempferol cannot act as inducer with the nodD gene product of R.meliloti, while the flavones luteolin and apigenin are active. Only luteolin reaches an induction level which is comparable with that of V.hirsuta exudate. The second difference is the maximum level of induction. The induction with the nodD gene of R.leguminosarum, cloned in an IncP plasmid (pMP100), is comparable with the induction of a strain with an intact Sym plasmid (RBL5560), but surprisingly the induction of the p.nodA of R.leguminosarum is higher with the nodD gene product of R.trifolii (pRt032). The induction of this promoter by the nodD gene product of R.meliloti is much lower.

THE EXUDATES

The third component of the induction system of the nod genes of Rhizobium is the inducing substance present in exudate. As differences in the nodD gene product of fast-growing Rhizobia were detected, we considered the possibility that also exudates differ in their inducing capacity. Therefore we compared the induction of strains with different nodD gene products using several exudates. The results (Table 2) show that the inducing capacity of the various exudates is not identical. V.hirsuta and T.repens exudates induce all three strains rather well. The induction by V.sativa exudate is very low in the presence of the nodD gene product of R.meliloti, whereas T.subterraneum exudate can induce to a rather high level only with the nodD gene product of R.trifolii.

DISCUSSION

The induction of the nod genes of R.leguminosarum requires both the nodD gene product and an inducing substance present in root exudate of leguminous plants (Rossen et al 1985; Shearman et al 1986; Spaink et al, this volume). We found that the main inducing substance in V.sativa root exudate is probably a flavanone. To study the expression, regulation and promoter structure of the nod genes of R.leguminosarum, it is important to have a pure inducing substance at one's disposal. The results of the present paper show that naringenin is an excellent substitute for root exudate (Table 1). Maximal induction of the nod promoter was observed at very low concentrations, and no significant differences were found between naringenin and root exudate concerning the requirements for synthesis, the kinetics of induction and of the requirements for induction of the Tsr factor synthesis.

The induction systems for R.leguminosarum, R.trifolii and R.meliloti consist of three components: (i) the promoter of the nod operon including the nod box, (ii) the nodD gene product and (iii) an inducing substance of root exudate. The three nod promoters of R.leguminosarum (p.nodA, p.nodF and p.nodH) and the nodA promoters of R.trifolii and R.meliloti are inducible in combination with the nodD gene product of R.leguminosarum (RBL5560) by both Vicia exudates as well as by several flavonoids, but the maximal level of induction (varying between 5000 and 18000 units of β-galactosidase) is dependent on the promoter. Whether these differences play a role in the nodulation process and whether these differences are caused by the 200 bp between the nod box and the translational start of the first nod gene, is still uncertain.

The second component of the induction system is the nodD gene product. We found that the background level of β-galactosidase of pMP154 (p.nodA-lacZ) is dependent on the bacterial background (Table 2). In a cured strain or in a strain with the nodD gene product of R.leguminosarum this level is 300 β-galactosidase units, while the background level of the same plasmid in strains with the nodD gene product of either R.trifolii or R.meliloti is more than 1000 units. The most likely explanation of this increase seems to be the constitutive expression of the promoter of nodA at a low level in a nodD dependent way (see also Rolfe et al, this volume).

The homologous combination pMP100 (nodD) and pMP154 (p.nodA) of R.leguminosarum is induced at rather high levels (maximal 20000 units) by most flavonoids and exudates. Only with T.subterraneum the induction is markedly reduced, while with kaempferol hardly any induction was observed. The heterologous combination pRt032 (nodD of R.trifolii) with pMP154 is induced by all flavonoids and exudates at a level markedly higher than the induction by the nodD gene product of R.leguminosarum, with the exception of V.sativa exudate, which induces to the same level, and eriodictyol, which induces to a lower level. Both flavones and flavanones can act as inducers in the presence of a nodD gene product of R.leguminosarum or R.trifolii. In contrast the induction of pMP154 by the nodD gene product of R.meliloti requires a flavone, and for induction at a level comparable to the best exudate also hydroxylation of the 3' position is required, as is the case in luteolin. The natural inducer present in alfalfa seed exudate is luteolin (Long et al, this volume). The main inducer in V.sativa root exudate is probably a flavanone, and therefore it is not surprising that this exudate is a bad inducer in the strain with a nodD gene product of R.meliloti. Redmond et al (this volume) describe the presence of two efficient inducers in T.repens seed exudate, DHF (7,4'-dihydroxyflavone) and geraldone (7,4'-dihydroxy-3'-methoxyflavone). T.repens root exudate induces pMP154 in strains with the nodD gene product of R.meliloti better than apigenin (5,7,4'-trihydroxyflavone) does, suggesting that a methoxy-group at the 3' position is as efficient as a hydroxyl-group at that place, as is found in luteolin.

The observed differences in induction by exudates in strains with the nodD gene product of R.leguminosarum, R.trifolii or R.meliloti point to a certain host preference in the induction system. In addition, the inducing substances in exudates from plants from the same cross-inoculation group seem not to be identical (e.g. V.sativa and V.hirsuta exudate, and T.repens and T.subterraneum exudate in Table 2).

From the results described in this paper it is clear that all three components of the natural induction system of Rhizobium harbour differences between R.leguminosarum, R.trifolii and R.meliloti. How far these differences are more general for the induction system in the various cross-inoculation groups and whether the differences play a role throughout nodule development is subject of our present investigations.

ACKNOWLEDGEMENTS

The investigations were partly supported by the Foundation of Fundamental Biological Research, which is subsidized by the Netherlands Organization for the Advancement of Pure Research.

REFERENCES

Djordjevic MA, Schofield PR, Ridge RW, Bassam BJ, Plazinski J, Watson JM and Rolfe BG (1985a) Rhizobium nodulation genes involved in root hair curling (Hac) are functionally conserved. Plant Mol Biol 4: 147-160

Djordjevic MA, Schofield PR and Rolfe BG (1985b) Tn5 mutagenesis of R.trifolii host-specific nodulation genes result in mutants with altered host-range ability. Mol Gen Genet 200: 463-471

Downie JA, Knight CD, Johnston AWB and Rossen L (1985) Identification of genes and gene products involved in the nodulation of peas by Rhizobium leguminosarum. Mol Gen Genet 198: 255-262

Egelhoff TT, Fisher RF, Jacobs JW, Mulligan JT and Long SR (1985) Nucleotide sequence of Rhizobium meliloti 1021 nodulation genes: nodD is read divergently from nodABC. DNA 4: 241-248

Fisher RF, Tu JK and Long SR (1985) Conserved nodulation genes in Rhizobium meliloti and Rhizobium trifolii. Appl Env Microbiol 49: 1432-1435

Hooykaas PJJ, van Brussel AAN, den Dulk-Ras H, van Slogteren GMS and Schilperoort RA (1981) Sym plasmid of Rhizobium trifolii expressed in different rhizobial species and Agrobacterium tumefaciens. Nature (London) 291: 351-353

Innes RW, Kuempel PL, Plazinski J, Canter-Cremers H, Rolfe BG and Djordjevic MA (1985) Plant factors induce expression of nodulation and host-range genes in R.trifolii. Mol Gen Genet 201: 426-432

Long SR, Buikema WJ and Ausubel FM (1982) Cloning of Rhizobium meliloti nodulation genes by direct complementation of nod$^-$ mutants. Nature (London) 298: 485-488

Mabry TJ, Markham KR and Thomas MB (1970) The systematic identification of flavonoids. Springer Verlag, Berlin, Heidelberg, New York

Mulligan JT and Long SR (1985) Induction of Rhizobium meliloti nodC expression by plant exudate requires nodD. Proc Natl Acad Sci USA 82: 6609-6613

Rossen LR, Johnston AWB and Downie JA (1984) DNA sequence of the Rhizobium leguminosarum nodulation genes nodA,B and C required for root hair curling. Nucl Acids Res 12: 9497-9508

Rossen L, Shearman CA, Johnston AWB and Downie JA (1985) The nodD gene of Rhizobium leguminosarum is autoregulatory and in the presence of plant exudate induces the nodA,B,C genes. EMBO J 4: 3369-3373

Rostas K, Kondorosi E, Horvath B, Simoncsits A and Kondorosi A (1986) Conservation of extended promoter regions of nodulation genes in Rhizobium. Proc Natl Acad Sci USA 83: 1757-1761

Schofield PR, Ridge WR, Rolfe BG, Shine J and Watson JM (1984) Host-specific nodulation is encoded on a 14 kb DNA fragment in Rhizobium trifolii. Plant Mol Biol 3: 3-11

Schofield PR and Watson JM (1986) DNA sequence of Rhizobium trifolii nodulation genes reveals a reiterated and potentially regulatory sequence preceding nodABC and nodFE. Nucleic Acids Res 14: 2891-2903

Scott KF (1986) Conserved nodulation genes from the non-legume symbiont Bradyrhizobium sp. (Parasponia) Nucleic Acids Res 14:

2905-2919

Shearman CA, Rossen L, Johnston AWB and Downie JA (1986) The Rhizobium leguminosarum nodulation gene nodF encodes a polypeptide similar to acyl-carrier protein and is regulated by nodD plus a factor in pea root exudate. EMBO J 5: 647-652

Török I, Kondorosi E, Stepkowski T, Pósfai J and Kondorosi A (1984) Nucleotide sequence of Rhizobium meliloti nodulation genes. Nucl Acids Res 12: 9509-9524

Van Brussel AAN, Zaat SAJ, Canter-Cremers HCJ, Wijffelman CA, Pees E, Tak T and Lugtenberg BJJ (1986) Role of plant root exudate and Sym plasmid-localized nodulation genes in the synthesis by Rhizobium leguminosarum of Tsr factor, which causes thick and short roots on common vetch. J Bacteriol 165: 517-522

Vincent JM (1980) In: WE Newton, WH Orme-Johnson (eds) Factors controlling the Legume-Rhizobium symbiosis. Nitrogen Fixation Vol. II. Univ. Park Press, Baltimore p. 103-129

Wijffelman CA, Pees E, van Brussel AAN, Okker RJH and Lugtenberg BJJ (1985) Genetic and functional analysis of the nodulation region of the Rhizobium leguminosarum Sym plasmid pRL1JI. Arch Microbiol 143: 225-232

2505-2519

[illegible]

INDUCTION OF *RHIZOBIUM LEGUMINOSARUM NOD* GENES BY FLAVONOIDS IS ANTAGONIZED BY OTHER PLANT-SPECIFIED AROMATIC COMPOUNDS

J L Firmin, K E Wilson, L Rossen, and A W B Johnston
John Innes Institute,
Colney Lane,
NORWICH, NR4 7UH
United Kingdom

ABSTRACT

Chromatographic and chemical analyses indicated that the compounds present in peas which activated *nodD*-dependent transcription of the *R. leguminosarum nodFE* and *nodABCIJ* operons were flavonoids. This was confirmed by showing that three flavones, three flavonones and one flavone glucoside could induce transcription when present at concentrations as low as 10nM. Extracts of alfafa and *Vigna* also activated *R. leguminosarum nod* gene transcription. In addition to identifying molecules that activated the transcription of *nodFE* and *nodABCIJ*, several compounds were found to antagonize this induction. These anti-inducers included some phenolics which had previously been shown to activate transcription of the *vir* genes of *Agrobacterium tumefaciens*. Also, certain flavonols and isoflavonoids acted as potent anti-inducers.

INTRODUCTION

Rhizobium nodulates legumes because it has nodulation (*nod*) genes and these have been the subject of detailed physical and genetic analyses in several laboratories. The three species whose *nod* genes have been described in the greatest detail are the symbionts of peas, alfafa and clover; respectively these are *R. leguminosarum*, *R. meliloti* and *R. trifolii*. Several observations have emerged from these studies which show that the location, structure, function and regulation of *nod* genes in

NATO ASI Series, Vol. H 4
Recognition in Microbe-Plant Symbiotic and Pathogenic Interactions. Edited by B. Lugtenberg

these different Rhizobium species are in fact quite similar. In all three, genes for the early stages of nodulation and host-range specificity are on large 'symbiotic' plasmids (see Long, 1984). Secondly, comparison of the sequences has shown that there are highly conserved corresponding nod genes in the different species (Rossen et al 1984; Torok et al 1984; Egelhoff et al 1985; Schofield & Watson 1986; Shearman et al 1986); genetic studies showed that a given nod gene from one species is functionally equivalent to that of a different species - for example nodC mutants of R. meliloti, which are defective in the nodulation of alfafa, were corrected for nodulation of this host by the nodC gene from R. trifolii, (Fisher et al 1985). Thirdly, the fine scale location of particular nod genes relative to each other is conserved, though not identical, in these three Rhizobium species (Rossen et al 1984; Evans & Downie 1986; Shearman et al 1986 - for R. leguminosarum); (Schofield & Watson 1986 - for R. trifolii); (Torok et al 1984; Egelhoff et al 1985; A. Kondorosi, personal communication - for R. meliloti). Lastly, the regulation of the expression of the nod genes is similar but not identical in these Rhizobium species.

In this paper, recent data on the regulation of nod genes on the R. leguminosarum symbiotic plasmid pRLIJ1 will be presented. The region of pRLIJ1 to be described contains 9 kb of DNA which, when transferred to other species of Rhizobium, lacking a resident symbiotic plasmid, confers the ability to form pea nodules which contain infection threads and bacteroids and which, for the first 15 days following inoculation, are essentially indistinguishable from those induced by wild-type strains (Downie et al 1983). Genetic and physical analyses of the DNA in this nodulation region of pRLIJ1 showed it contained eight genes termed nod EFDABCIJ (Downie et al 1985; Rossen et al 1984; Shearman et al 1986; Evans & Downie 1986 and see Fig. 1). Mutations in nod D, A, B and C abolished nodulation whereas mutations in the other four nod genes delayed nodule onset and reduced the number of nodules formed. In R. trifolii a very similar arrangement of nod genes was found (Schofield & Watson 1986). In R. meliloti nodD also resides upstream from nodABCI but in this species nodF and nodE are in a region of the symbiotic plasmid which is 10 kb downstream of the nodC gene (Torok et al 1984; Egelhoff et al 1985; A. Kondorosi, personal communication).

Rossen et al (1984) showed that the nodABC genes of the R. leguminosarum pRLIJ1 were sufficient for root hair curling, an early step in the infection process. However, mutations in the nodD gene abolished root hair curling and nodulation and it was suggested (Rossen et al 1984) that nodD might be regulatory, being required for the activation of transcription of the other nod genes. In both R. meliloti and R. leguminosarum this prediction has been borne out (Rossen et al 1985; Shearman et al 1986; Mulligan & Long 1985).

Studies on nod gene regulation have been done using translational (Mulligan & Long 1985; Rossen et al 1985) or transcriptional (Innes et al 1985) fusions between the E.coli lacZ gene lacking its promoter (and for the translational fusions its ribosomal binding site) to individual nod genes. Figure 1 shows the dimension of the nodC-lacZ and nodF-lacZ fusion plasmids that have been used for the analysis of nod gene regulation in R. leguminosarum. The following was deduced about the regulation of nod gene expression in this species (Rossen et al 1985; Shearman et al 1986; I.J. Evans, personal communication).

(1) These 8 genes are in three transcriptional units, nodABCIJ, nodFE and nodD.

(2) nodD is expressed constitutively, but in normal media nodABCIJ and nodFE are transcribed at very low levels, if at all.

(3) nodD has at least two regulatory functions; in media which either contains or lacks pea root exudate it represses its own transcription and it activates high-level expression (70-fold increase over background) of the nodFE and the nodABCIJ genes when the cells were exposed to root exudate from peas.

(4) Also, DNA sequence analysis revealed a conserved sequence (the 'nod-box') upstream of nodFE and nodABC which appears to be involved in their activation by nodD plus factors from root exudate. This same (or very closely related) sequence is found in the same position relative to the start of nodF and nodA in R.

trifolii (Schofield & Watson 1986) and R. meliloti (Torok et al 1984; Rostas et al 1986).

In R. meliloti (Mulligan & Long 1985) and R. trifolii (Innes et al 1985) nodD is also transcribed constitutively and transcription of the other nod genes required factors from the exudate of alfafa and clover respectively. For R. meliloti the activation of nodABC was shown to depend upon the presence of nodD but, unlike R. leguminosarum, R. meliloti nodD was not autoregulated.

In this paper, we show that certain flavonoids can, in the presence of nodD, activate transcription of nodABCIJ and nodFE of R. leguminosarum. Also, we introduce a novel phenomenon, the identification of compounds which, at low concentration (1-10 μM) prevent or greatly diminish the activation of nodFE and nodABC1J by nodD plus flavonoids.

RESULTS

Previously, a nodC-lacZ fusion (pIJ1477) and a nodF-lacZ fusion (pIJ1521) where constructed in which these two nod genes plus their regulatory upstream sequences were fused, in-frame to the lacZ gene of E.coli which had been cloned in the wide host-range P-1 group fusion plasmid pIJ1363 (Rossen et al 1985; Shearman et al 1986; see Fig. 1).

The effects of different compounds on the expression of these two nod-lac fusions were measured using strains containing either pIJ1477 or pIJ1521. These strains lacked a symbiotic plasmid but contained a second recombinant plasmid, pIJ1518 (Rossen et al 1985, Fig. 1), which has the pRL1JI nodD gene cloned in the Q-group vector pKT230.

There are at least four activating compounds in pea exudate

It was found that induction of nodF-lacZ and nodC-lacZ transcription was obtained with extracts of roots, seeds (both dried and germinating) and callus of peas. Further, extracts of seeds of alfalfa, Vigna and clover, three legumes not nodulated by R. leguminosarum also had activity.

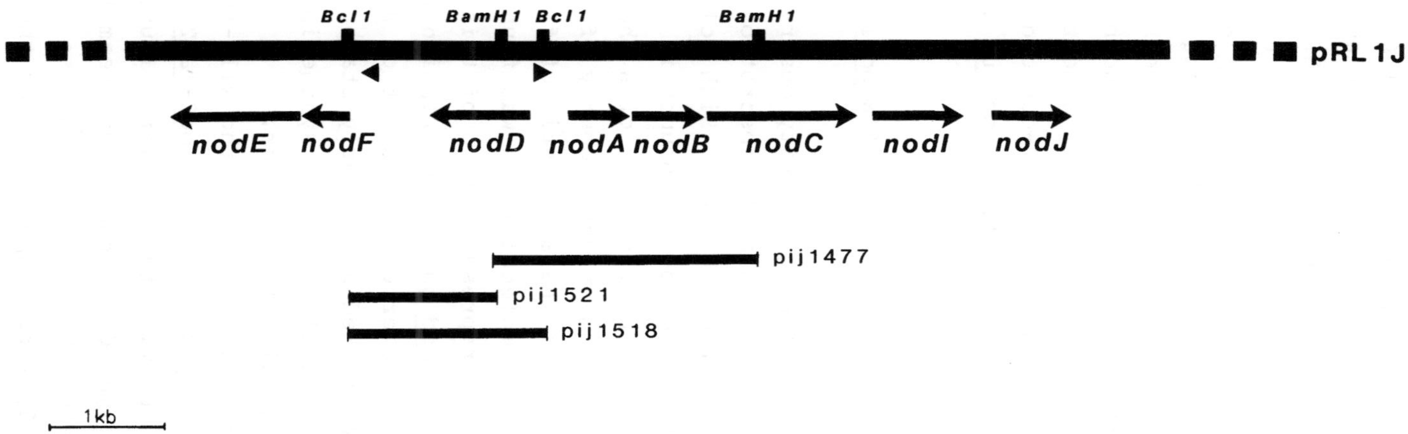

Figure 1. Location of eight nod genes of pRL1JI.

The dimensions and direction of transcription of the eight nod genes are shown. Restriction sites used for the construction of the two fusions pIJ1477 (nodC-lacZ), and pIJ1521 (nodF-lacZ) are indicated. pIJ1518 contains the nodD gene, as a BclI fragment cloned in the BamH1 site of pKT230. The closed triangles show the locations and the orientations of the conserved sequences (nod-boxes) that are associated with upstream sequences of nodFE and nodABCIJ that may be involved in nodD-dependent activation of transcription.

To characterize in more detail the nature of the activation molecule(s) in peas, seeds (variety Wisconsin Perfection) were soaked for three days in water, the liquor was concentrated and following a preliminary clean up by reverse phase HPLC was analysed further by analytical HPLC. Fractions of 0.25 ml were assayed for their ability to activate the nodF-lacZ and the nodC-lacZ fusions. Four different peaks of inducing ability were found one of which had the greatest activity. A peak of activity with the same retention time as this peak was found both in a methanol-extract of dried peas and in the root exudate of 10-day old pea seedlings. Extracts of germinating 8-day old alfafa seeds also induced transcription of R. leguminosarum nodF-lacZ and nodC-lacZ.

Induction of nod gene expression by selected flavonoids

The major peak of nod gene-inducing activity observed following HPLC could also be detected by UV absorbtion, suggesting the presence of an aromatic ring. High voltage paper electrophoresis showed that at high pH, compounds in this peak carried a weak negative charge characteristic of a phenolic hydroxyl group. Electron-impact mass spectrometry of the material in this peak showed ions characteristic of flavonoid fragmentation although no clear molecular ion was found.

To determine whether or not representatives of this class of compound were able to activate nod gave transcription we took advantage of our access to a flower colour mutant of Antirrinhum majus, which is genetically blocked at an early stage in flavonoid biosynthesis, and therefore cannot synthesise any of the flavonoids present in the wild-type flower.

A methanolic extract from the flowers of the wild-type and the mutant nivea lines was bioassayed using the nodC-lacZ fusion. Activity was only found in the wild-type flower. Thus, compound(s) produced by biosynthetic steps downstream from the nivea block (which inactivates the chalcone synthase group) (Spiribille & Forkmann 1982) were responsible for the nod gene-inducing activity found in the wild-type flower.

Several commercially available flavonoids, naringinin,

eriodictyol and luteolin, which are present in wild-type Antirrhinum flowers, but absent in the mutant strains, activated the nodF-lacZ and nodC-lacZ fusions (Table 1). Analogues of these compounds were also tested for nod gene-inducing ability; active compounds are listed in Table 1.

TABLE 1
Induction of nod gene expression in R. leguminosarum by selected flavonoids[a]

Compound	β-galactosidase activities			
FLAVANONES	100μM	1μM	100nM	10nM
4', 5,7,-trihydroxy-flavanone(naringinin)	4552	3154	100	58
3',5,7,-trihydroxy-4'methoxy-flavanone (hesperitin)	9207	6409	1036	250
3', 4',5,7,-tetra-hydroxy-flavanone (eriodictyol)	7144	5066	523	75
FLAVONE GLUCOSIDE				
4',5,7,-trihydroxy flavone 7-0-glucoside (apigenin 7-0-glucoside)	5964	3642	239	55
FLAVONES				
4',5,7,-trihydroxy-flavone(apigenin)	5163	4649	541	59
3',4',7,-trihydroxy-flavone	6814	1069	109	64
3',4',7,-tetra-hydroxy-flavone(luteolin)	5120	1437	77	73

[a]The strain of Rhizobium containing the nodC-lacZ fusion plasmid pIJ1477 plus the nodD gene cloned in pIJ1518 was grown in minimal medium containing different concentrations of the flavonoids shown. β-galactosidase activities were measured as described by Rossen et al (1985). The control values (i.e. no inducer added) were approximately 60 units (see Rossen et al 1985)

In general, flavones and flavonones with hydroxylated A and B rings appear to be active; however all flavonols and isoflavanoids tested were inactive. Two of the active flavonoids (eriodyctiol and apigenin 7-O-glucoside) had identical retention time (on HPLC) to that of the major inducing peak from pea exudate. Acid hydrolysis (2N HCL/100° (1 hr)) reduced the biological activity associated with this peak by c.50% and results in an additional biologically active peak with the same retention time as apigenin. Thus the major inducing peak from pea seed exudate contains at least two biologically active compounds one of which is indistinguishable from a glycoside of apigenin.

Identification of phenolics and isoflavonoids that inhibit transcription of nodFE and nodABCIJ

Having established that certain flavones and flavonones could activate *nod* gene expression we tested related compounds for their abilities to enhance or to inhibit the *nodD*-dependent induction of *nodFE* and *nodABCIJ* in the presence of pea exudate (Table 2).

TABLE 2
Antagonism of transcription of *nodC-lacZ* fusion by phenolic compounds[a]

COMPOUND	Concentration of antagonist (μM) 1000	500	100	50	10
Syringic acid	83	39	31	29	23
Syringaldehyde	88	37	22	23	23
Acetovanillone	95	95	91	85	20
Sinapinic acid	89	60	28	15	22
Acetosyringone	96	94	95	46	5
3',5-dimethoxyacetophenone	92	69	14	21	19
4-hydroxyacetophenone	94	95	94	91	41
Kaempfeirol	ND	78	80	78	63
Genistein	ND	100	100	100	68

[a]*Rhizobium* cells containing the *nodC-lacZ* fusion plasmid pIJ1477 plus the cloned *nodD* gene (in pIJ1518) were grown overnight in minimal medium containing *nod* gene inducer from pea seeds plus various concentrations of the 'anti-inducers' shown. The figures refer to the % age inhibition of β-galactosidase activities compared to the control containing no antagonists. ND, not determined.

Several inhibitors, but no enhancers were found. The most potent inhibitor that was identified was the isoflavonoid genestein but the flavonol kaempferol and various simple phenolics were also inhibitory at low (1-10μM) concentrations. (Table 2) Interestingly, some of these phenolic 'anti-inducers' included compounds (e.g. acetosyringone, acetovanillone and 4-hydroxyacetophenone which had been shown by Stachell et al (1985) to be active in inducing transcription of the vir genes of A. tumefaciens.

DISCUSSION

The results in this paper clearly show that several different flavones, flavonones and one flavonone-glucoside are very potent at inducing nodD-dependent transcription of the R. leguminosarum nodFE and nodABC1J operons. The seeds of alfafa also exuded molecules that could activate the transcription of these R. leguminosarum nod genes. S.R. Long and her colleagues (personal communication) found that the flavanone luteolin was the most active compound extracted from alfafa seeds which could induce transcription of the R. meliloti nodABC genes. However, authentic luteolin did not have the same retention time on HPLC as the major bioactive compounds present in alfafa seedlings when assayed with the R. leguminosarum nodABC1J or nodFE fusions. It is possible that the compounds excuded by different varieties of alfafa are not the same. Alternatively, it may be that, although nodD genes of different species of Rhizobium are similar (Egelhoff et al 1985; Shearman et al 1986; Schofield & Watson 1986), there may be quantitative differences in the interactions (either direct or indirect) between different activator molecules and the nodD gene products of a particular Rhizobium species.

The distribution of flavones and flavonones is widespread in the plant kingdom and indeed the data presented here showed that activating molecules were present in the flowers of Antirrhinum. The reasons why no inducing activities were detected in the roots and seeds of several non-legumes (e.g. millet, Datura, barley, Zinnia) indicates that legumes may be unusual in that they excude flavones and flavonones into the surrounding medium whereas non-legumes do not. Alternatively, non-legumes may exude more active 'anti-inducers'.

Several flavones and flavonones acted as inducers, the flavonone hesperetin being the most active of these tested. The means by which anti-inducers work is not clear. An examination of their structures shows that they have similarities to those of the flavones or flavonone inducers. Perhaps by competitive (or non-competitive inhibition) they interfere with the uptake of the inducer or the interaction between the inducer molecules and the nodD gene product.

It will be of interest to determine if these anti-inducers are of importance in the normal nodulation process. Are they exuded into the rhizosphere and, if so, does their concentration increase in response to infection to provide a negative control on the expression of the Rhizobium nod genes. In this connection, Stachel et al (1985) showed that the phenolics that activated vir gene transcription in A. tumefaciens, are made at much higher concentrations in wounded than unstressed plants. It may be relevant to note that the most active anti-inducer that we identified was an isoflavonoid, a class of molecule whose distribution is largely confined to legumes. Thus, although the inducing molecules may be widespread in the plant kingdom, compounds with potent anti-inducing behaviour appear to be confined to legumes. A provoking thought on which to end.

Note added in proof

In the present volume, Wijffelman et al, Redmond et al and Long et al have also shown that various flavones and/or flavanones induce the nodulation genes of R. leguminosarum, R. trifolii and R. meliloti respectively.

REFERENCES

Downie JA, Hombrecher G, Ma Q-S, Knight CD, Wells B, Johnston AWB (1983a) Cloned nodulation genes of Rhizobium leguminosarum determine host range specificity. Mol Gen Genet 190: 359-365

Downie JA, Knight CD, Johnston AWB, Rossen L (1985) Identification of genes and gene products involved in the nodulation of peas by Rhizobium leguminosarum. Mol Gen Genet 198: 255-262

Egelhoff TT, Fisher RF, Jacobs TW, Mulligan JT, Long SR (1985) Nucleotide sequence of Rhizobium meliloti 1021 nodulation genes: nodD is transcribed divergently from nodABC. DNA 4: 241-248

Evans IJ, Downie JA (1986) The nodI gene product of Rhizobium leguminosarum is closely related to ATP-binding bacterial transport proteins; nucleotide sequence analysis of the nodI and nodJ genes. Gene (in press)

Fisher RT, Tu JK, Long SR (1985) Conserved nodulation genes in Rhizobium meliloti and Rhizobium trifolii. Appl Env Microbiol 49: 1432-1435

Innes RW, Kuempel PL, Plazinski J, Canter-Cremers H, Rolf BG, Djordjevic MA (1985) Plant factors induce expression of nodulation and host-range genes in Rhizobium trifolii. Mol Gen Genet 201: 426-432

John M, Schmidt J, Wieneke V, Kondorosi E, Kondorosi A, Schell J (1985) Expression of the nodulation gene nodC of Rhizobium meliloti in Escherichia coli: role of the nodC gene product in nodulation. EMBO J 4: 2425-2430

Long SR (1984) Genetics of Rhizobium nodulation. In: Kosage T, Nester EW (eds) Plant Microbe Interactions Vol. 1, Molecular and Genetic Aspects. Macmillan, New York

Mulligan JT, Long SR (1985) Induction of Rhizobium meliloti nodC expression by plant exudate requires nodD. Proc Natl Acad Sci USA 82: 6609-6613

Rossen L, Johnston AWB, Downie JA (1984) DNA sequence of the Rhizobium leguminosarum nodulation genes nodAB and C required for root hair curling. Nucl Acids Res 12: 9497-9508

Rossen L, Shearman CA, Johnston AWB, Downie JA (1985) The nodD of Rhizobium leguminosarum is autoregulatory and in the presence of plant exudate induces the nodABC genes. EMBO J 13: 3369-3373

Rostas K, Kondorosi E, Horvarth B, Semoncsits A, Kondorosi A (1986) Conservating extended promoter regions of nodulation genes in Rhizobium. Proc Natl Acad Sci USA 83: 1757-1761

Schofield PR, Watson JM (1986) DNA sequence of Rhizobium trifolii nodulation genes reveals a reiterated and potentially regulatory sequence preceding nodABC and nodFE. Nucl Acid Res 14: 2891-2903

Shearman CA, Rossen L, Johnston AWB, Downie JA (1986) The Rhizobium leguminosarum nodulation gene nodF encodes a polypeptide similar to acyl-carrier protein and is regulated by nodD plus a factor in pea root exudate. EMBO J 5: 647-652

Spiribille R, Forkmann G (1982) Genetic control of chalcone synthase in flowers of Antirrhinum majus. Phytochemistry 21:

2231-2234
Stachel ES, Gynhcung A, Flores C, Nester EW (1985) A Tn3 *lacZ* transposon for the random generation of β-galactosidase gene fusions: application to the analysis of gene expression in *Agrobacterium*. EMBO J 4: 891-898
Torok I, Kondorosi E, Slepkowski T, Pasfai J, Kondorosi A (1984) Nucleotide sequence of *Rhizobium meliloti* nodulation genes. Nucl Acids Res 12: 9509-9524

FRACTIONATION OF RHIZOBIUM LEGUMINOSARUM CELLS INTO OUTER MEMBRANE, CYTOPLASMIC MEMBRANE, PERIPLASMIC AND CYTOPLASMIC COMPONENTS

Ruud A. de Maagd, Carel A. Wijffelman and Ben Lugtenberg
Department of Plant Molecular Biology, University of Leiden,
Botanical Laboratory, Nonnensteeg 3, 2311 VJ Leiden,
The Netherlands

INTRODUCTION

The intensive studies on the genetics of Rhizobium have revealed a number of genes involved in nodulation of Leguminosae by Rhizobia (see other papers in this volume). Little is known about the functioning of the nod gene products at the molecular level. An important step toward the elucidation of these functions may be the subcellular localization of the gene products. We have therefore developed methods to fractionate cultured Rhizobium cells into their four cellular fractions.

MEMBRANE SEPARATION

For the separation of the cytoplasmic and outer membranes we have modified the original methods of Osborn et al (1972) and of Hancock and Nikaido (1978). Rhizobium leguminosarum strain 248 (Josey et al 1979) was grown in 500 ml tryptone yeast extract (TY) medium to mid-logarithmic phase (A_{620} from 0.5 - 0.6) at 28°C. All further parts of the procedure were carried out at 0 - 4°C. Cells were harvested, resuspended in 10 ml of a buffer containing 50 mM Tris pH 8.5, - 20% (w/v) sucrose - 0.2 mM dithiothreitol (DTT) and 0.2 mg/ml of both ribonuclease A and deoxyribonuclease I, and broken by three passages through a French pressure cell at 25,000 p.s.i. Subsequently, the suspension was incubated with 0.2 mg/ml lysozyme in ice for 30 min, diluted with 2 volumes of 50 mM Tris, pH 8.5 and centrifuged

NATO ASI Series, Vol. H 4
Recognition in Microbe-Plant Symbiotic and
Pathogenic Interactions. Edited by B. Lugtenberg

for 20 min at 900 x g to pellet unbroken cells. After addition of KCl (2 M) to a final concentration of 0.2 M, cell envelopes were harvested from the supernatant by centrifugation at 262,000 x g for 2 hrs in a Sorvall T865 rotor. Cell envelopes were carefully resuspended in 0.5 to1.0 ml 5 mM EDTA, - 0.2 mM DTT - 10% (w/w) sucrose, pH 7.5 with a syringe and injection needle. The resuspended cell envelopes were fractionated in a density gradient consisting of 0.5 ml 60%, 0.5 ml 55%, 1.5 ml 50%, 3.0 ml 45%, 3.0 ml 40%, 3.0 ml 35%, 3.0 ml 30% and 1.5 ml 25% (w/w) sucrose in 5 mM EDTA - 0.2 mM DTT, pH 7.5. After centrifugation two clearly visible bands were observed in the gradient. After fractionation of the gradient, fractions were assayed for protein content, buoyant density and distribution of the outer membrane marker 2-keto-3-deoxyoctonic acid (KDO) and the cytoplasmic membrane marker β-nicotinamide adenine dinucleotide (NADH)-oxidase.

The lower band with a buoyant density of 1.243 g/cm^3, had a mean KDO/protein ratio of 34 μg/mg and a mean specific NADH--oxidase activity of 17 μmol/min/mg. For the upper band with a buoyant density of 1.147 g/cm^3, these values were 2 μg/mg and 74 μmol/min/mg, respectively. These bands were therefore designated as outer membrane and cytoplasmic membrane, respectively.

ISOLATION OF PERIPLASMIC AND CYTOPLASMIC PROTEIN FRACTIONS

For the isolation of periplasmic proteins a modification of the method of Glenn and Dilworth (1979) was used. Cells of strain RBL1-AP4, an alkaline phosphatase constitutive mutant of wild type strain RBL1 was used for assays of alkaline phosphatase release. Cells were grown at 28°C in TY medium to an A_{620} of 0.4 - 0.6. After harvesting, cells were resuspended in 50 mM Tris, pH 8.0 - 20% (w/v) sucrose - 2 mM EDTA - 0.2 mg/ml lysozyme and incubated for 30 min at 20°C. This released 88% of the total activity of alkaline phosphatase, a periplasmic marker and only 4% of the activity of malate dehydrogenase, a cytoplasmic marker, into the supernatant. The remaining cells were pelleted by centrifugation, and broken after resuspending in 50 mM Tris, pH 8.0

by sonication. Subsequently, cell envelopes were removed by centrifugation and the final supernatant was shown to contain 96% of the activity of malate dehydrogenase as well as the remaining 12% of the activity of alkaline phosphatase.

PROTEIN PROFILES OF SUBCELLULAR FRACTIONS

Comparison of the proteins of the four cellular fractions by SDS-polyacrylamide gel electrophoresis showed that the fractions have unique protein patterns as will be published in detail elsewhere (De Maagd and Lugtenberg 1986).

RELATIVE PROTEIN CONTENTS OF THE CELLULAR COMPARTMENTS

To determine the relative protein contents of the isolated fractions, bacteria were grown at 28°C to an A_{620} of 0.4 - 0.6 in TY medium, supplemented with 0.2 μCi/ml L-[^{35}S]-methionine. After fractionation as described for the isolation of periplasmic and cytoplasmic components, macromolecules of the various fractions, including the culture supernatant with 0.1 mg/ml Bovine serum albumin added as carrier, were precipitated with 5% (w/v) trichloroacetic acid. Precipitates were collected on glass fibre filters, washed with cold 5% TCA and ethanol, and their radioactivity was determined. Results were averaged and corrected for cross-contaminations. Of the total recovered radioactivity the supernatant fluid contained 1.0 ± 0.2%, the periplasm 15.9 ± 4.2%, the cytoplasm 45.4 ± 4% and the cell envelope fraction 37.7 ± 2.6%. Based on the amounts of protein recovered in the separated membranes of unlabelled cells, it could be calculated that outer membrane and cytoplasmic membrane contain 20.4% and 17.3% of the total cellular protein, respectively.

REFERENCES

De Maagd RA, Lugtenberg B (1986) Fractionation of Rhizobium leguminosarum cells into outer membrane, cytoplasmic membrane, periplasmic and cytoplasmic components. J Bacteriol (in press)

Glenn AR, Dilworth MJ (1979) An examination of Rhizobium leguminosarum for the production of extracellular and periplasmic proteins. J Gen Microbiol 112: 405-409

Hancock REW, Nikaido H (1978) Outer membrane of Gram-negative bacteria XIX. Isolation from Pseudomonas aeruginosa PAO1 and use in reconstitution of the permeability barrier. J Bacteriol 136: 381-390

Josey DP, Beynon JL, Johnston AWB, Beringer JE (1979) Strain identification in Rhizobium using intrinsic antibiotic resistance. J Appl Microbiol 46: 343-350

Osborn MJ, Gander JE, Parisi E, Carson J (1972) Mechanism of assembly of the outer membrane of Salmonella typhimurium. J Biol Chem 247: 3962-3972

A STUDY OF SURFACE INTERACTIONS BETWEEN RHIZOBIUM BACTEROIDS AND THE PERIBACTEROID MEMBRANE USING MONOCLONAL ANTIBODIES

N.J. Brewin, D.J. Bradley, E.A. Wood, G. Galfre* and G.W. Butcher*

John Innes Institute and *Monoclonal Antibody Centre
Colney Lane
NORWICH NR4 7UH, UK

AFRC Institute of Animal Physiology
Babraham, Cambridge CB2 4AT, UK

INTRODUCTION

As a way of supplementing the other techniques for research into the molecular biology of the Rhizobium-legume symbiosis, we have chosen to apply the technology of monoclonal antibodies to the problem of surface interactions between the endosymbiotic bacteroid and the encircling plant-derived peribacteroid membrane. Using monoclonal antibodies in conjunction with the recently developed techniques for immunogold electron microscopy (Robertson et al 1985), there are three kinds of biological questions that we want to ask. Firstly, how is this interaction set up? In other words, how does the Rhizobium cell become taken up by the plant cell plasma membrane; what governs subsequent divisions of the intracellular bacterium and the concomitant division of the encircling peribacteroid membrane (Robertson & Lyttleton 1984); and what physical interactions occur between bacteroid and peribacteroid membrane surfaces which might control morphogenesis? A second class of question relates to membrane transport phenomena - what kinds of molecules are transported across the bacteroid outer membrane and the peribacteroid membrane and how is this achieved? The third class of question relates to the ultimate breakdown of the symbiosis - nodule senescence - which is in some way related to the nitrogen status of the whole plant.

In theory, monoclonal antibodies (McAb) could provide a set of specific affinity probes for each of the molecular components of the bacteroid surface, the peribacteroid fluid

NATO ASI Series, Vol. H 4
Recognition in Microbe-Plant Symbiotic and Pathogenic Interactions. Edited by B. Lugtenberg

and the peribacteroid membrane. In practice, however, we would be interested in the most abundant antigens, because these may define the physical chemistry of the interaction and the nodule-specific antigens (bacteroid or plant) because these are presumed to be 'biologically meaningful'. The attractive feature about making and using monoclonal antibodies is that from relatively small quantities of unpurified and uncharacterised biological material used as immunogen (in this case either bacteroids or the peribacteroid material), it is possible to obtain virtually unlimited quantities of a monoclonal antibody reacting with one specific component of the system. This McAb can then be used for subsequent analysis of the antigen by a wide range of immunochemical techniques. For example, the cytological distribution of the antigen in time and space can be examined by immunogold electron microscopy, and the biochemical nature of the antigen can be ascertained by immunopurification and conventional biochemical analysis. The rest of this paper will describe the isolation and use of monoclonal antibodies reacting with the bacteroid surface and the peribacteroid membrane (summarised in Table 1). It will illustrate some of the advantages of working with McAb, but also some of the pit-falls of this experimental approach.

FRACTIONATION OF PEA NODULE HOMOGENATES

The starting material for all biochemical and immunochemical procedures was a preparation of membrane-enclosed bacteroids isolated by sucrose gradient fractionation from extracts of pea nodules that had been homogenised in the presence of 0.5M sucrose (Brewin et al 1985; Bisseling et al 1984). In this preparation, more than 85% of the isolated bacteroids were still enclosed by peribacteroid membrane, which could be ruptured by resuspension of the cells in 50mM Tris-HCl buffer without sucrose. Naked bacteroids were then removed by centrifugation and the peribacteroid membranes (pbm) and peribacteroid fluid (pbf) could, if necessary, be further fractionated by ultracentrifugation on sucrose gradients (Robertson et al 1978). Subsequently either sonicated

Table 1. Isolation of monoclonal antibodies reacting with bacteroids peribacteroid membrane (pbm) or peribacteroid fluid (pbf)

Immunisation	No of McAb Isolated	Antigen Type
I. Sonicated bacteroids	24	Bacterial LPS
	1	Bacteroid Peptidoglycan
II. pbm + pbf	34	Bacterial LPS
	10	pbm glycoprotein
	3	Plant glycoprotein
	1	Bacterial protein
	1	Bacterial protein
III. Antigen-depleted pbm + pbf	16	pbm glycoprotein
	1	Bacterial protein
	1	Bacterial protein
	1	Bacterial protein
IV. Deglycosylated pbm + pbf	23	Bacterial protein

bacteroids or the unfractionated peribacteroid material (pbm + pbf) were used as antigens for rats so that, after fusion to a myeloma cell line, monoclonal antibody producing hybridoma cell lines could be isolated.

McAb FROM SONICATED BACTEROIDS

In the first fusion experiments rats were immunised with a supernatant fraction from sonicated bacteroids of strain 3841 and strain B155 (Brewin et al 1986).

The results of 6 fusion experiments (involving 3 different immunised rats) were that from 217 hybridoma lines, 25 were identified that reacted with antigens present in strain B155 bacteroids (Table 1). Subsequent biochemical analysis indicated that 24 of the monoclonal antibodies recognised components of bacteroid lipopolysaccharide which in all cases were also represented in cultures of the corresponding free-living bacteria (Brewin et al 1986). The one exception was a McAb (produced by cell line AFRC MAC 90) which reacted with bacteroids of strain B155 and 3841 but not with the corresponding free-living cultures. Immunogold localisation of nodule thin sections revealed that, as for LPS, the bacteroid specific antigen was a component of the bacteroid cell wall. Preliminary biochemical evidence indicates that it may form part of the peptidoglycan layer - although solubilised by sonication, the antigen was not released from intact bacteroids by extraction with boiling 10% SDS, by a mixture of phenol, chloroform, and petroleum ether, or by 1M NaOH.

McAb FROM PERIBACTEROID MATERIAL

In order to obtain McAb reacting with components of the peribacteroid membrane and peribacteroid fluid, the peribacteroid material was used to inoculate rats (Brewin et al 1985 and unpublished results). From four fusion experiments (Table 1) the majority of McAb isolated reacted with bacteroid lipopolysaccharide; the next most abundant class reacted with a glycoprotein or group of glycoproteins of disperse molecular weight which were components of the peribacteroid membrane (Brewin et al 1985); then there was a class of McAb that reacted with another periodate sensitive antigen (?glycoprotein) which accumulated in the matrix of infection droplets; finally there were two McAb which reacted with components of both the plant and bacteroid cytoplasms - and, since these McAb also reacted with components present in free-living bacteria, it is believed that the antigens originate in the bacteria and are transported to the plant cytoplasm. In the case of MAC 201, it is interesting to note that the pattern of antigen bands seen by western blotting is different for nodule cytoplasmic, peribacteroid, bacteroid and free-living bacterial fractions (Fig. 1).

From these fusion experiments, a number of generalisations can be made about the classes of McAb that have been isolated.

1. McAb against certain antigens are strongly represented in

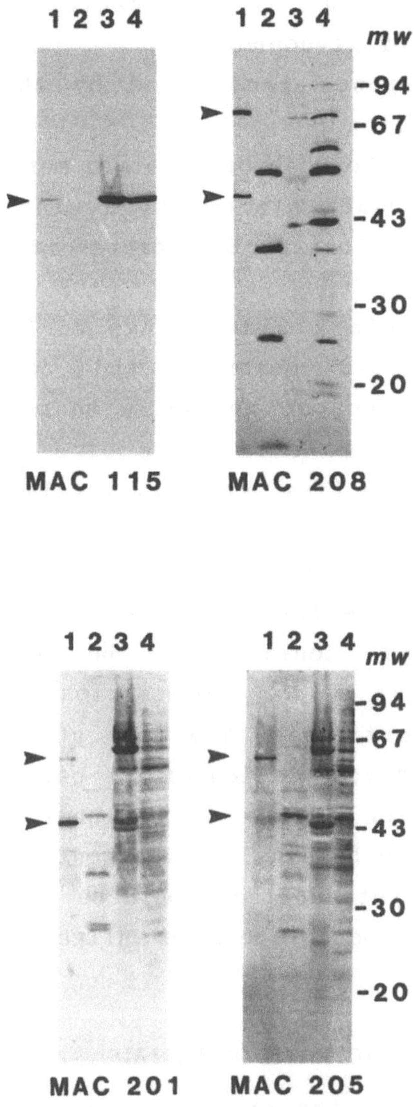

Figure 1. Monoclonal antibodies that recognise epitopes present in both the nodule cytoplasmic fraction (arrowheads) and the free-living Rhizobium strain. Fractions run on SDS-PAGE gels were:- 1, nodule cytoplasmic supernatant; 2, peribacteroid material; 3, bacteroids; 4, free-living bacteria. After electro-blotting to nitrocellulose sheets, the antigens were incubated with rat monoclonal antibodies (a) MAC 115, (b) MAC 208, (c) MAC 201, (d) MAC 205, and subsequently immunostained using goat anti-rat phosphatase, nitroblue tetrazolium, and bromochloroindoyl phosphate. Note that MAC 201 and MAC 205 are epitope variants which both react with the same bands in the nodule supernatant fraction (arrowheads), but these are stained with differing intensities.

the final collection, e.g. to bacteroid LPS or to one particular group of glycoproteins;

2. most of the epitopes recognised by these McAb are to carbohydrate functions;
3. in western blots, the McAb often stain more than one band;
4. the pattern of bands stained by a single McAb is often different for peribacteroid, bacteroid and free-living bacterial fractions.

In order to obtain a wider range of McAb to bacteroid surfaces, peribacteroid membrane and components of the peribacteroid fluid, it was necessary to modify the material used as antigen and two strategies were adopted to try to achieve this goal. The first strategy involved a process known as cascade selection, i.e. the construction of immuno-affinity columns bearing representatives of the most abundant classes of McAb in order to deplete the respective antigens. LPS and the glycoprotein antigen could thus be removed from the peribacteroid material before immunisation of rats, so that a different range of antibodies might be stimulated. A second strategy involved deglycosylation of the peribacteroid material used for immunisation so as to eliminate the carbohydrate groups which appear to be highly immunogenic in this system. A third strategy, not yet adopted, would be to obtain a fraction that is enriched for nodule-specific antigens.

ANTIGEN DEPLETED PERIBACTEROID MATERIAL

In order to remove the antigens giving the most abundant classes of McAb, immunoaffinity columns were made using MAC 57, a McAb reacting with bacterial LPS, and MAC 64, a McAb reacting with the predominant plant glycoprotein antigen. Peribacteroid material, solubilised by the addition of NP40 (0.5% v/v), was passed successively through each of these immunoaffinity columns, resulting in at least a 40-fold depletion of the respective antigens. The column eluate was concentrated and used to immunise rats. After fusion of spleen cells with the myeloma line, 240 cell lines were screened for antibodies reacting with peribacteroid material and 21 lines were isolated for further investigation. The McAb could be grouped into 4 different families on the basis of immuno-staining patterns for western blots of nodule supernatant, peribacteroid material, bacteroids and free-living bacteria. The largest group (16 examples) reacted with the same glycoproteins as MAC 64, implying that the immunoaffinity column had not been successful in depleting this antigen to a sufficient degree before immunisation of rats. On the other hand, no McAb were isolated that reacted with bacteroid LPS, and so the MAC 57 immuno-affinity column was operationally effective. The three other classes of McAb that were isolated all gave complex staining patterns on Western blots (Fig. 1), and cross-reacted with bands present in free-living cultures of Rhizobium as well as peribacteroid material and the plant cytoplasmic fraction. It is presumed that these antigens are bacterial products which appear in the peribacteroid fraction and the plant cytoplasm as a result of release from the periplasmic space of bacteroids. These possibilities merit further investigation because they could identify a transmembrane transport process that may be operating between the bacteroid and the host plant symbiont.

DEGLYCOSYLATED PERIBACTEROID MATERIAL

As another way of increasing the range of McAb isolated from a fusion experiment, we have deglycosylated a preparation of peribacteroid material using trifluoromethane sulphonic acid (Edge et al 1978). After deglycosylation, the peribacteroid material did not react with MAC 57 or MAC 64 antibodies (which both recognise carbohydrate epitopes), although a number of protein bands were still visible after SDS-PAGE and silver-staining. Rats were immunised with this deglycosylated material and, after the fusion and screening, 23 reactive McAb were isolated from a total of 424 cell lines. Unfortunately, western blotting indicated that all 23 McAb were similar in their antigen binding properties. The McAb also reacted with free-living bacteria and again showed evidence of transport and modification between different compartments of the infected nodule cell, similar to the other McAb illustrated in Fig. 1.

CONCLUSIONS

Using the range of McAb that have been isolated already, there is a great deal of scope for the analysis of surface recognition effects in the Rhizobium-legume interaction (Bradley et al 1986), for the analysis of the transport of macromolecules between bacteroid and plant compartments within the nodule and the analysis of the structure and development of particular macromolecules, e.g. LPS (Brewin et al 1986) and the plant glycoprotein from the peribacteroid membrane (Brewin et al 1985). However, it is obvious that the range of McAb that has been obtained so far is not at all comprehensive. In future, it will have to be extended either by focussing more specifically on the contents of purified peribacteroid membranes or by confining the analysis to those antigens of plant or bacterial origin that are only expressed within the legume root nodule.

REFERENCES

Bradley DJ, Butcher GW, Galfre G, Wood EA, Brewin NJ. (1986) Physical association between the peribacteroid membrane and lipopolysaccharide from the bacteroid outer membrane in Rhizobium-infected pea root nodule cells. J Cell Biol. In the press

Brewin NJ, Robertson JG, Wood EA, Wells B, Larkins AP, Galfre G, Butcher GW. (1985) Monoclonal antibodies to antigens in the peribacteroid membrane from Rhizobium-induced root nodules of pea cross-react with plasma membranes and Golgi bodies. EMBO J 4:605-611

Brewin NJ, Wood EA, Larkins AP, Galfre G, Butcher GW. (1986) Analysis of lipopolysaccharide from root nodule bacteroids of R. leguminosarum using monoclonal antibodies. J Gen Microbiol 132 (in press)

Bisseling T, Been C, Klugkist J, van Kammen A, Nadler K. (1983) Nodule-specific host proteins in effective and ineffective root nodules of Pisum sativum. EMBO J 2:961-966

Edge ASB, Faltynek CR, Hof L, Reichert LE, Weber P. (1981) Deglycosylation of glycoporteins by trifluoromethanesulfonic acid. Anal Biochem 118:131-137

Robertson JG, Warburton MP, Lyttleton P, Fordyce AM, Bullivant S. (1978b) Preparation and properties of peribacteroid membranes and bacteroid envelope inner membranes from developing lupin nodules. J Cell Sci 30:151-174

Robertson JG, Lyttleton P. (1984) Division of peribacteroid membranes in root nodules of white clover. J Cell Sci 69: 147-157

Robertson JG, Wells B, Brewin NJ, Williams MA. (1985). Immunogold localisation of cellular consituents in the legume-Rhizobium symbiosis. In: BJ Mifiln (ed) Oxford Reviews of Plant Molecular and Cell Biology, Vol. 2, pp.69-89

NODULINS INVOLVED IN EARLY STAGES OF PEA ROOT NODULE DEVELOPMENT

Ton Bisseling, Henk Franssen, Ton Gloudemans, Francine Govers,
Marja Moerman, Jan-Peter Nap and Albert van Kammen
Agricultural University
Department of Molecular Biology
Transitorium, De Dreijen 11
6703 BC Wageningen
The Netherlands

INTRODUCTION

In nodules formed on the roots of pea plants at least 25 plant genes are specifically expressed, the nodulin genes. The majority of these genes are first expressed 13 days after sowing and inoculation with Rhizobium leguminosarum (Govers et al 1985). Histological examination of longitudinal sections of infected roots of 8, 10 and 13 day old plants showed that a complete nodule structure is already present at day 10 (Figure 1A). From the apical meristem infected as well as uninfected cells, the two cell types that are present in the symbiotic zone of the mature nodule, have been formed (Figure 1A, B). Vascular bundles are present at the periphery of the nodule, an uninfected cortex surrounds the central part of the nodule and in the zone with the differentiating cells a great number of infection threads is present. At day 8 this differentiation process has started, but the difference between infected and uninfected cells is not yet clear (Figure 1B), and at 13 days the infected cells are fully packed with rhizobia (Figure 1B). Since the complete nodule structure is already present before the majority of the nodulin genes is expressed these nodulin genes cannot be involved in the morphogenesis of this new organ.
To study the part of the developmental program that leads to the formation of a root nodule structure we have searched for nodulin genes that are expressed during the early stages of root nodule formation, isolated cDNA clones that represent early nodulin genes, and we have studied the expression of these nodulin genes in nodules that are blocked in their development.

NATO ASI Series, Vol. H 4
Recognition in Microbe-Plant Symbiotic and
Pathogenic Interactions. Edited by B. Lugtenberg

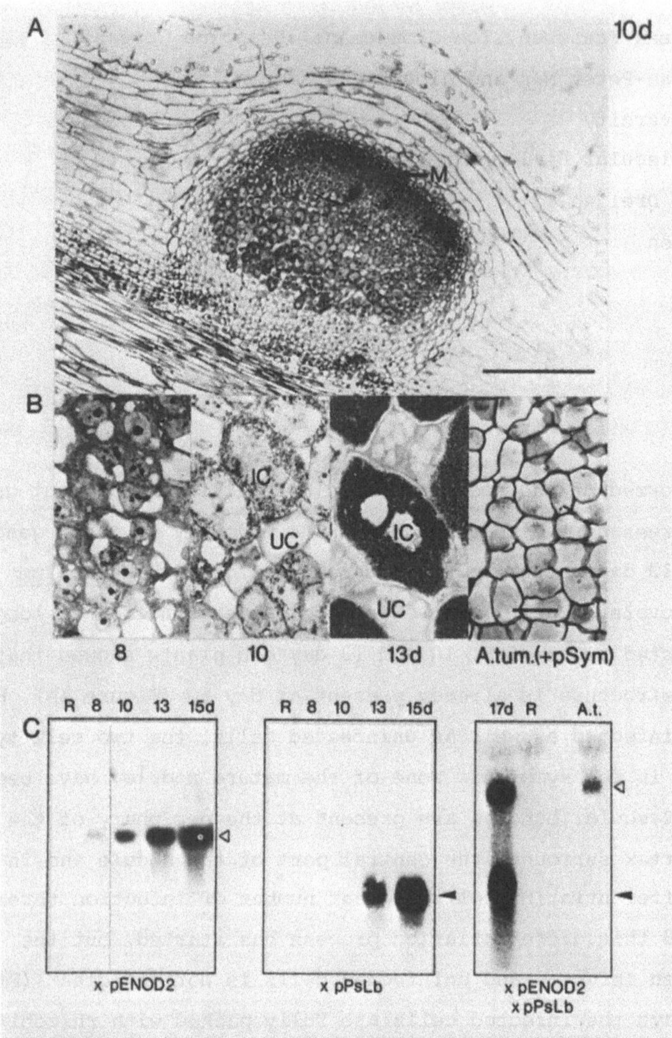

Figure 1. Cytology of pea nodules and the expression of the ENOD2 and Lb genes during root nodule development.

(A) Light micrograph of a longitudinal section through an infected pea root (10d) stained with toluidine blue. (M) nodule meristem. Bar = 300 µm.

(B) Light micrographs of parts of the nodule tissue containing differentiated cells from 8, 10 and 13 day old pea nodules and pea nodules formed by the Agrobacterium transconjugant (LBA2712) carrying a Sym plasmid from R.leguminosarum. (IC) infected cell, (UC) uninfected cells. Bar = 600 µm.

(C) Autoradiographs of Northern blots. The two left blots contain RNA from 8 day old uninfected roots (R) and nodules from 8, 10, 13 and 15 day old pea nodules and they were hybridized with pENOD2, the soybean early nodulin cDNA clone, and pPsLb102, a pea Lb cDNA clone, respectively. The Northern blot on the right contains RNA from wild type pea root nodules (17d), 8 day uninfected pea roots (R) and "empty" nodules induced by the Agrobacterium transconjugant LBA2712 (28d) and was hybridized with both cDNA clones.

IDENTIFICATION OF EARLY NODULINS

We surveyed the expression of nodulin genes by two methods. In the first place, by in vitro translation of nodule and root RNAs followed by two-dimensional gelelectrophoresis and secondly by isolation of nodulin clones from cDNA-libraries.

Analysis of the in vitro translation products from pea root nodules and uninfected roots on two-dimensional gels showed the presence of 21 nodulin mRNAs (Govers et al 1985). The majority of these RNAs appeared concomitantly with the leghemoglobin (Lb) mRNAs at day 13. An example of this group, N-40, is shown in Figure 2.

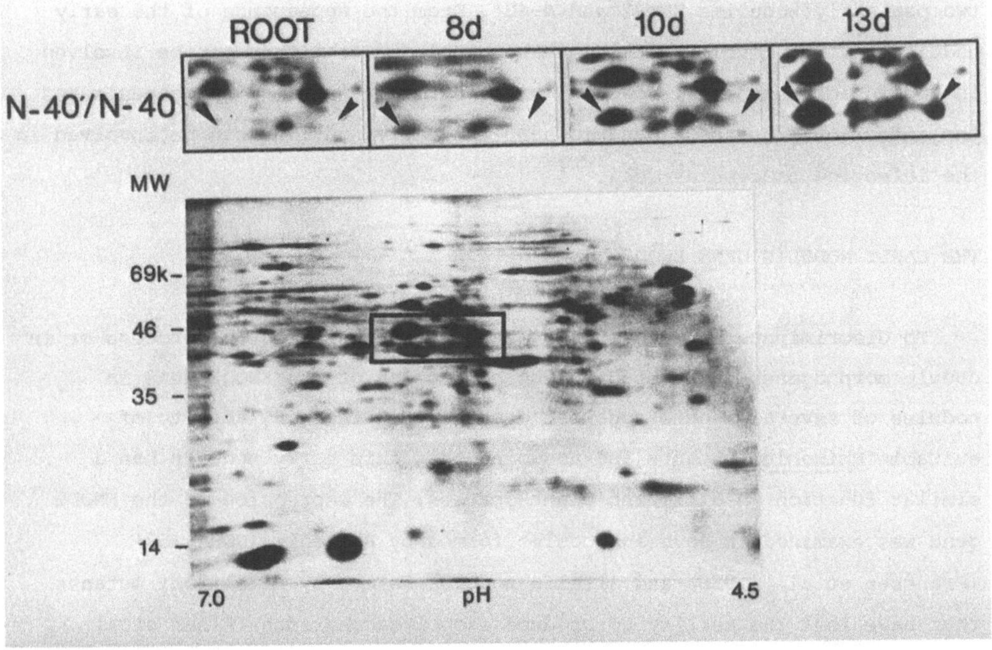

Figure 2. Detection of the N-40' RNA during pea root nodule development (A) Fluorograph of a two dimensional gel of in vitro translation products from RNA isolated from effective pea root nodules 17 days after sowing and inoculation with R.leguminosarum PRE. In (B) only the part of the gel within the square indicated in (A) is shown (from uninfected roots and 8, 10 and 13 day old root nodules). The early nodulin N-40' and the nodulin N-40 that appears at day 13 are indicated with arrowheads. N-40' is the spot on the left.

However, two nodulins were already present in infected roots of 8 day old pea plants, and they will be called early nodulins. These two early nodulins have apparent molecular weights of 80.000 and 40.000, respectively, and are named N-80 and N-40' in accordance with the previously established nomenclature (Van Kammen 1984). The appearence of N-40' during root nodule formation is shown in Figure 2. From a soybean cDNA library we isolated 3 cDNA clones that represent early nodulin genes (Franssen et al 1986). One of these clones, pENOD2, shows strong homology to pea, alfalfa and Vicia sativa nodulin mRNA on Northern blots. During pea root nodule development the ENOD2 RNA is first detectable in root segments from 8 day old plants and it accumulates in older nodules, while the Lb mRNA is first detectable 5 days later at day 13 (Figure 1C). The pENOD2 hybrid selected pea RNA encodes a nodulin with an apparent molecular weight of 80.000. Hence, pENOD2 might represent N-80.

In the following two paragraphs we will discuss the characteristics of the two pea early nodulins ENOD2 and N-40'. From the appearance of the early nodulins during development it can be concluded that they can be involved in nodule morphogenesis. However, since normally nodule morphogenesis and the infection process go hand in hand the early nodulins can be involved in the infection process as well.

THE EARLY NODULIN GENE ENOD2

To discriminate between its involvement in the infection process or in nodule morphogenesis, we studied the expression of the ENOD2 gene in nodules of several legume species, depending on the availability of suitable Rhizobium mutants and assuming that this early nodulin has a similar function in different plant species. The expression of the ENOD2 gene was examined in soybean nodules formed by R.fredii (USDA 257) (Franssen et al 1986) and alfalfa nodules formed by R.meliloti mutants that have lost the ability to produce exopolysaccharides (Finan et al 1985). Both types of nodules lack intracellular bacteria as well as infection threads, but in both nodules ENOD2 RNA was present as shown by Northern blot hybridization. Since on the other hand ENOD2 RNA is not detected in root hairs from infected plants (Franssen et al 1986) we conclude that ENOD2 will not be involved in the infection process, but more likely in nodule morphogenesis.

Studies of Calvert et al (1984) have shown that in soybean roots many more infection loci are formed than the final number of nodules indicates. Up to the stage where the nodule emerges through the epidermis it can be aborted, but when it reaches the "emergence" stage the nodule will continue to develop. In soybean root nodule development the first expression of the ENOD2 gene is detectable when the nodule reaches this "emergence" stage. Therefore ENOD2 may be related to the developmental decision of abortion or continuation.

THE EARLY NODULIN N-40'

Since no cDNA clone for N-40' is available yet, the analysis of the characteristics of the second early nodulin is confined to two dimensional gel analysis. An Agrobacterium transconjugant carrying a Sym plasmid from R.leguminosarum is able to form small nodules on the roots of pea plants. These nodules lack intracellular bacteria (Figure 1B), although infection thread-like structures are present (Govers et al 1986). These nodules have the same organisation as normal nodules: a central tissue surrounded by uninfected cortex cells and peripherically located vascular bundles. However, the central tissue contains only one cell type. This consists of relatively small non-dividing cells that seem to represent neither the infected nor the uninfected cells that are found in normal root nodules (Figure 1B). In these "empty" nodules we observed the expression of the ENOD2 gene (Figure 1C) but the other early nodulin N-40' was not detectable. So the two early nodulin genes we have identified may represent different steps in the developmental program. The nodulins that appear concomitantly with Lb during normal root nodule development were also absent in these nodules. The absence of N-40' in these "empty" nodules suggests that also N-40' is not essential for the infection process since infection threads are present. Moreover, N-40' is not involved in the formation of the nodule structure that is formed by this Agrobacterium transconjugant.

On Vicia sativa the same Agrobacterium strain forms nodules that do contain infected and uninfected cells. In these nodules the early nodulin N-40 (V.sativa) is present. This early nodulin has the same molecular weight and a similar isoelectric point as the pea N-40' suggesting that N-40(V.sativa) and N-40' are similar nodulins. Assuming that this is the case, we can conclude that the expression of the N-40' gene is only observed when the bacteria are released from the infection threads. This expression might be related to the part of the differentiation process involved in the formation of the two cell types found in the symbiotic zone of a root nodule, infected and uninfected cells.

RHIZOBIUM GENES

The analysis of the patterns of expression of the early nodulin genes in nodules elicited by engineered Agrobacterium strains not only facilitates the unravelling of steps in the plant differentiation process, but also addresses to the part of the rhizobial genome that is responsible for the induction of the expression of early nodulin genes.
It was shown previously that Rhizobium strains cured of their own Sym plasmid regain the ability to form root nodules after the introduction of cloned DNA containing the nod region (Downie et al 1983). Downie et al (1983) showed by deduction that 10 kb of the Sym plasmid, in a R.phaseoli chromosomal background, is sufficient for the induction of nodules on pea roots. Using the same Rhizobium strains we have shown that in these nodules all nodulin genes are expressed (Govers et al 1986). Hence it was concluded that besides the nod genes all other genes encoded by the Sym plasmid cannot be involved in the induction of the expression of nodulin genes. This result indicates that the nod genes are involved in the induction of nodulin genes, although it does not exclude a role of the Rhizobium chromosome. However, in an Agrobacterium chromosomal background the total Sym plasmid gives rise to nodules on pea in which we have demonstrated the exclusive expression of the ENOD2 gene (Figure 1C) (Govers et al 1986). Since A.tumefaciens is unable to induce the expression of the ENOD2 gene in tumors formed on the stem of pea plants the Agrobacterium genome itself does not harbour genes involved in the induction of the expression of the ENOD2 gene. Hence these genes are most likely located on the Sym plasmid and since only 10 kb nod region of this Sym plasmid is essential, it can be deduced that the Rhizobium genes involved in the induction of expression of the ENOD2 genes have to be located in that nod region of the Sym plasmid (Govers et al 1986).
Since the other early nodulin N-40' is not expressed in the "empty" nodules on pea, one might argue that the nod region is not sufficient for the induction of this nodulin gene. However, the very similar early nodulin gene N-40(V.sativa) is expressed in nodules formed on V.sativa by the same Agrobacterium transconjugant. It is not found in tumors on V.sativa, so the

expression of N-40(V.sativa) is also regulated by the nod genes. Assuming that N-40(V.sativa) and N-40' are similar, we speculate that the expression of the N-40' gene is regulated by the nod genes as well, but its expression is obscured. Since the outside of an Agrobacterium differs from Rhizobium it can be anticipated that the Agrobacterium is recognized by the plant in a different way than Rhizobium and the Agrobacterium might be subject to plant defence mechanisms.

The striking differences between the V.sativa and pea nodules formed by the Agrobacterium transconjugant indicate that one has to be extremely careful with conclusions about the genetic potentials of such a strain with respect to nodule development. The presence of genetic information for the induction of a part of the developmental program is apparently no guarantee for the actual realization of that induction. The failure to induce a part of the developmental program can be due not only to the absence of the required genetic information, but also to the blockade of that program by the defence response of the host plant.

REFERENCES

Calvert HE, Pence M, Pierce M, Malik NSA, Bauer WD (1984) Anatomical analysis of the development and distribution of Rhizobium infections in soybean roots. Can J Bot 62: 2375

Downie JA, Ma QS, Knight CD, Hombrecher G, Johnson AWB (1983) Cloning of the symbiotic region of Rhizobium leguminosarum: the nodulation genes are between the nitrogenase genes and a nifA-like gene. The EMBO J 2: 947-952

Finan TM, Hirsch AM, Leigh JA, Johansen E, Kuldan GA, Deegan S, Walker GC, Signer ER (1985) Symbiotic mutants of Rhizobium meliloti that uncouple plant from bacterial differentiation. Cell 40: 869-877

Franssen HJ, Gloudemans T, Stiekema W, Van Dam H, Govers F, Louwerse J, Bisseling T (1986) Molecular cloning of cDNA for Nodulin-75: a gene product involved in early stages of nodule development. Submitted

Govers F, Gloudemans T, Moerman M, Van Kammen A, Bisseling T (1985) Expression of plant genes during the development of pea root nodules. The EMBO J 4: 861-867

Govers F, Moerman M, Downie JA, Hooykaas P, Franssen HJ, Louwerse J, Van Kammen A, Bisseling T (1986) Rhizobium nodulation genes are involved in expression of an early nodulin gene. Submitted

Van Kammen A (1984) Plant genes involved in nodulation. In: Veeger C, Newton WE (eds) Advances in Nitrogen Fixation Research. Nijhoff, The Hague, p 587

PATHOGENIC STRATEGIES OF FUNGI

N. T. Keen
University of California
Department of Plant Pathology
Riverside, California 92521, U.S.A.

INTRODUCTION

Fungi are the most important group of pathogens on plants from an economic point of view. This results in part from the fact that fungi are extremely varied in the ways they attack plants and actively move through host tissues, forcibly breaching barriers by enzymatic and/or physical mechanisms. In addition, fungal pathogens exhibit a complex array of tropic responses which may contribute to complex infection processes and are important in determining host range. Fungal interactions with plants are based on parasitism, a nutritional relationship, but the economic and aesthetic damage which they cause may result from the operation of other factors that constitute pathogenicity. While these two terms are frequently used interchangeably, they are distinct; indeed, it is possible to have pathogenesis without parasitism and vice versa. Although fungal diseases have been studied for over a century, we are only beginning to understand the molecular factors which determine whether a certain fungus is pathogenic on a higher plant. In this paper, I will review some of the progress that has been made, point out deficiencies in our knowledge and speculate on the future, especially on ways in which the powerful machinery of molecular genetics is being brought to bear on fungal pathogens. Because of the survey nature of the paper, references are exemplary only.

The basic strategies of fungal pathogenicity fall into two general classes, which may be analogized with certain of the criminals that afflict human civilizations (Table 1). One class is the thugs. These are pathogens which damage the host through production of toxins and/or enzymes and then utilize the food base of the dead and incapacitated host cells. Such pathogens often elicit host defense mechanisms, but are not always inhibited because they may be equipped with enzymes or other devices that inactivate and/or avoid the defenses. The other class of fungal pathogens is the confidence men (Table 1). They do not seriously damage their hosts in

NATO ASI Series, Vol. H 4
Recognition in Microbe-Plant Symbiotic and Pathogenic Interactions. Edited by B. Lugtenberg

the early stages of the interaction and avoid non-specific initiation of host defense mechanisms. Instead, these pathogens co-exist with their hosts while utilizing nutrients and accumulating fungal biomass. Such pathogens frequently convince hosts that their infection sites are normal sinks for the import of carbon assimilates translocated from distant plant sources. When the infected host tissue begins to senesce, the pathogen quickly produces and releases large quantities of spores, and in the process contributes to the decline of the host tissue by accelerating the turnover of proteins and starch.

Table 1. Characteristics of the two basic approaches of fungal pathogenicity in plants

Group I - thugs	Group II - confidence men
Parasite is easily cultured	Parasite is not easily cultured
Disease generally involves premature killing or senescence of host tissue	No early killing of host tissue; instead a local stimulation of metabolism often occurs
Often copiously produce enzymes and toxins	Restricted production of enzymes, toxins
Little or no directed translocation to infection site	Long distance, directed translocation of nutrients to infection site
Generally do not have complex interactions with host and little if any intimate contact of host and parasite cells occurs	Generally have several developmental stages during host infection and exhibit intimate contact of host and parasite protoplasm (e.g., haustoria)
Little stimulation of host protein and nucleic acid biosynthesis occurs	Considerable stimulation of host protein and nucleic acid biosynthesis
Often elicit plant defence reactions non-specifically (e.g., phytoalexins) but may contain degrading enzymes or other escape devices	Generally do not elicit defense reactions non-specifically; if elicited, defense reactions are effective

Toxins and host-damaging enzymes produced by fungal pathogens have preoccupied plant pathologists for most of the last century, but pathogenicity is doubtless more complex than this, especially with the 'confidence men'. As shown in Table 2, other factors such as the ability to complete certain developmental or tropic requirements in the infection process are essential to the success of certain fungi, noteably obligate parasites.

Another area that has not received enough attention is the necessity for effective nutrient uptake systems. While this may be a relatively simple matter for 'thugs', the 'confidence men' clearly go to considerable trouble to obtain host nutrients without killing or seriously damaging host cells. Finally, successful pathogens require mechanisms to handle host defenses. While enzymes may be important in overcoming physical barriers to pathogen ingress, pathogens must also deal with inducible plant defenses such as the hypersensitive response. For instance, the pathogen must in some way control the red-ox environment of the host tissue to prevent the generation of activated oxygen species which are associated with invocation of the hypersensitive reaction. Successful 'thugs' may also require mechanisms for detoxifying or otherwise evading host phytoalexins. Information in all of these latter areas is limited and accordingly they present major challenges to disease physiology over the next decade.

Table 2. Some of the factors contributing to fungal pathogenicity
Ability to complete required developmental stages on host plant
Toxins
Enzymes
Nutrient uptake systems
Red-ox control mechanisms
Suppressors and phytoalexin degrading mechanisms

BEHAVIORAL AND TROPIC TRAITS AND THE COMPLEX EVENTS LEADING TO HOST INFECTION

We are only beginning to understand the processes involved in host infection, especially with obligate fungal parasites. Largely as a consequence of the work of Staples, Wynn and their associates (e.g., Wynn and Staples 1981; Staples et al 1985), it is clear that many genes contribute to germination of spores on the host surface and in the development of germ tubes properly appressed to the cuticle surface. In addition, many genes are responsible for appresorium formation, infection hyphum production, sub-stomatal vesicles, haustorial mother cells, etc. Unfortunately, all of these processes are poorly understood genetically and biochemically. A few inroads are being made, however, as illustrated by the recent work of Epstein et al (1985). These workers used proteolytic enzymes to demonstrate that one or more cell surface proteins on bean rust germ tubes may mediate a thigmotropic response which leads to appresorium initiation when the germ tube encounters a surface discontinuity such as a stomate on a leaf or a scratch on a piece of glass or plastic. This finding could eventually result in an understanding of the specific gene(s) which encode the cell surface protein(s) and accounts for the thigmotropic response.

Another interesting recent line of work has been the demonstration that melanin biosynthesis is required for appresorial penetration of plants by Pyricularia oryzae (Woloshuk et al 1983) and Colletotrichum lagenarium (Kubo et al 1985). Tolousbalides and Sisler (1978) first demonstrated that control of the rice blast disease by tricyclazole was not due to antifungal activity but instead to a specific inhibition of melanin production by P. oryzae. Perhaps because their structural integrity is compromised, melanin-deficient appresoria are incapable of penetrating the host cuticle. This suggestion of a physical requirement for host cell penetration complements work done on cutinase enzymes, discussed in the next section.

ENZYME PRODUCTION

Pathogens which utilize the 'thug' approach to pathogenesis frequently produce batteries of enzymes that attack one or more structural components of the host cell. These may include one to several endopectic enzymes, cellulases, phosphatidases, proteases, galactosidases, xylanases, arabanases, cutinases and nucleases. A large literature is concerned with the possible role of these enzymes in pathogenesis (see Collmer and Keen 1986; Cooper 1984, for reviews), but we do not have conclusive proof for the involvement of a single cell-wall degrading enzyme with a fungal disease. A strong circumstantial case has been made for the importance of pectic enzymes in certain diseases; however, studies with enzyme-less mutants have sometimes yielded results which do not support a pathogenic role for the pectic enzyme(s) (e.g., Howell 1976). Unfortunately, fungal genes encoding cell-wall degrading enzymes have not yet been cloned and gene replacement experiments performed in order to critically test the role of a particular enzyme in pathogenicity.

Kolattukudy's group has tested the role of cutinase in the pathogenicity of _Fusarium_ _solani_ f.sp. _pisi_ on pea seedlings and much of this work has been recently reviewed (Kolattukudy 1985). Cutinase deficient mutants were observed to be non-pathogenic but their pathogenicity was restored by the addition of purified cutinase to the inoculated mutant spores. Specific cutinase inhibitors and cutinase antibodies also eliminated the pathogenicity of wild-type isolates of the fungus. The Fusarium cutinase gene has recently been cloned and characterized (Soliday et al 1984), but attempts have not yet been made to _in vitro_ mutagenize the gene and use it to construct a defined cutinase deficient mutant strain via a gene replacement experiment. This would constitute a definitive and conclusive test for the role of the Fusarium cutinase and its structural gene in pathogenicity.

Several enzymes are produced by class I fungal pathogens which degrade host phytoalexins that otherwise would inhibit the pathogen. For example, Smith et al (1984) showed that wild type isolates of _Fusarium_ _solani_ f.sp. _phaseoli_ actively degrade kievitone, a major phytoalexin produced by the host plant, bean. Several variant strains, however, exhibited both reduced degradation of kievitone and pathogenicity. Van Etten's group has also studied the role of pisatin demethylases produced by _Fusarium_ _solani_ f.sp. _pisi_ which degrade the pea phytoalexin pisatin to a less toxic product. The demethylases were found to be produced in larger quantity and in an inducible manner by highly pathogenic field isolates of the fungus but not by field isolates of lower pathogenicity. In order to critically test this suggested importance of the demethylases in high pathogenicity, Van Etten's group crossed isolates of high and low pathogenicity and examined the segregating ascospore progeny. These experiments showed that high, inducible production of pisatin demethylase activity invariably segregated in association with high pathogenicity and low pisatin demethylase activity invariably segregated with low pathogenicity (Kistler and VanEtten 1984; Tegtmeier and VanEtten 1982). Thus, a convincing case has been made for importance of the pisatin demethylase activity in pathogenicity. Finally, Van Etten's group has recently completed the molecular cloning of one of the Fusarium pisatin demethylase genes and expressed it in _Aspergillus_ _nidulans_ (unpublished data). This work should permit examination of the gene structure of the cytochrome P-450 enzyme accounting for pisatin demethylation. In summary, Van Etten's work constitutes a convincing case for the physiologic importance of the pisatin demethylases in the pathogenicity of _F._ _solani_ f. sp. _pisi_. An important corollary of the genetic experiments is to prove the importance of pisatin as a defense system in pea against Fusarium isolates which cannot degrade it. The work is therefore consistent with the general role of phytoalexins as a defense system.

TOXINS AS WEAPONS OF THE PATHOGEN

Class I pathogens frequently produce toxins that damage plant cells. As with pathogen enzymes, some uncertainty exists regarding the physiologic importance of several pathogen-produced toxins. However, the greater use of mutational and genetic approaches in this field as well as the fact that several toxins are host-selective have resulted in considerable support for the role of certain toxins in pathogenesis (see Durbin 1981; Scheffer and Livingston 1984). Since toxins have been adequately discussed in these reviews, I will mention only a few examples of recent studies germane to the question of toxins and their role in pathogenicity. One of them involves the host-selective toxin victorin, produced by *Helminthosporium victoriae*. While it has been the most studied of the host-selective toxins, there is now serious question whether victorin is a toxin in the classic sense or whether it in fact functions by eliciting a plant defense mechanism, the hypersensitive response. Victorin has now been isolated by several groups (Macko et al 1985; Gloer et al 1986; Mayama et al 1986; Walton and Earle 1984) and landmark work by Macko and Arigoni and their associates has resulted in a proposed structure for the major victorin-active peptide (Fig. 1) (Wolpert et al 1985).

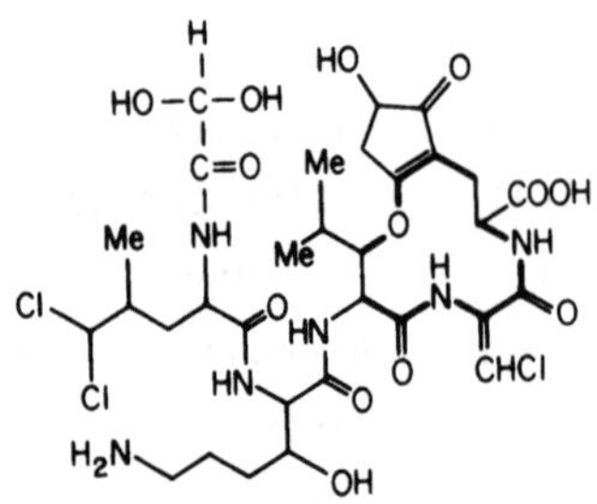

Fig. 1. Proposed structure of Victorin

Victorin affects the *Pc-2* genotype of oat plants but neither other known oat genotypes nor any other higher plant. This selectivity offers strong *a priori* support for the importance of the toxin in the pathogenicity of *H.*

victoriae. Of considerable interest is the fact that the Pc-2 oat gene was discovered because it confers resistance to the fungus pathogen, Puccinia coronata. This suggested to Litzenberger (1949) that victorin was not a toxin in the classic sense but might instead elicit a hypersensitive response in oat tissues infected with H. victoriae and that this fungus was able to withstand the defense reaction and continue colonization of the tissue. Mayama et al (1982) later showed that plants carrying Pc-2 produced phytoalexins in response to inoculation with incompatible but not compatible races of P. coronata. In more recent work, Mayama et al (1986) observed that victorin indeed in an efficient elicitor of phytoalexin accumulation in Pc-2 oat leaves but not in pc-2 leaves. They also noted that H. victoriae is relatively insensitive to the antibiotic effects of the oat phytoalexins, unlike Puccinia coronata. These facts therefore support the original Litzenberger (1949) suggestion that victorin is an elicitor of the hypersensitive reaction conditioned by the Pc-2 allele. Since the structures of the victorin-active peptides are now known, it should be possible to address the possibility that victorin has a binding site on oat cells which is encoded by the Pc-2 resistance gene. Such experimentation could also ultimately result in the molecular cloning of the Pc-2 gene.

Nachmias et al (1985, 1986) have shown that low molecular weight peptide toxins from Verticillium dahliae exhibit host-selective effects on tomato plants containing or lacking the Ve gene for Verticillium tolerance. Race 1 isolates of the fungus cause disease symptoms on ve tomatoes, but not Ve plants. On the other hand, race 2 isolates give severe symptoms on both genotypes. These symptoms and their relative specificity were reproduced by purified preparations of the peptide toxins isolated from the respective races (Nachmias et al 1986). Surprisingly, colonization of tomato plants by the two fungus races appears to be similar; thus, the appearance of disease symptoms may be a consequence of plant sensitivity to the fungal peptide toxins. The difference in sensitivity of ve and Ve gene tomatoes to the race 1 peptide was only ca. 10x, unlike the 10^6 difference observed with victorin on Pc-2 and pc-2 oats. Although additional work is required to characterize the Verticillium peptide toxins, preliminary results suggest that they may be cyclic (Nachmias, Burstein and Buchner, personal communication). It will be of interest to deduce the basis of their specificity and mode of action on sensitive tomato plants.

PLANT DISEASE DEFENSE MECHANISMS

Plants exhibit several types of resistance to potential plant pathogens. General resistance implies the resistance of an entire plant species to an entire pathogen species. On the other hand, specific resistance is exhibited by only certain genotypes of a plant species to some but not all biotypes or races of a pathogen taxon. In its most defined sense, the occurrence of specific resistance is regulated by single dominant genetic alleles in both the plant and pathogen, and their interaction is referred to as gene-for-gene complementarity (Keen 1982; Ellingboe 1985).

I will not discuss in detail the various mechanisms which contribute to general and specific plant disease resistance, but a brief consideration is necessary because successful pathogens must have evolved the means to deal with any and all preformed and inducible defense mechanisms of a host plant. Relatively little research has been done on preformed defense mechanisms, but these clearly function to restrict the number of pathogenic agents which attack a particular plant. Such preformed mechanisms can be categorized either as structural/morphologic features of a plant or chemical factors. Although these mechanisms are suspected to play a wide-spread role in preventing potential pathogens from gaining entry into plants, critical evidence is generally lacking which proves that a particular preformed barrier is physiologically functional in conferring disease resistance. It should be noted, however, that Kolattukudy's cutinase work discussed earlier has as a corollary the tenet that pea cuticle functions as a barrier which effects resistance against *Fusarium solani* unless the fungus secretes an active cutinase to degrade it.

Among preformed chemical factors, several compounds with antibiotic activity are present in plants which may prevent infection by a potentially pathogenic fungus, but very little conclusive experimental evidence supports this role. Two recently studied systems are the *Fusarium*-tomato system and the *Colletotrichum*-avocado system. In the former system, Defago et al (1983) demonstrated that the resistance of green tomato fruits to *F. solani* was due to their content of the preformed antifungal compound, tomatine (Fig. 2). Tomatine resistant mutants were selected and found to be pathogenic on green fruits. One of these mutants was then crossed with a tomatine-sensitive, non-pathogenic wildtype isolate. Analysis of the resulting cross progeny showed that tomatine insensitivity and pathogenicity always segregated together. The experiments therefore make a strong case for the role of tomatine in the resistance of green tomato fruits to wild-type isolates of F. solani.

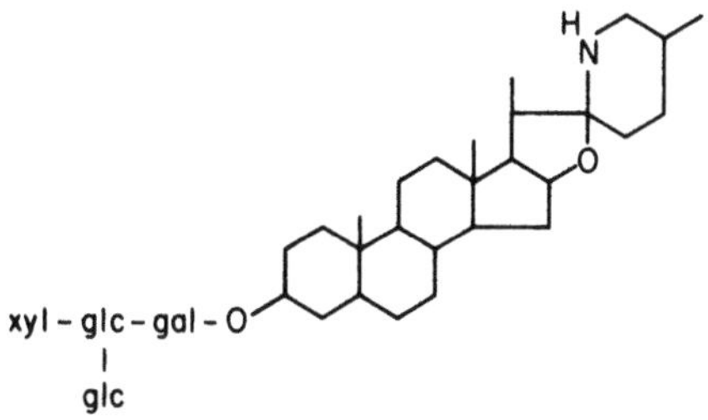

Fig. 2. Structure of tomatine

Prusky et al (1982) showed that levels of a preformed antifungal diene compound in the peel of avocado fruits bearing latent Colletotrichum gloeosporioides lesions correlated with the observed resistance of unripe fruits to colonization by the fungus. During ripening, however, levels of the diene were observed to decrease and resistance was lost. The antifungal factor was identified as a diene compound with considerable similarity to linoleic acid (Fig. 3). Since the latter compound is a well known substrate for plant lipoxygenases, the antifungal diene was tested and found to be efficiently detoxified by avocado and other lipoxygenases.

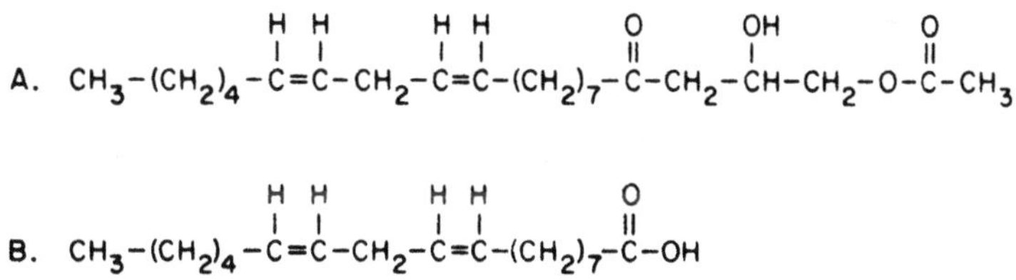

Fig. 3. Antifungal diene (A) and linoleic acid (B)

Activated lipoxygenase activity also was observed during avocado fruit ripening. This was believed responsible for the decrease of the diene compound because antioxidants and other lipoxygenase inhibitors retarded its disappearance and fruits remained resistant. Recently, Prusky et al (1985) observed that the level of epicatechin, a natural antioxidant in avocado peel, appears to modulate the activity of lipoxygenase, which in turn determines the rate of degradation of the antifungal diene and disease resistance. The data therefore suggest that the preformed antifungal diene compound is a resistance factor in unripe avocado fruits against C. gloeosporioides and that control of its concentration is complex.

Inducible plant defense mechanisms can be grouped into inducibly-produced antibiotic chemicals (phytoalexins) and inducibly-formed structural barriers. These topics have been reviewed several times in recent years (e.g. Brooks and Watson 1985; Darvill and Albersheim 1984; Keen 1986; Kuc and Rush 1985; Vance et al 1980) and considerable evidence suggests that they are major factors leading to the inhibition of pathogen growth in hypersensitive and other defense reactions such as adult plant resistance (Douglas et al 1984). Hypersensitive responses are typified by the rapid death of host cells surrounding the pathogen infection site and the arrested movement and multiplication of the pathogen. Although little is known about the mechanism(s) by which restriction of pathogen development occurs, the current perspective is that several biochemical changes which occur in the hypersensitive tissue may antagonize pathogen development. With viruses, the deposition of inducible barriers may restrict the movement of virus particles to other parts of the plant, while the production of inducible structural barriers (Hammerschmidt et al 1985) and phytoalexins may antagonize microbial pathogens. The emerging view, therefore, is that expression of the hypersensitive reaction may vary between plants and includes a complex array of factors such as the accumulation of lignin, callose and hydroxyproline-rich glycoprotein barriers on the one hand and chemical antagonists such as phytoalexins, chitinase and glucanases on the other.

Recent work from several laboratories has shown that expression of the hypersensitive response involves the early generation of reactive oxygen species which would be expected to account for the subsequently observed lipid peroxidation and loss of membrane integrity and the death of host cells (Doke and Chai 1985; Keppler and Novacky 1986). In an unknown manner, these early changes lead to the increased expression of a relatively small number of plant genes (Bell et al 1984; Chappell and Hahlbrock 1984; Cramer et al 1985) among which are those encoding enzymes in phytoalexin biosynthetic pathways and hydroxyprotein-rich glycoproteins (Showalter et al 1985).

ELICITORS, SUPPRESSORS AND PATHOGEN ESCAPE MECHANISMS

Elicitors are components of a pathogen which may initiate the invocation of one or more active host defense systems. It is suspected that elicitors are detected by a plant cell surveillance mechanism involving specific receptors, but little definitive evidence is yet available to test this point. Most fungi seem to contain elicitor-active fractions; the recognized fungal elicitors include low molecular weight peptides, unsaturated fatty acids, certain carbohydrate polymers such as glucans and chitin or chitosan, and the carbohydrate side chains of glycoconjugates. Oligogalacturonides released from plant cell walls or from sodium polypectate by fungal or bacterial enzymes are elicitors in some plants (Bruce and West 1982; Davis et al 1986). However, there is as yet no proof that these elicitors are of significance in determining the outcome of a plant-pathogen interaction. Similarly, despite the high activity of a well characterized hepta-glucan elicitor isolated by Albersheim's group from *Phytophthora megasperma* f.sp. *glycinea* (Sharp et al 1984), it is uncertain whether the elicitor is important in plant-pathogen specificity. My suspicion is that widely conserved fungal elicitors such as β-glucans and chitin are probably of significance as determinants of general resistance but are not involved in eliciting race-specific resistance. Thus, if a fungus displays cell wall carbohydrates to a host plant, it will not be a pathogen unless it has the additional capability to deal with the host resistance reaction.

Pathogens have evolved several means to escape non-specifically elicited host defenses and these may operate at various stages of the plant-

Table 3. Mechanisms used by successful plant pathogens to escape active general and specific plant defense mechanisms

1. Pathogen does not display effective elicitors to the host
2. Pathogen contains elicitors which are recognized by the plant but also produces suppressors that block invocation of an active defense system
3. Pathogen elicits defense reactions (e.g. phytoalexins) but produces enzymes that degrade them
4. Pathogen elicits defense system, but is able to tolerate the reaction (e.g. by incorporating host components which decrease sensitivity to a plant-produced phytoalexin)

pathogen interaction (Table 3). The simplest method for a pathogen to avoid an active host defense response is not to display an effective elicitor to the host cells. Note that the fungus can contain one or more elicitors, but they may not be presented to host cells in a recognizable way. This mechanism may largely account for the differential recognition associated with gene-for-gene plant-pathogen systems (Keen 1982). There is now substantial evidence for race-specific elicitors associated with the occurrence of distinct avirulence genes in fungal pathogens (Anderson 1980; Anderson and Tepfer, unpublished; DeWit et al 1985; Higgins and DeWit 1985; Keen and Legrand 1980; Keen et al 1983). Bacterial avirulence genes have been cloned which confer race specific host defense reactions (Gabriel 1984; Staskawicz et al 1984), but corresponding genes have not been cloned from fungal pathogens. It will be interesting when such cloned genes are available to test their relationship with the observed race specific elicitors.

Another mechanism used by successful pathogens to confound host defenses is the production of suppressor substances. In these cases the pathogen does display an effective elicitor to host cells but also produces a second substance which in some way inhibits the occurrence of the active defense mechanism. Evidence has been obtained suggesting that such an elicitor-suppressor mechanism may account for the determination of race specificity in the Phytophthora-potato system (Doke and Tomiyama 1980; Kuc et al 1984), but its occurrence has not yet been firmly established. On the other hand, certain pathogens which do not occur in gene-for-gene interactions with their hosts appear to utilize defense suppressors. For instance, Shiraishi et al (1978) isolated a low molecular weight peptide suppressor from Mycosphaerella pinodes which blocked the production of pisatin by infected pea plants. Significantly, the suppressor also rendered pea plants susceptible to attack by a Stemphylium species not normally pathogenic on peas. Similarly, Tietjen and Matern (1984) have also shown that a low molecular weight suppressor produced by Alternaria carthami blocks the ability of safflower plants to produce phytoalexins, thus presumably leading to the success of the pathogen.

I have already reviewed the work of Smith et al (1984) and that of Van Etten's group dealing with pathogen-mediated degradation of host phytoalexins. In such systems, the pathogen contains and displays elicitors to the host which then undergoes the phytoalexin response. The pathogen, however, produces enzymes which detoxify the phytoalexins to less antibiotic adducts.

A final mechanism for pathogen escape of a phytoalexin host defense response is exemplified by the Aphanomyces euteiches-pea system. Pueppke and VanEtten (1976) showed that infected pea tissue accumulated high amounts of pisatin, but growth of the pathogen was not inhibited. This observation was subsequently used to question the general role of phytoalexins in disease defense (Pueppke 1978). However, more recent work by Sweigard and VanEtten (1984) showed that growth of A. euteiches in medium supplemented with extracts from pea plants led to considerable fungal tolerance to the growth inhibitory effects of pisatin, unlike the case on culture medium without pea additives. In contrast to Fusaria, Aphanomyces did not appear to enzymatically degrade pisatin. Thus, the work of Sweigard and VanEtten suggests that Aphanomyces has developed a tolerance for pisatin based on the utilization of polar lipids derived from the host plant which permit the fungus to escape the growth inhibitory properties of pisatin. Thus, instead of arguing against the role of phytoalexins in plant disease defense, the evidence garnered from the Aphanomyces-pea system constitutes a case where the pathogen has developed a clever mechanism to escape the antibiotic property of a host phytoalexin.

PROSPECTS AND REQUIREMENTS FOR FUTURE WORK

Despite the ponderous literature on fungal pathogen-host interactions, we still do not know a great deal about the molecular mechanisms which dictate fungal pathogenicity and host defense to these organisms. In part this has resulted from the general unavailability of suitable genetic systems in the various pathogens to permit incisive genetic tests of candidate mechanisms. The recent advent of molecular cloning, transformation and gene replacement techniques for filamentous fungi, however, promises to revolutionize the study of fungal plant pathogens. For instance, it should be possible to molecularly clone the genes encoding certain enzymes or peptide toxins that are candidate pathogenicity factors. The in vitro mutation of these genes and their reintroduction into the wild-type organism by a gene replacement technique would then provide a near-isogenic mutant strain which should have lost some or all pathogenicity if the candidate gene is indeed important. The wider development of mutant strains by conventional techniques should also aid in elucidation of relevant pathogenicity mechanisms, especially if these mutants can be complemented by a gene library of wild-type fungal DNA in order to see if pathogenicity is then restored.

The relatively large genome size of fungi will be a major impediment to the routine use of molecular genetic techniques, especially if gene cloning must be done by function. On the other hand, if it is possible to isolate individual chromosomes by fluorescence activated sorting techniques or pulsed electrophoresis, these may be libraried or used in attempts to transform an entire chromosome into a recipient fungal strain. Finally, it would be useful if a transposeable element system could be developed for fungal pathogens. The yeast Ty elements (Boeke et al 1985) might be a lucrative candidate for this role if they could be satisfactorily engineered to function in filamentous fungal pathogens. The resulting mutational tagging of genes suspected to be involved in pathogenesis would permit their cloning and facilitate critical genetic tests of their importance.

The author's research is supported by grants from the National Science Foundation.

REFERENCES

Anderson AJ (1980) Differences in the biochemical compositions and elicitor activity of extracellular components produced by three races of a fungal plant pathogen, Colletotrichum lindemuthianum. Can J Microbiol. 26: 1473-1479

Bell JN, Dixon RA, Bailey JA, Rowell PM, Lamb CJ (1984) Differential induction of chalcone synthase mRNA activity at the onset of phytoalexin accumulation in compatible and incompatible plant-pathogen interactions. Proc Nat Acad Sci USA 81: 3384-3388

Boeke JP, Garfinkel DJ, Styles CA, Fink CR (1985) Ty elements transpose through an RNA intermediate. Cell 40: 494-500

Bruce RJ, West CA (1982) Elicitation of casbene synthetase activity in castor bean. The role of pectic fragments of the plant cell wall in elicitation by a fungal endopolygalacturonase. Plant Physiol 69: 1181-1188

Brooks CJW, Watson DG (1985) Phytoalexins. Nat Prod Reports 2: 427-459

Chappell J, Hahlbrock K (1984) Transcription of plant defense genes in response to UV light or fungal elicitor. Nature 311: 76-78

Collmer A, Keen NT (1986) The role of pectic enzymes in plant pathogenesis. Annu Rev Phytopathol 24: 383-409

Cooper RM (1984) The role of cell wall-degrading enzymes in infection and damage. In: Wood RKS, Jellis GJ (eds) Plant Diseases: infection, damage and loss, Blackwell Scientific Pub, Oxford, p. 13-27

Cramer CL, Ryder TB, Bell JN, Lamb CJ (1985) Rapid switching of plant gene expression induced by fungal elicitor. Science 227: 1240-1243

Darvill AG, Albersheim P (1984) Phytoalexins and their elicitors--a defense against microbial infection in plants. Annu Rev Plant Physiol 35: 243-275

Davis KR, Darvill AG, Albersheim P, Dell A (1986) Host-pathogen interactions. XXIX. Oligogalacturonides released from sodium polypectate by endopolygalacturonic acid lyase are elicitors of phytoalexins in soybean. Plant Physiol (in press)

Défago G, Kern H, Sedlar L (1983) Genetic analysis of tomatine insensitivity, sterol content and pathogenicity for green tomato fruits in mutants of _Fusarium solani_. Physiol Plant Pathol 22: 39-43

DeWit PJGM, Hofman AE, Velthuis GCM, Kuc JA (1985) Isolation and characterization of an elicitor of necrosis isolated from intercellular fluids of compatible interactions of _Cladosporium fulvum_ (syn. _Fulvia fulva_) and tomato. Plant Physiol 77: 642-647

Doke N, Chai HB (1985) Activation of superoxide generation an enhancement of resistance against compatible races of _Phytophthora infestans_ in potato plants treated with digitonin. Physiol Plant Pathol 27: 323-334

Doke N, Tomiyama K (1980) Suppression of the hypersensitive response of potato tuber protoplasts to hyphal wall components by water soluble glucans isolated from _Phytophthora infestans_. Physiol Plant Pathol 16: 177-186

Douglas SM, Sherwood RT, Lukezic FL (1984) Effect of adult plant resistance on primary penetration of oats by _Erysiphe graminis_ f. sp. _avenae_. Physiol Plant Pathol 25: 219-228

Durbin RD (1981) Toxins in Plant Disease. Academic Press, New York. 515 p

Ellingboe AH (1985) Prospects for using recombinant DNA technology to study race-specific interactions between host and parasite. In: Groth JV, Bushnell WR (eds), Genetic basis of biochemical mechanisms of plant disease, APS Press, St. Paul, Minn, p 103-125

Epstein L, Laccetti L, Staples RC, Hoch HC, Hoose WA (1985) Extracellular proteins associated with induction of differentiation in bean rust uredospore germlings. Phytopathology 75: 1073-1076

Gabriel DW (1984) Molecular cloning of specific avirulence genes from Xanthomonas malvacearum. Proc 2nd International Symp on the Mol Gen of the Bacterial-Plant Interaction

Gloer JB, Meinwald J, Walton JD, Earle ED (1985) Studies on the fungal phytotoxin victorin: structures of three novel amino acids from the acid hydrolyzate. Experientia 41: 1370-1374

Hammerschmidt R, Bonne AM, Bergstrom GC, Baker KK (1985) Association of epidermal lignification with nonhost resistance of cucurbits to fungi. Can J Bot 63: 2393-2398

Higgins VJ, DeWit PJGM, (1985) Use of race- and cultivar-specific elicitors from intercellular fluid for characterizing races of Cladosporium fulvum and resistant tomato cultivars. Phytopathology 75: 695-699

Howell C (1976) Use of enzyme-deficient mutants of Verticillium dahliae to assess the importance of pectolytic enzymes in symptom expression of Verticillium wilt of cotton. Physiol Plant Pathol 9: 279-283

Keen NT (1982) Specific recognition in gene-for-gene host-parasite systems. Adv Plant Pathol 1: 35-82

Keen NT (1986) Phytoalexins and their involvement in plant disease resistance. Iowa State J Sci (in press)

Keen NT, Legrand M (1980) Surface glycoproteins: evidence that they may function as the race specific phytoalexin elicitors of Phytophthora megasperma f. sp. glycinea. Physiol Plant Pathol 17: 175-192

Keen NT, Yoshikawa M, Wang MC (1983) Phytoalexin elicitor activity of carbohydrates from Phytophthora megasperma f. sp. glycinea and other sources. Plant Physiol 71: 466-471

Keppler LD, Novacky A (1986) Involvement of membrane lipid peroxidation in the development of a bacterially induced hypersensitive reaction. Phytopathology 76: 104-108

Kistler HC, VanEtten HD (1984) Three non-allelic genes for pisatin demethylation in the fungus Nectria haematococca. J Gen Microbiol 130: 2595-2603

Kolattukudy PE (1985) Enzymatic penetration of the plant cuticle by fungal pathogens. Annu Rev Phytopathol 23: 223-250

Kubo Y, Suzuki K, Furusawa I, Yamamato M (1985) Melanin biosynthesis as a prerequisite for penetration by appresoria of Colletotrichum lagenarium: site of inhibition by melanin-inhibiting fungicides and their action on appresaria. Pest Biochem Physiol 23: 47-55

Kuc J, Rush JS (1985) Phytoalexins. Arch Biochem Biophys 236: 455-472

Kuc J, Tjamos E, Bostock R (1984) Metabolic regulation of terpenoid accumulation and disease resistance in potato. In: Nes W, Fuller G, Tsai L (eds), Isopentenoids in plants, biochemistry and function, Mercel Dekker, New York, p 103-126

Litzenberger SC (1949) Nature of susceptibility to Helminthosporium victoriae and resistance to Puccinia coronata in Victoria oats. Phytopathology 39: 300-318

Macko V, Wolpert TJ, Acklin W, Juan B, Seibl J, Meili J, Arigoni D (1985) Characterization of victorin C, the major host-selective toxin from Cochliobolus victoriae: structure of degradation products. Experientia 41: 1366-1370

Mayama S, Matsuura Y, Iida H, Tani T (1982) The role of avenalumin in the resistance of oat to crown rust, Puccinia coronata f.sp. avenae. Physiol Plant Pathol 20: 189-199

Mayama S, Tani T, Ueno T, Midland SL, Sims JJ, Keen NT (1986) The purification of victorin and its phytoalexin elicitor activity in oat leaves. Physiol Plant Pathol (in press)

Nachmias A, Buchner V, Burstein Y (1985) Biological and immunochemical characterization of a low molecular weight phytotoxin isolated from a protein-lipopolysaccharide complex produced by a potato isolate of Verticillium dahliae Kleb. Physiol Plant Pathol 26: 43-55

Nachmias A, Buchner V, Tsror L, Burstein Y, Keen NT (1986) Differential phytotoxicity of peptides from culture fluids of Verticillium dahliae races 1 and 2 and their relationship to pathogenicity. Phytopathology (in press)

Prusky D, Keen NT, Sims JJ, Midland SL (1982) Possible involvement of an antifungal compound in the latency of Colletotrichum gloeosporioides on unripe avocado fruits. Phytopathology 72: 1578-1582

Prusky D, Kobiler I, Jacoby B, Sims JJ, Midland SL (1985) Inhibitors of avocado lipoxygenase: their possible relationship with the latency of Colletotrichum gloeosporioides. Physiol Plant Pathol 27: 269-279

Pueppke SG (1978) Some intriguing aspects of the putative role of isoflavonoid phytoalexins in plant disease. Mycopathologia 65: 115-119

Pueppke SG, VanEtten HD (1976) The relation between pisatin and the development of Aphanomyces euteiches in diseased Pisum sativum. Phytopathology 66: 1174-1185

Scheffer RP, Livingstone RS (1984) Host-selective toxins and their role in plant diseases. Science 223: 17-21

Sharp JK, McNeil M, Albersheim P (1984) The primary structures of one elicitor-active and seven elicitor-inactive hexa(β-D-glucopyranosyl)-D-glucitols isolated from the mycelial walls of Phytophthora megasperma f.sp. glycinea. J Biol Chem 259: 11321-11336

Shiraishi T, Oku H, Yamashita M, Ouchi S (1978) Elicitor and suppressor of pisatin induction in spore germination fluid of pea pathogen, Mycosphaerella pinodes. Ann Phytopathol Soc Japan 44: 659-665

Showalter AM, Bell JN, Cramer CL, Bailey JA, Varner JE, Lamb CJ (1985) Accumulation of hydroxyproline-rich glycoprotein mRNAs in response to fungal elicitor and infection. Proc Nat Acad Sci USA 82: 6551-6555

Smith DA, Wheeler HE, Banks SW, Cleveland TE (1984) Association between lowered kievitone hydratase activity and reduced virulence to bean in variants of Fusarium solani f.sp. phaseoli. Physiol Plant Pathol 25: 135-147

Soliday CL, Flurkey WH, Okita TW, Kolattukudy P (1984) Cloning and structure determination of cDNA for cutinase, an enzyme involved in fungal penetration of plants. Proc Nat Acad Sci USA 81: 3939-3943

Staples RC, Hoch HC, Epstein L, Laccetti L, Harsouna S (1985) Recognition of host morphology by rust fungi: responses and mechanisms. Can J Plant Pathol 7: 314-322

Staskawicz BJ, Dahlbeck D, Keen NT (1984) Cloned avirulence gene of Pseudomonas syringae pv. glycinea determines race-specific incompatibility on Glycine max (L.) Merr. Proc Natl Acad Sci USA 81: 6024-6028

Sweigard JA, VanEtten HD (1984) Polar lipids from pea reduce the pisatin sensitivity of Aphanomyces euteiches. Phytopathology 74: 833 (Abstr.)

Tegtmeier KJ, VanEtten HD (1982) The role of pisatin tolerance and degradation in the virulence of Nectria haematococca on peas: a genetic analysis. Phytopathology 72: 608-612

Tietjen KG, Matern U (1984) Induction and suppression of phytoalexin biosynthesis in cultured cells of sunflower, Carthamus tinctorius L., by metabolites of Alternaria carthami Cnowdhury. Arch Biochem Biophys 229: 136-144

Tolousbalides MC, Sisler HD (1978) Effect of tricyclazole on growth and secondary metabolism in Pyricularia oryzae. Pesticide Biochem Physiol 8: 26-32

Vance CP, Kirk TK, Sherwood RT (1980) Lignification as a mechanism of disease resistance. Annu Rev Phytopathol 18: 259-288

Walton JD, Earle ED (1984) Characterization of the host-specific phytotoxin victorin by high-pressure liquid chromatography. Plant Sci Lett 34: 231-238

Woloshuk CP, Sisler HD, Vigil EL (1983) Action of the antipenetrant, tricyclazole, on appresoria of Pyricularia oryzae. Physiol Plant Pathol 22: 245-259

Wolpert TJ, Macko V, Acklin W, Juan B, Seibl J, Meili J, Arigoni D (1985) Structure of victorin C, the major host-selective toxin from Cochliobolus victoriae. Experientia 41: 1524-1529

Wynn, WK, Staples RC (1981) Tropisms of fungi in host recognition. In: Staples RC, Toenniessen GH (eds) Plant Disease Control: Resistance and Susceptibility. Wiley-Interscience, New York, p 45-69

THE MOLECULAR BASIS OF THE AGROBACTERIUM-PLANT INTERACTION

- Characteristics of Agrobacterium virulence genes and their possible occurrence in other plant-associated bacteria -

Paul J.J. Hooykaas and Rob A. Schilperoort
Leiden University,
Dept. Plant Molecular Biology,
Biochemistry Lab,
Wassenaarseweg 64,
2333 AL Leiden,
The Netherlands

INTRODUCTION

Bacterial species of the genus Agrobacterium can either induce tumours or adventitious roots from infected wound sites on a wide range of dicotyledonous plant species (Fig.1). While A.tumefaciens induces tumours which are called crown galls (Smith and Townsend 1907), A.rhizogenes (Riker 1930) provokes hairy root formation on plant species such as apple, tobacco, clover, carrot or tumour formation on plants such as sunflower and pea. Most Agrobacterium strains have a wide host range, but certain strains - usually grapevine isolates - have a limited host range, which is restricted to grapevine and a number of Nicotiana species (Panagopoulos and Psallidas 1973). There is variation among plant species (De Cleene and De Ley 1976) and even between cultivars of one particular species (Owens and Cress 1985) in their susceptibility to Agrobacterium infection, while also various parts and organs of the same plant may respond differently to this bacterium.

Three major events can be recognized in the interaction between Agrobacterium and plants, namely: a) establishment of a host-bacterium complex, b) signal communication, and c) host response. During the last decade many data have been acquired regarding the bacterial genes involved in these three events. A number of these genes are located in the chromosome of Agrobacterium, but most are present in a large, about 200kbp plasmid, which is called Ti (tumour inducing) plasmid in the case of A.tumefaciens and Ri (root inducing) plasmid in the case of A.rhizogenes. The chromosomal virulence or chv-genes play a role in the establishment of a host-bacterium complex. They mediate

NATO ASI Series, Vol. H 4
Recognition in Microbe-Plant Symbiotic and Pathogenic Interactions. Edited by B. Lugtenberg

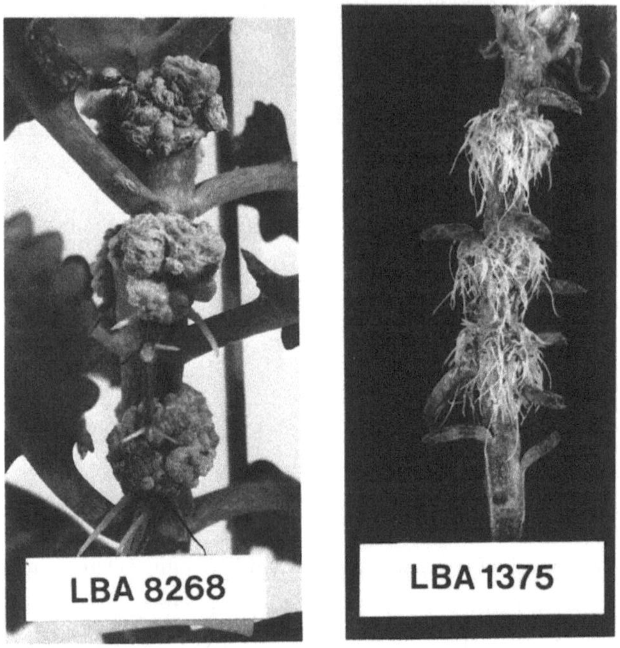

Fig.1. Tumour (left) and hairy root (right) formation on kalanchoe as a result of infection with A.tumefaciens and A.rhizogenes, respectively.

attachment of the bacterium to the plant cell walls (Douglas et al 1982) as well as the formation of bacterial aggregates via cellulose fibrils (Matthijsse 1983). The Ti and Ri plasmid genes necessary for tumor induction are clustered in two adjacent regions, which are called the transferred (T-) region and the virulence (Vir)-region, respectively (Fig.2).

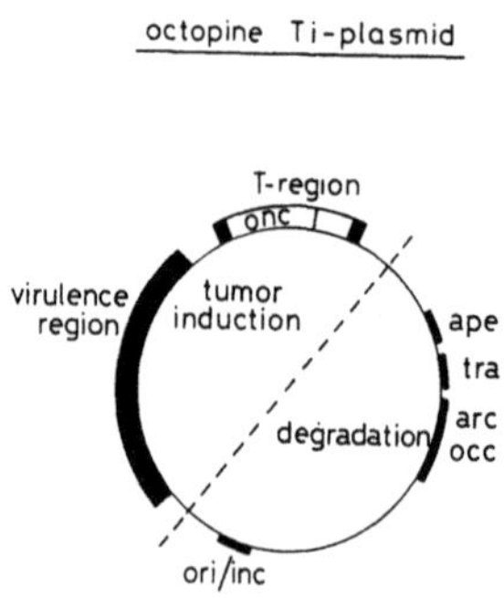

Fig.2. Schematic representation of the genetic map of an octopine Ti plasmid.

During tumour induction the T-region becomes integrated in the nuclear genome of plant cells (Reviewed by Bevan and Chilton 1982). The process of T-region transfer is mediated in an unknown way by products determined by vir-genes located in the Vir-region. It was found by Okker et al (1984) and Stachel et al (1985) that the vir-genes are silent until they are induced by particular plant compounds of low molecular weight (signal communication).

The T-region contains a number of genes, which are expressed in the transformed plant cells (Willmitzer et al 1982). The expression of some of these genes, which are onc-genes, leads to the formation of a tumour in dicotyledonous plants (host response). As a result of the presence of other T-DNA genes the tumour cells produce (Tempé and Goldmann 1982) and excrete (Messens et al 1985) a number of unusual amino acid derivatives and phosphorylated sugars which together are called opines. The opines can be catabolized by virulent Agrobacterium strains, but not by most other bacteria (Petit et al 1970). The catabolic genes involved in this are located on the Ti plasmid (Fig.2 : occ, arc). Interestingly, these genes as well as those involved in the conjugative transfer of the Ti plasmid, are inducible, but only by particular opines (signal communication) (Klapwijk et al 1978; Petit and Tempé 1978). A number of reviews on the Agrobacterium-plant interaction have appeared (e.g. Nester and Kosuge 1981; Van Montagu and Schell 1982; Hooykaas and Schilperoort 1984). In this paper we shall describe the genes in A.tumefaciens that are necessary for its tumorigenic properties, and discuss their (possible) functions as well as their (possible) occurrence in other fytopathogenic and rhizosphere bacteria.

THE ONC-GENES OF THE T-DNA

The T-region is flanked by 24 bp imperfect, direct repeats in the Ti plasmid (Yadav et al 1982). It is likely that these repeats act as recognition signals for the Vir-system by which the T-region is transferred to plant cells. Plant cells may receive one or more copies of the T-DNA, and also truncated or internally deleted T-DNA copies may be present in tumour cells. According to Peerbolte (1986) irregularities in T-DNA structure may arise at four stages namely: 1) in the bacteria during processing of the T-region (See chapter by Van Haaren et al); 2) in the plant cells before integration, whereby T-DNA may even come linked to certain plant DNA fragments; 3) after integration in a process that might act to stabilize the chromatin structure; 4) during prolonged tissue culture via somaclonal variation (at a very low frequency). Although the T-DNA is of bacterial origin, the genes it carries are expressed in an eukaryotic system, namely when present in the nucleus of plant cells. This is due to the fact that T-DNA genes are surrounded by regulatory sequences which are recognized by plant factors involved in transcription and translation.

The functions determined by the three most important onc-genes in the Ti T-DNA are known now. The gene called cyt or tmr determines an isopentenyl-transferase, which converts AMP into the cytokinin isopentenyl-AMP (Akiyoshi et al 1984; Barry et al 1984), while the genes called aux-1 or tms-1 (Thomashow et al 1986; Van Onckelen et al 1986) and aux-2 or tms-2 (Schröder et al 1984; Thomashow et al 1984) code for the enzymes tryptophan monooxygenase and indoleacetamide hydrolase, respectively. The latter two enzymes together mediate the production of the auxin indole acetic acid (IAA) from tryptophan. The production of an auxin and a cytokinin by the transformed plant cells explains why they are tumorous and can grow as phytohormone autotrophs (Braun 1958). Agrobacteria do not form tumours on monocotyledonous plants (De Cleene and De Ley 1976) and this was long taken as evidence that T-DNA transfer does not occur to these plants.

However, this assumption turned out not to be true. Data that were acquired by us from opine assays showed unequivocally that T-DNA transfer is possible to monocots, but that this does not lead to tumour formation (Hooykaas-Van Slogteren et al 1984). The reason for the latter may be that monocots are not very responsive to the auxins and cytokinins produced via the T-DNA onc-genes.

The bacterium A.rhizogenes provokes the hairy root disease in plants. However, in certain plant species this bacterium induces tumours just like A.tumefaciens. Mutants of A.tumefaciens with a mutation in one of the aux-genes can complement mutants with a mutation in the cyt-gene for tumour induction in coinoculation experiments (Ooms et al 1982). We have similarly studied whether agropine type A.rhizogenes strains could complement aux- or cyt mutants of A.tumefaciens in order to find out whether the T-DNA of the Ri plasmid contained genes with a similar function. Our experiments showed that the presence of an agropine type A.rhizogenes strain led to the complementation of aux-, but not of cyt-mutants of A.tumefaciens. Further work led to the identification of two genes in the T_R-region of the Ri plasmid, which determined similar functions as the Ti aux-1 and aux-2 genes, respectively

(Offringa et al 1986). These Ri genes turned out to be important for the initiation of the hairy root disease (Offringa et al 1986), but appeared not essential for the phytohormone independent growth of established hairy root cultures (Peerbolte 1986). Although neither the T_L- nor the T_R-region of the Ri plasmid carried a gene that could complement Ti cyt mutants, we observed that the Ri T_L-DNA contained a gene, which mediated a suppression of root formation, a property in common with the Ti cyt-gene.

The bacterium Pseudomonas syringae pv. savastanoi is the causal agent of gall formation on olive, oleander and privet (Wilson 1935). This bacterium excretes the auxin IAA and the cytokinin trans-zeatin (Schmidt and Kosuge 1978; Powell and Morris 1986). Mutants which are not able to produce IAA are avirulent (Schmidt and Kosuge 1978), which shows that IAA-production is a virulence factor in this bacterium. The genes involved in the formation of IAA, called iaaH and iaaM (Yamada et al 1985) and trans-zeatin, called tzp (Powell and Morris 1986) have been identified and sequenced. They turned out to be very similar to the aux- and cyt-genes of the Agrobacterium Ti plasmid, but in contrast to the Agrobacterium genes, which are well expressed only in plant cells, the Ps.syringae pv. savastanoi genes are surrounded by regulatory sequences that guarantee good expression within a bacterial background.

Infection of plants by symbiotic and pathogenic bacteria is often accompanied by the formation of structures such as root nodules, galls, hairy roots or (teratomatous) shoots. It is evident that plant growth regulators must play an important role in this. The finding that such widely separated bacteria as Agrobacterium and Rhizobium use similar genes for the production of phytohormones makes it likely that many more plant-associated bacteria contain similar genes. The regulatory sequences connected to the coding sequences, may, however, be different in each case, since they must be adapted to the expression system of the particular host.

THE VIR-REGION

During tumour induction the T-region, but none of the rest of the Ti plasmid is introduced into the plant genome. Transfer of the T-region from bacterium to plant cell even occurs when the T-region is physically separate from the rest of Ti present on a compatible plasmid (De Framond et al 1983; Hoekema et al 1983) or in the chromosome of A.tumefaciens (Hoekema et al

1984a). Strains with such binary systems are equally virulent as wild-type strains. This shows that the T-region as such is recognized by the transfer-apparatus of Agrobacterium. Direct repeats of 24 bp which border the T-region at both extremities probably form the recognition signals for the transfer apparatus of Agrobacterium. The transfer apparatus itself is encoded by genes that are located outside the T-region, since none of the T-region genes is important for the transfer process (Leemans et al 1982). It is most likely that genes located in the Vir-region are involved in transfer. The 45 kbp Vir-region is adjacent to the left border of the T-region, and contains 7 loci most of which contain a number of genes (Hille et al 1982; Klee et al 1983; Hooykaas et al 1984; Lundquist et al 1984). Different names have been given to the different vir-operons, but at the First International Congress of Plant Molecular Biology (Savannah, USA, 1985) it was agreed to use the following nomenclature (going in the direction of the left border of the T-region): virA, virB, virG, virC (previously virO, bak), virD (previously virC, hdv), virE and virF. With the exception of virA, which is constitutive, the vir-genes are induced by certain low mol. wt. compounds present in root exudate and wound juice (Okker et al 1984; Stachel et al 1986). The compounds that induce these vir-genes include acetosyringone, sinapinic acid, syringaldehyde and acetovanillone (Stachel et al 1985). A number of vir-loci (virA, virB, virG, virD) contain genes that are essential for tumour induction on all plant species tested, but mutations at other vir-loci (virC, virE, virF) only lead to a reduction in the host range for tumour induction (Garfinkel and Nester 1980; Hille et al 1984; Hooykaas et al 1984). The DNA-sequence of most of the Vir-region has been determined and this has shown that this region determines at least 21 different proteins. The functions of these proteins are unknown, but it is obvious that some may be involved in the regulation of expression of the vir-regulon, while other may play a role in the processing of the T-region during tumour induction. Among the latter there may be proteins that recognize, bind and/or nick the Ti plasmid at the 24 bp border repeats.

During hairy root induction A.rhizogenes transfers the T-region of its Ri plasmid to plant cells (Reviewed by Bevan and Chilton 1982). It was therefore of interest to find out whether the Vir-systems of A.tumefaciens and A.rhizogenes are related. In fact adjacent to the left border of the T-region of the Ri plasmid a region with homology to the Ti Vir-region was detected (White et al 1980; Risuleo et al 1982). In order to find out whether there was also functional homology complementation experiments were done between Ri and Ti vir-mutants. The Ti plasmid belongs to a different incompatibility group as the Ri plasmid and both can therefore be stably maintained in one cell (Hooykaas 1979; Costantino et al 1980). The results of the complementation experiments showed that Ti vir-genes were interchangeable with Ri vir-genes (Hooykaas et al 1984; unpublished; Hoekema et al 1984b).

During root nodule induction on legumes by Rhizobium there is no DNA-transfer to plant cells. Therefore, it was not expected that Rhizobium would contain a Vir-system comparable to that of Agrobacterium. Indeed only limited homology is detected when the Ti Vir-region is used as a probe to find homologous sequences in Rhizobium Sym plasmids. Moreover, the small areas which share homology probably do not represent vir-genes, but are located between certain vir-operons and might be transposable elements or remnants thereof (Prakash et al 1982). In order to find out whether Sym plasmid carrying R.trifolii or R.leguminosarum strains could provide a system that could substitute for the Ti Vir-system, a plasmid with the Ti T-region, but lacking the Ti Vir-region was introduced into these strains. The resulting strains remained avirulent (Hoekema et al 1984b). Since it is already known for many years that R.trifolii and R.leguminosarum become tumorigenic after receipt of a complete Ti plasmid (Hooykaas et al 1977; Hooykaas 1979), the results obtained by Hoekema et al (1984b) show that Rhizobium cannot provide a functional Vir-system. There is, however, one caveat that we must consider. This concerns the possibility that hypothetical Rhizobium vir-genes are only expressed in the presence of certain plant exudate factors that are specific for its leguminous hosts, and therefore were not expressed during the tumour assays on tomato, tobacco, kalanchoe.

CHROMOSOMAL VIRULENCE GENES

Besides the Ti plasmid encoded vir-genes the Agrobacterium chromosome also contains a number of genes (chv), which are required for virulence. The absence of one or more of these chv-genes in other bacterial species than Agrobacterium, Rhizobium trifolii and R.leguminosarum may be the main reason why these bacteria, which include for instance R.meliloti and R.parasponia (Hooykaas 1979; unpublished) do not become tumorigenic upon receiving a Ti or Ri plasmid.

Some of the chv-genes are host range determinants (Hooykaas et al 1984), while other determine the production of cellulose fibrils (Matthysse 1983). These fibrils are not absolutely essential for tumour induction, but they cause an aggregation of the agrobacteria, which leads to an enhanced ability for tumour induction under certain conditions (Matthysse 1983). Certain chv-genes determine the ability of Agrobacterium to attach to plant cell walls, which is an initial and essential step in plant tumorigenesis (Douglas et al 1982; See also chapter by Matthysse). Two of these "attachment" genes, which are called chvA and chvB, have been partially characterized. They are located in the Agrobacterium chromosome in the vicinity of the trp-13 marker (Vanderleyden 1985) on the map of Hooykaas et al (1982) and form constitutive, convergently transcribed genes (Douglas et al 1985). Mutations in chvA or chvB not only lead to a loss of virulence and an inability to attach to plant cells, but also result in a loss of bacterial motility (Bradley et al 1984). Agrobacterium chvB mutants lack 1,2-β-glucans, which are present in the periplasm of wild-type Agrobacterium strains (Puvanesarajah et al 1985). Although is has been reported that chvA and chvB mutants are avirulent on plant species such as kalanchoe, sunflower and tomato (Garfinkel and Nester 1980) and it was implied that these genes are essential for tumorigenicity on all plant species (Douglas et al 1985), we found that these mutants are still virulent on a few plant species (Solanum tuberosum, Nicotiana debneyi). It is difficult to explain this result; however, it might be that the chvA and chvB genes are regulatory rather than structural, or one might even speculate that Agrobacterium can use a different mode of infection in the case of the two plant species mentioned above.

We were interested to know whether the chvA and chvB genes had homologous counterparts in other plant-associated bacteria. Therefore, we used the chv-genes as probe in Southern blotting experiments with total DNAs of a number of bacteria. In this way we found homologous sequences in bacteria such as Rhizobium, Azospirillum, Azotomonas, Pseudomonas putida and Enterobacter agglomerans (R.J.M. van Veen, R.A.S. and P.J.J.H., unpublished). Similarly, homology was detected in the genome of several different Azospirillum strains by J. Vanderleyden (pers. commun.). Since chv-like sequences are present in Rhizobium species, we analysed their role in the process of root nodule induction (Van Veen et al 1986). Therefore, we introduced the Sym plasmid of R.phaseoli into Agrobacterium strains, which were either wild-type for the chvA- and chv-B-genes or had a chvA or chvB mutation. We used the

R.phaseoli plasmid pSym9 in these experiments, since we already knew that this plasmid renders Agrobacterium wild-type strains capable of inducing root nodules on Phaseolus vulgaris abundantly (Fig.3; Hooykaas et al 1985). We discovered that the introduction of pSym9 into the chvA and chvB mutants did not result in nodulating Agrobacterium strains (Fig.3). In spite of the presence of pSym9 no nodules were formed at all by the chvB mutants on P.vulgaris, while the Sym plasmid carrying chvA mutants only induced a few tiny, non-fixing root nodules on the same plant species (Van Veen et al 1986). These results showed that the chvA and chvB genes are important not only for tumour induction, but also for root nodule formation on plants. Since the available data indicate that these chv-genes mediate a non-host-specific attachment of Agrobacterium to plant cell walls, it can be assumed that a non-host-specific adhesion is a first step not only in the process of crown gall induction, but also in that of root nodule formation. This conclusion is in disagreement with the lectin-hypothesis, which states that the initial step in root nodulation consists of a host-specific attachment of Rhizobium to the root hairs via a plant species specific lectin. (Dazzo et al 1984). In order to reconcile all the available data it can be hypothesized that lectins play an important role as a specificity factor in the process of root nodule induction, but at a later stage than the initial non-host-specific adhesion step.

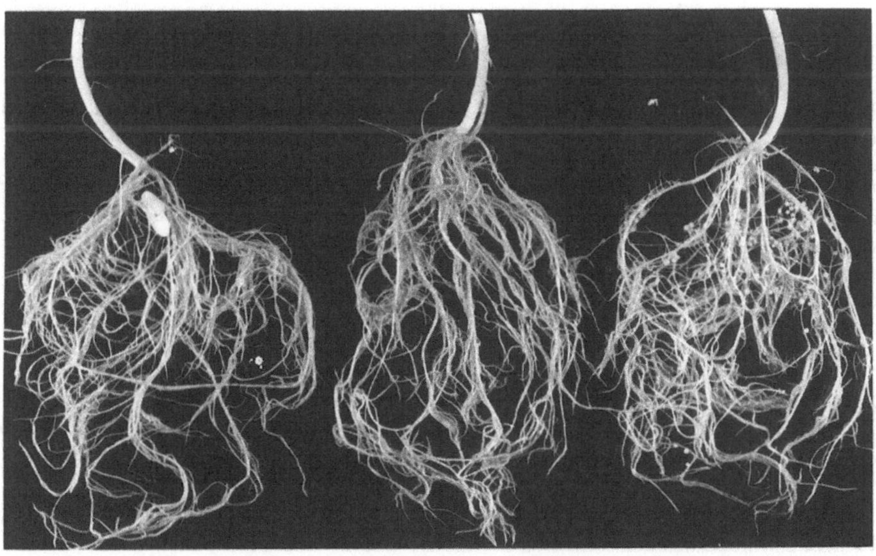

Fig.3. Root nodule induction on Phaseolus vulgaris by Agrobacterium strains (from left to right): wild-type (no Sym plasmid); chvB mutant (with pSym9); wild-type (with pSym9).

CONCLUDING REMARKS

Many, but probably not all, Agrobacterium genes that are involved in crown gall induction have been identified and characterized in some detail. Unfortunately, the functions which most of these genes determine in the establishment of a host-bacterium complex, in signal communication between host and bacterium or in provoking a host response, are largely unknown. However, some of the virulence genes of Agrobacterium have been shown to be involved in the attachment of bacterium to plant cell, and others in the production of plant growth regulators which trigger a plant response.
It is striking that genes involved in the latter two processes seem to have counterparts in other plant-associated bacteria. We suspect, therefore, that some of the principles used by A.tumefaciens to colonize and parasitize on plants are also used by many other soil microorganisms for similar purposes.

REFERENCES

Akiyoshi DE, Klee H, Amasino RM, Nester EW, Gordon MP (1984) T-DNA of Agrobacterium tumefaciens encodes an enzyme of cytokinin biosynthesis. Proc Natl Acad Sci USA 81:5994-5998

Barry GF, Rogers SG, Fraley RT, Brand L (1984) Identification of a cloned cytokinin biosynthetic gene. Proc Natl Acad Sci USA 81:4776-4780

Bevan MW, Chilton MD (1982) T-DNA of the Agrobacterium Ti and Ri plasmids. Ann Rev Genet 16:357-384

Bradley DE, Douglas CJ, Peschon J (1984) Flagella-specific bacteriophages of Agrobacterium tumefaciens: demonstration of virulence of nonmotile mutants. Can J Microbiol 30:676-681

Braun AC (1958) A physiological basis for autonomous growth of crown gall tumor cell. Proc Natl Acad Sci USA 44:344-349

Costantino P, Hooykaas PJJ, Den Dulk-Ras H, Schilperoort RA (1980) Tumor formation and rhizogenicity of Agrobacterium rhizogenes carrying Ti plasmids. Gene 11:79-87

Dazzo FB, Truchet GL, Sherwood JE, Hrabak EM, Abe M, Pankratz SH (1984) Specific phases of root hair attachment in the Rhizobium trifolii-clover symbiosis. Appl Env Microbiol 48:1140-1150

De Cleene M, DeLey J (1976) The host range of crown gall. Botan Rev 42: 389-466

De Framond AJ, Barton KA, Chilton MD (1983) Mini-Ti: a new vector strategy for plant genetic engineering. Bio/Technology 1:262-269

Douglas CJ, Halperin W, Nester EW (1982) Agrobacterium tumefaciens mutants affected in attachment to plant cells. J Bacteriol 152:1265-1275

Douglas CJ, Staneloni RJ, Rubin RA, Nester EW (1985) Identification and genetic analysis of an Agrobacterium tumefaciens chromosomal virulence region. J Bacteriol 161:850-860

Garfinkel DJ, Nester EW (1980) Agrobacterium tumefaciens mutants affected in crown gall tumorigenesis and octopine catabolism. J Bacteriol 144:732-743

Hille J, Klasen I, Schilperoort R (1982) Construction and application of Rprime plasmids, carrying different segments of an octopine Ti plasmid from Agrobacterium tumefaciens for complementation of vir genes. Plasmid 7:107-118

Hille J, Van Kan J, Schilperoort R (1984) Trans-acting virulence functions of the octopine Ti plasmid from Agrobacterium tumefaciens. J Bacteriol 158:754-756

Hoekema A, Hirsch PR, Hooykaas PJJ, Schilperoort RA (1983) A binary plant vector strategy based on separation of vir and T-region of the Agrobacterium tumefaciens Ti plasmid. Nature 303:179-180

Hoekema A, Roelvink PW, Hooykaas PJJ, Schilperoort RA (1984a) Delivery of T-DNA from the Agrobacterium tumefaciens chromosome into plant cells. EMBO J 3:2485-2490

Hoekema A, Hooykaas PJJ, Schilperoort RA (1984b) Transfer of the octopine T-DNA segment to plant cells mediated by different types of Agrobacterium tumor-or root inducing plasmids: generality of virulence systems. J Bacteriol 158:383-385

Hooykaas PJJ (1979) The role of plasmid determined functions in the interactions of Rhizobiaceae with plant cells. A genetic approach. Thesis, Leiden, The Netherlands

Hooykaas PJJ, Schilperoort RA (1984) The molecular genetics of crown gall tumorigenesis. In: Scandalios JG (ed) Molecular Genetics of Plants. Advances in Genetics, vol 22. Academic Press, Orlando, USA, pp 209-283

Hooykaas PJJ, Klapwijk PM, Nuti MP, Schilperoort RA, Rörsch A (1977) Transfer of the Agrobacterium tumefaciens Ti plasmid to avirulent agrobacteria and to Rhizobium ex planta. J Gen Microbiol 98:477-484

Hooykaas PJJ, Peerbolte R, Regensburg-Tuïnk AJG, De Vries P, Schilperoort RA (1982) A chromosomal linkage map of Agrobacterium tumefaciens and a comparison with the maps of Rhizobium spp. Mol Gen Genet 188:12-17

Hooykaas PJJ, Hofker M, Den Dulk-Ras H, Schilperoort RA (1984) A comparison of virulence determinants in an octopine Ti plasmid, a nopaline Ti plasmid, and an Ri plasmid by complementation analysis of Agrobacterium tumefaciens mutants. Plasmid 11:195-205

Hooykaas PJJ, Den Dulk-Ras H, Regensburg-Tuïnk AJG, Van Brussel AAN, Schilperoort RA (1985) Expression of a Rhizobium phaseoli Sym plasmid in R.trifolii and Agrobacterium tumefaciens: incompatibility with a R.trifolii Sym plasmid. Plasmid 14:47-52

Hooykaas-Van Slogteren GMS, Hooykaas PJJ, Schilperoort RA (1984) Expression of Ti plasmid genes in monocotyledonous plants infected with Agrobacterium tumefaciens. Nature 311:763-764

Klapwijk PM, Scheulderman T, Schilperoort RA (1978) Coordinated regulation of octopine degradation and conjugative transfer of Ti plasmid in Agrobacterium tumefaciens: evidence for a common regulatory gene and separate operons. J Bacteriol 136:775-785

Klee HJ, White FF, Iyer VN, Gordon MP, Nester EW (1983) Mutational analysis of the virulence region of an Agrobacterium tumefaciens Ti-plasmid. J Bacteriol 153:878-883

Leemans J, Deblaere R, Willmitzer L, De Greve H, Hernalsteens JP, Van Montagu M, Schell J (1982) Genetic identification of functions of T_L-DNA transcripts in octopine crown galls. EMBO J 1:147-152

Lundquist RC, Close TJ, Kado CI (1984) Genetic complementation of Agrobacterium tumefaciens Ti plasmid mutants in the virulence region. Mol Gen Genet 193: 1-7

Matthysse AG (1983) Role of bacterial cellulose fibrils in Agrobacterium tumefaciens infection. J Bacteriol 154:906-915

Messens E, Lenaerts A, Van Montagu M, Hedges RW (1985) Genetic basis for opine secretion from crown gall tumour cells. Mol Gen Genet 199:344-348

Nester EW, Kosuge T (1981) Plasmids specifying plant hyperplasias. Ann Rev Microbiol 35:531-565

Offringa IA, Melchers LS, Regensburg-Tuink AJG, Costantino P, Schilperoort RA, Hooykaas PJJ (1986) Complementation of Agrobacterium tumefaciens Ti aux mutants by genes from the T_R-region of the Ri plasmid of Agrobacterium rhizogenes. Proc Natl Acad Sci USA 83:

Okker RJH, Spaink H, Hille J, Van Brussel TAN, Lugtenberg B, Schilperoort RA (1984) Plant-inducible virulence promoter of the Agrobacterium tumefaciens Ti plasmid. Nature 312:564-566

Ooms G, Molendijk L, Schilperoort RA (1982) Double infection of tobacco plants by two complementing octopine T-region mutants of Agrobacterium tumefaciens. Plant Mol Biol 1:217-226

Owens LD, Cress DE (1985) Genotypic variability of soybean response to Agrobacterium strains harboring the Ti or Ri plasmid. Plant Physiol 77:87-94

Panagopoulos CG, Psallidas PG (1973) Characteristics of Greek isolates of Agrobacterium tumefaciens (Smith and Townsend) Conn. J Appl Bact 36: 233-240

Peerbolte R (1986) The fate of T-DNA during vegetative and generative propagation - crown gall and hairy root tissues of Nicotiana spp - Thesis, Leiden, The Netherlands

Petit A, Tempé J (1978) Isolation of Agrobacterium Ti-plasmid regulatory mutants. Mol Gen Genet 167:147-155

Petit A, Delhaye S, Tempé J, Morel G (1970) Recherches sur les guanidines des tissues de crown-gall. Mise en évidence d'une relation biochimique spécifique entre les souches d'Agrobacterium tumefaciens et les tumeurs qu'elles induisent. Physiol Végét 8:205-213

Powell GK, Morris RO (1986) Nucleotide sequence and expression of a Pseudomonas savastanoi cytokinin biosynthetic gene: homology with Agrobacterium tumefaciens tmr and tzs loci. Nucl Ac Res 14:2555-2565

Prakash RK, Schilperoort RA (1982) Relationship between nif plasmids of fast-growing Rhizobium species and Ti plasmids of Agrobacterium tumefaciens. J Bacteriol 149:1129-1134

Puvanesarajah V, Schell FM, Stacey G, Douglas CJ, Nester EW (1985) Role for 2-linked-β-d-glucan in the virulence of Agrobacterium tumefaciens. J Bacteriol 164:102-106

Riker AJ (1930) Studies on infectious hairy root of nursery apple trees. J Agr Res 41:507-540

Risuleo G, Battistoni P, Costantino P (1982) Regions of homology between tumorigenic plasmids from Agrobacterium rhizogenes and Agrobacterium tumefaciens. Plasmid 7:45-51

Schröder G, Waffenschmidt S, Weiler EW, Schröder J (1984) The T-region of Ti-plasmids codes for an enzyme synthesizing indole-3-acetic acid. Eur J Biochem 138:387-391

Smidt M, Kosuge T (1978) The role of indole-3-acetic acid accumulation by alpha-methyl tryptophan-resistant mutants of Pseudomonas savastanoi in gall formation on oleanders. Physiol Plant Pathol 13:203-214

Smith EF, Townsend CO (1907) A plant tumor of bacterial origin. Science 25:671-673

Stachel SE, Messens E, Van Montagu M, Zambryski P (1985) Identification of the signal molecules produced by wounded plant cells that activate T-DNA transfer in Agrobacterium tumefaciens. Nature 318:624-629

Stachel SE, Nester EW, Zambryski PC (1986) A plant cell factor induces Agrobacterium tumefaciens vir gene expression. Proc Natl Acad Sci USA 83:379-383

Tempé J, Goldmann A (1982) Occurence and biosynthesis of opines. In: Kahl G, Schell J (eds) Molecular Biology of Plant Tumors, Acad Press New York, pp 427-449

Thomashow LS, Reeves S, Thomashow MF (1984) Crown gall oncogenesis: evidence that a T-DNA gene from the Agrobacterium Ti plasmid pTiA6 encodes an enzyme that catalyzes synthesis of indoleacetic acid. Proc Natl Acad Sci USA 81:5071-5075

Thomashow MF, Hugly S, Buchholz WG, Thomashow LS (1986) Molecular basis for the auxin-independent phenotype of crown gall tumor tissues. Science 231: 616-618

Vanderleyden JP, Verdickt KM, Waelkens FM, Van Gool AP, Mergeay M (1985) Construction and characterization of R-prime plasmid carrying a chromosomal virulence region of Agrobacterium tumefaciens. Abstract First Intern Congress on Plant Mol Biol p 161

Van Montagu M, Schell J (1982) The Ti plasmids of Agrobacterium. Curr Topics 96:237-254

Van Onckelen H, Prinsen E, Inzé D, Rüdelsheim P, Van Lijsebettens M, Follin A, Schell J, Van Montagu M, De Greef J (1986) Agrobacterium T-DNA gene 1 codes for tryptophan 2-monooxygenase activity in tobacco crown gall cells. FEBS Lett 198:357-360

Van Veen RJM, Den Dulk-Ras H, Hooykaas PJJ, Schilperoort RA (1986) Chromosomal nodulation genes: Sym-plasmid containing Agrobacterium strains need chromosomal virulence genes (chvA and chvB) for nodulation. J Bacteriol

White FF, Nester EW (1980) Relationship of plasmids responsible for hairy root and crown gall tumorigenicity. J Bacteriol 144:710-720

Willmitzer L, Simons G, Schell J (1982) The T_L-DNA in octopine crown-gall tumours codes for seven well-defined polyadenylated transcripts. EMBO J 1:139-146

Wilson EE (1935) The olive knot disease: its inception, development, and control.Hilgardia 9:231-264

Yadav NS, Vanderleyden J, Bennett DR, Barnes WM, Chilton MD (1982) Short direct repeats flank the T-DNA on a nopaline Ti plasmid. Proc Natl Acad Sci USA 79:6322-6326

Yamada T, Palm CJ, Brooks B, Kosuge T (1985) Nucleotide sequences of the Pseudomonas savastanoi indoleacetic acid genes show homology with Agrobacterium tumefaciens T-DNA. Proc Natl Acad Sci USA 82:6522-6526

T-REGION TRANSFER FROM <u>AGROBACTERIUM TUMEFACIENS</u> TO PLANT CELLS: FUNCTIONAL CHARACTERIZATION OF BORDER REPEATS

Mark J.J. van Haaren, Jack T. Pronk, Rob A. Schilperoort and Paul J.J. Hooykaas
Leiden University,
Department of Plant Molecular Biology,
Biochemistry Laboratory,
Wassenaarseweg 64,
2333 AL Leiden,
The Netherlands

INTRODUCTION

Oncogenic *Agrobacterium tumefaciens* strains can transfer a specific part of their large tumor inducing (Ti) plasmid to the plant cell and thereby cause the plant disease called Crown Gall (Smith and Townsend 1907). The transferred DNA (T-DNA) is stably integrated into the genome of the plant cell (Chilton et al 1977; Thomashow et al 1980). Expression of genes present on this T-DNA results in the production of the plant hormones cytokinin and auxin, which leads to tumor formation on a wide range of dicotyledonous plants (Thomashow et al 1984; Inzé et al 1984; Barry et al 1984; Akiyoshi et al 1984; Buchmann et al 1985). Furthermore unusual amino acid derivatives (opines) are synthesized in the tumor cells as a consequence of the expression of certain T-DNA genes. Which opines are synthesized in the tumors is dependent of the particular Ti-plasmid that was present in the tumor inducing *A.tumefaciens* strain (Bomhoff et al 1976; Guyon et al 1980). Research has focussed on the socalled octopine and nopaline Ti-plasmids.

The T-region of nopaline Ti-plasmids consists of one continuous DNA fragment, whereas the T-region of octopine Ti-plasmids is organized in two separate segments, which are called the T_L- and T_R-region, respectively. The T_L-region of octopine Ti-plasmids harbors all the genes involved in oncogenicity (*onc*), i.e. for Crown Gall development and maintenance (Ooms et al 1981; 1982). Sequence data of different Ti-plasmids revealed that the T-regions are always flanked by imperfect 24bp direct repeats (Simpson

NATO ASI Series, Vol. H 4
Recognition in Microbe-Plant Symbiotic and
Pathogenic Interactions. Edited by B. Lugtenberg

et al 1982; Yadav et al 1982; Barker et al 1983).
Studies on the T-region by transposon insertion and deletion mutagenesis revealed that none of the T-region genes is required for the transfer of this DNA to the plant cell (Leemans et al 1982; Hille et al 1983), but that the flanking regions containing the 24bp repeats are essential in this process (Ooms et al 1982).

The genes involved in the transfer of the T-region to the plant cells are located in the virulence (Vir) region of the Ti-plasmid, which is located adjacent to the left end of the T-region (Ooms et al 1980; Garfinkel and Nester 1980). The Vir-region embraces at least seven _vir_-operons, some of which contain _vir_-genes involved in T-region transfer. Certain other _vir_-operons carry host range determinants (Klee et al 1983; Hille et al 1984; Hooykaas et al 1984). In the chromosome of _A.tumefaciens_ a few _chv_ (chromosomal virulence) genes have been identified (Douglas et al 1985). Some of these _chv_-genes mediate the attachment of _Agrobacterium_ to the plant cell walls, which is an essential step in the process of plant tumor induction.

A number of different plant vector systems based on the Ti-plasmid of _A.tumefaciens_ have been developed and have been used for the introduction of foreign genes into several dicotyledonous plant species (Hoekema et al 1983; Zambryski et al 1983; Fraley et al 1985). In spite of this fact, the mechanism by which DNA is transferred and integrated into the plant genome is still largely unknown. The possibility that the whole Ti-plasmid is transferred to the plant cell is not likely, since a physical separation of the T-region from the rest of the Ti-plasmid, and even a chromosomal location of the T-region in _Agrobacterium_, does not lead to a reduction in virulence (Hoekema et al 1983; 1984).

This suggests that the T-region is "processed" in the bacterium before transfer to the plant cell. This processing might be mediated by products from the Vir-region (hereinafter called _vir_-products), which might recognize the border repeats flanking the T-region, bind there and then induce excision or replication of the T-region. Possible T-DNA intermediates were recently detected in Agrobacteria that had been cocultured with plant cells (Koukolikova-Nicola et al 1985). It is possible that certain vir-products bind to such T-region intermediates in order to assure their safe transport to the plant nucleus and their integration into the plant genome.

BORDER REPEATS

Studies on the T-DNA structure in transformed tissues have shown that integration of T-DNA into the plant genome often occurs near or precisely at the 24bp repeats, which suggests that the border sequences were used as the integration points (Zambryski et al 1982; Simpson et al 1982; Holsters et al 1983; Kwok et al 1985; Slightom et al 1985). However when DNA, harboring the border repeats, is introduced into plant cells by other methods than via A.tumefaciens transformation, for instance via PEG/Ca2+ transformation or microinjection, the border repeats appear not to be used for integration of this DNA into the plant genome (Krens et al 1985; Crossway et al 1986). This indicates that most likely the border repeats are of special importance within Agrobacterium, e.g. to act as recognition signals for some sort of processing by vir-products.

Plant transformation experiments with strains containing Ti-plasmids from which either one or both of the border repeats had been deleted have been used to understand the importance of these repeats in the T-region transfer process. It was found that deletion of the right border repeats from octopine or nopaline Ti-plasmids resulted in an almost complete loss of oncogenicity (Ooms et al 1982; Shaw et al 1984; Wang et al 1984; Hepburn and White 1985; Rubin 1985), whereas deletion of left border repeats did not have any detectable effect on virulence (Hille et al 1983). This result suggests either a difference between the functions of a left and right border repeat in the transfer process, or a strictly directed transfer of the T-region starting from the right border repeat.

Following the second idea a mechanism can be imagined in which vir-products first attach to the right border repeat and from there start processing the T-region, until at a left border repeat or at a similar socalled "pseudoborder" the excision or replicative process is terminated. When the sequence of 24 bp border repeats of different origin are compared (Table I), no significant difference in sequence between left and right border repeats became apparent (Barker et al 1983; Slightom et al 1985). It is however conceivable that sequences surrounding the 24bp repeats rather than the repeats themselves make the distinction between left and right borders.

TABLE I. SEQUENCE OF THE 24bp BORDER REPEATS IN DIFFERENT PLASMIDS

PLASMID	BORDER REPEAT	SEQUENCE
oct. Ti	LB (T_L-region)	GGCAGGATATATTCAATTGTAAAC
oct. Ti	RB (T_L-region)	GGCAGGATATATACCGTTGTAATT
oct. Ti	LB (T_R-region)	GGCAGGATATATCGAGGTGTAAAA
oct. Ti	RB (T_R-region)	GGCAGGATATATGCGGTTGTAATT
nop. Ti	LB (T-region)	GGCAGGATATATTGTGGTGTAAAC
nop. Ti	RB (T-region)	GACAGGATATATTGGCGGGTAAAC
Ri	RB (T_L-region)	GACAGGATATATGTTCCTGTCATG
	Derived consensus	GGCAGGATATATxxxxxTGTAAxx

Sequence data from: Yadav et al 1982; Barker et al 1983; Slightom et al 1985.

FUNCTIONAL ANALYSIS OF BORDER FRAGMENTS

We performed experiments to obtain information on the capability of different Ti-plasmid fragments to act as border fragments. In these experiments we used a deletion mutant of the octopine Ti-plasmid pTiAch5 called pAL4417, which misses the entire T_R-region as well as the right border repeat of the T_L-region. Consequently strains containing pAL4417 are almost completely non-oncogenic (Ooms et al 1982). Using a shuttle vector system, DNA fragments containing or lacking border repeats were introduced at the right side of the T-region in pAL4417 and the resulting strains were tested for oncogenicity. If a functional border fragment is introduced into pAL4417 in this way, the resulting Agrobacterium strain is expected to become normally oncogenic. In agreement with this rationale we found that the introduction of a large restriction fragment, called RBE, carrying the deleted right border region restored oncogenicity completely (Table II, Fig.1). As a next step we tested whether the orientation with which the border region was inserted was important for activity, and whether left border fragments could replace the deleted right border when inserted in its place at the right end of the T-region. Insertion of the above mentioned restriction fragment RBE in pAL4417 in an inverted orientation (relative to the left border repeat) resulted in a strain that had become more virulent than the strain lacking this DNA-fragment, but the tumors formed were smaller than those induced by a wild-type strain (Table II, Fig.1).

TABLE II. Virulence of derivatives of pAL4417 with fragments inserted instead of the right border.
Abbreviations used for the description of border fragments are explained in the text.

Inserted border fragment	symbol	Virulence[2]
-[1]		-
LB		+
LB inverted		+/-
RB		+
RB inverted		+/-
LBE		+ + +
LBE inverted		+ +
RBE		+ + +
RBE inverted		+
E		-
E inverted		-

1) as a control the shuttle vector used was inserted alone into pAL4417.
2) symbols are as follows: + + + wild type virulence; + to ++ partial virulence; +/- very weak virulence; - avirulent

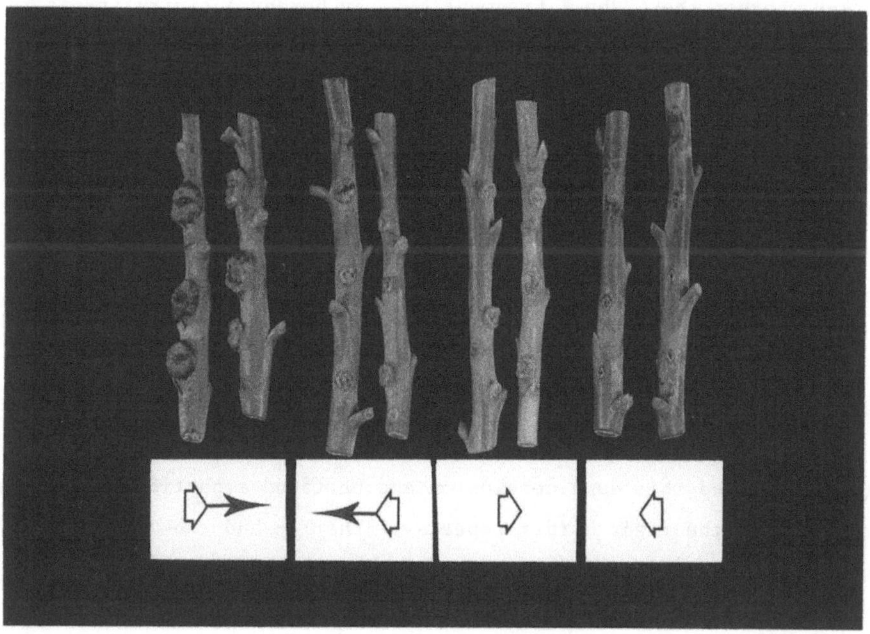

FIG. 1. Inoculation test of strains containing the border fragments as indicated, on N.glauca. Scored 3 weeks after inoculation.

The introduction of a restriction fragment with a left border repeat (LB) in pAL4417 resulted in a weakly virulent strain if the repeat was present in its native orientation, but the strain remained avirulent when the repeat was introduced in an inverted orientation. From these results we can conclude that the orientation of the border repeat containing fragment strongly influences its biological activity, and that right border fragments are much more active than left border fragments. Since right and left border repeats are almost identical, this latter finding suggested that sequences surrounding the border repeats might be important.

In order to investigate this in more detail we inserted a smaller 315bp fragment (RB) with the right border repeat instead of the 937bp fragment RBE in pAL4417. This did not result in a restoration of oncogenicity to wild-type level as was observed after introduction of the 937bp fragment. We concluded that the 622bp piece (E) of fragment (RBE), which was absent in fragment RB, contained an "enhancer-like" activity, and hypothesized that a sequence with this activity was absent from left border fragments. In order to prove the latter we introduced this 622bp fragment E together with a left border fragment LB into pAL4417 and examined the oncogenicity of the resulting strain. We found that oncogenicity had been restored to wild-type level. As a control we checked whether the 622bp E fragment had any border activity itself in the absence of a border repeat, but this was not the case. Therefore, these results show the presence of an enhancer sequence next to the right border repeat, which effectuates an enormously increased "border activity" of 24bp border repeats.

CONCLUDING REMARKS

In the experiments discussed here no functional difference was found between the right and left border 24bp repeats themselves. Surrounding sequences determine whether a repeat functions as a left or right border. Ream (1985) reached this same conclusion and he found a particular consensus sequence next to the right border repeats of the T_L- and the T_R-region of the octopine Ti plasmid, which might represent the enhancer sequence (Fig.2). However, only a sequence that is partially homologous with this consensus turned out to be located next to the right border repeat of the nopaline Ti-plasmid. Moreover, we could not find a sequence resembling the proposed enhancer sequence in the vicinity of the right border repeat of the T-region of the A.rhizogenes root-inducing (Ri) plasmid.

FIG. 2. POSSIBLE ENHANCER SEQUENCE

<right border T_L> TGAGCTCGTGTGAATAAGTCGCTGTGTATGTTTGTTTGATTGTTT

<right border T_R> CATTTTTATTGTCTAAATTTCTGTATTTGTTTGTTTGTTCGGTT

--- - ---- - ----------

CTGTTGGAGTGCAGCCCATTT

GTAAATTTTTTTGGAAGACCC

Sequences at the right hand site of the RB of the octopine T_L- and T_R-region. Underlined identical nucleotides according to Ream, 1985.

Whether a strictly defined nucleotide sequence is required for the enhancing effect or just some secondary structure of the DNA molecule is not yet clear. Possible roles of the enhancing sequence may lie in helping the vir-genes to recognize specifically the right border repeat. After recognition of the right border repeat an orientated processing of the T-region may start on this right border until a left border repeat is encountered. In systems different from the A.tumefaciens transfer system enhancer sequences are also found. No clear sequence homology can be found between the regions near the T-region right border repeats and the recombinational enhancers found in the G-inversion system of bacteriophage Mu (Kahmann et al 1985) or the Hin-mediated site-specific recombination in Salmonella (Johnson et al 1985).

In our experiments both a right and a left border repeat inserted in an inverted orientation in pAL4417 gave rise to no or very attenuated tumor responses on the tested plants but both were strongly enhanced when an enhancer-like sequence is present in the inserted fragment (Table II). This effect of the enhancer-like sequence appeared to be even stronger for the left border fragment (Table II, Fig.1.). It is possible that in these cases not the native border repeats but related sequences which are called pseudo-borders are used to direct transfer of the T-region to the plant cell. If these pseudo-borders are situated in direct repeat with the left border repeat available in pAL4417 they will be better suited to mediate the transfer of onc-genes to the plant than the inverted 25bp border repeats.

In recent publications in which detailed studies have been made on the T-DNA content of transformed plant tissues, it has been demonstrated that apart from the regular transfer and integration of the whole T-region into the plant genome, integration of highly truncated T-region segments may occur in transformed tissues (Peerbolte et al 1986; Lijsebettens et al 1986). These can be explained by assuming the participation of pseudo-border repeats in the transfer process. A list of possible pseudo-borders present in the octopine T-DNA is presented in Table III. Pseudo-borders that possibly have been used in the transfer of the truncated T-DNA's in the transformed tissues studied by Peerbolte et al (1986) are indicated.

TABLE III. PSEUDO BORDERS LOCATED IN OCTOPINE T-REGION

Position		Sequence	
consensus:		GGCAGGATATATxxxxxTGTAA	
100	:	GGCcGGATAgAT	catcc
2406	:	GtaAGGATATta	TGTAt
2992	:	caCAGGAaATAg	TtgAA
3310	:	GaCAaacTATAT	cTatA
3334	:	GGCgctATATAT	ccggg
3680	:	GGCAaatTATAT	ataAA
3783	:	cttAaGATATAT	aaatt
7345	:	aGCgGGATATcT	Ttatg
* 8207	:	GaCAtaATATga	TaTAA
12300	:	aGCAcaATATAT	Tcatt
12400	:	cGggtGATATAT	aGaAt
14345	:	GaacGcATATtT	TGcAA
18084	:	GaCAGtcTAgAT	TcTtc
20190	:	GGgAaGATATAa	aGcAA
20790	:	GcCAGGcTcTgT	gGcAA
*21425	:	aaCAGGATcgAT	aGTAA
23246	:	GcCcGGtaAaAT	TGgAA
23591	:	GtCAtGATtaAT	aGgAg

The positions are according to Barker et al (1983), whose sequence data were used to detect pseudoborders by computer analysis

* Possible pseudo-borders used in studies of Peerbolte et al 1986.

Studies on the T-DNA structure in tissues transformed by T-regions flanked by border repeats in inverted orientation or by mutated border repeats in the presence or absence of enhancer sequences may provide additional insight in the mechanism used by A.tumefaciens to transfer its T-region to the plant cell and the requirement for efficient border mediated DNA transfer.

REFERENCES

Akiyoshi DE, Klee H, Amasino RM, Nester EW, Gordon MP (1984) T-DNA of Agrobacterium tumefaciens encodes an enzyme of cytokinin biosynthesis. Proc Natl Acad Sci USA 81:5994-5998

Barker RF, Idler KD, Thompson DV, Kemp JD (1983) Nucleotide sequence of the T-DNA region from Agrobacterium tumefaciens octopine Ti plasmid pTi15955. Plant Mol Biol 2:335-350

Barry GF, Rogers SG, Fraley RT, Brand L (1984) Identification of a cloned cytokinin biosynthetic gene. Proc Natl Acad Sci USA 81:4776-4780

Bommhoff G, Klapwijk PM, Kester HCM, Schilperoort RA, Hernalsteens JP, Schell J (1976) Octopine and nopaline synthesis and breakdown genetically controlled by a plasmid of Agrobacterium tumefaciens. Mol Gen Genet 145:177-181

Buchmann I, Marner FJ, Schröder G, Waffenschmidt S, Schröder J (1985) Tumour genes in plants: T-DNA encoded cytokinin biosynthesis. EMBO J 4:853-859

Chilton MD, Drummond MH, Merlo DJ, Sciaky D, Montoya AL, Gordon MP, Nester EW (1977) Stable incorporation of plasmid DNA into higher plant cells: the molecular basis of crown gall tumorigenesis. Cell 11:263-271

Corssway A, Oakes JV, Irvine JM, Ward B, Knauf VC, Shewmaker CK (1986) Integration of foreign DNA following microinjection of tobacco mesophyll protoplasts. Mol Gen Genet 202:179-185

Douglas CJ, Staneloni RJ, Rubin RA, Nester EW (1985) Identification and genetic analysis of an Agrobacterium tumefaciens chromosomal virulence region J Bacteriol 161:850-860

Fraley RT, Rogers SG, Horsch RB, Eichholtz DA, Flick JS, Fink CL, Hoffmann NL, Sanders PR (1985) The SEV system: a new disarmed Ti plasmid vector system for plant transformation. Bio/Technology 3:629-635

Garfinkel DJ, Nester EW (1980) Agrobacterium tumefaciens mutants affected in crown gall tumorigenesis and octopine catabolism. J Bacteriol 144: 732-743.

Guyon P, Chilton MD, Petit A, Tempé J (1980) Agropine in "null-type" crown gall tumors: evidence for generality of the opine concept. Proc Natl Acad Sci USA 77:2693-1697

Hepburn AG, White J (1985) The effect of right terminal repeat deletion on the oncogenicity of the T-region of pTiT37. Plant Mol Biol 5:3-11

Hille J, Wullems G, Schilperoort R (1983) Non-oncogenic T-region mutants of Agrobacterium tumefaciens do transfer T-DNA into plant cells. Plant Mol Biol 2:155-163

Hille J, Van Kan J, Schilperoort R (1984) Trans-acting virulence functions of the octopine Ti plasmid from Agrobacterium tumefaciens. J Bacteriol 158:754-756

Hoekema A, Hirsch PR, Hooykaas PJJ, Schilperoort RA (1983) A binary plant vector strategy based on separation of vir and T-region of the Agrobacterium tumefaciens Ti-plasmid. Nature 303:179-180

Hoekema A, Roelvink PW, Hooykaas PJJ, Schilperoort RA (1984) Delivery of T-DNA from the Agrobacterium tumefaciens chromosome into plant cells. EMBO J 3:2485-2490

Holsters M, Villarroel R, Gielen J, Seurinck J, De Greve H, Van Montagu M, Schell J (1983) An analysis of the boundaries of the octopine T_L-DNA in tumors induced by Agrobacterium tumefaciens. Mol Gen Genet 190:35-41

Hooykaas PJJ, Hofker M, Den Dulk-Ras H, Schilperoort RA (1984) A comparison of virulence determinants in an octopine Ti plasmid, a nopaline Ti plasmid, and an Ri plasmid by complementation analysis of Agrobacterium tumefaciens mutants. Plasmid 11:195-205

Inzé D, Follin A, Van Lijsebettens M, Simoens C, Genetello C, Van Montagu M, Schell J (1984) Genetic analysis of the individual T-DNA genes of Agrobacterium tumefaciens; further evidence that two genes are involved in indole-3-acetic acid synthesis. Mol Gen Genet 194:265-274

Johnson RC, Simon MI (1985) Hin-mediated site-specific recombination requires two 26 bp recombination sites and a 60 bp recombinational enhancer. Cell 41:781-791

Kahmann R, Rudt F, Koch C, Mertens G (1985) G inversion in bacteriophage Mu DNA is stimulated by a site within the invertase gene and a host factor. Cell 41, 771-780

Klee HJ, White FF, Iyer VN, Gordon MP, Nester EW (1983) Mutational analysis of the virulence region of on Agrobacterium tumefaciens Ti-plasmid. J Bacteriol 153:878-883

Koukolíková-Nicola Z, Shillito RD, Hohn B, Van Montagu M, Zambryski P (1985) Involvement of circular intermediates in the transfer of T-DNA from Agrobacterium tumefaciens to plant cells. Nature 313 (1985) 191-196

Krens FA, Mans RWM, Van Slogteren TMS, Hoge JHC, Wullems GJ, Schilperoort RA (1985) Structure and expression of DNA transferred to tobacco via transformation of protoplasts with Ti-plasmid DNA: co-transfer of T-DNA and non-T-DNA sequences. Plant Mol Biol 5:223-234

Kwok WW, Nester EW, Gordon MP (1985) Unusual plasmid DNA organization in an octopine crown gall tumor. Nucl Ac Res 13:459-471

Leemans J, Deblare R, Willmitzer L, De Greve H, Hernalsteens JP, Van Montagu M, Schell J (1982) Genetic identification of functions of T_L-DNA transcripts in octopine crown galls. EMBO J 1:147-152

Ooms G, Klapwijk PM, Poulis JA, Schilperoort RA (1980) Characterization of Tn904 insertions in octopine Ti plasmid mutants of Agrobacterium tumefaciens. J Bacteriol 144:82-91

Ooms G, Hooykaas PJJ, Moolenaar G, Schilperoort RA (1981) Crown gall plant tumors of abnormal morphology, induced by Agrobacterium tumefaciens carrying mutated octopine Ti plasmids; analysis of T-DNA functions. Gene 14:33-50

Ooms G, Hooykaas PJJ, Van Veen RJM, Van Beelen P, Regensburg-Tuïnk AJG, Schilperoort RA (1982) Octopine Ti-plasmid deletion mutants of Agrobacterium tumefaciens with emphasis on the right side of the T-region. Plasmid 7:15-29

Peerbolte R, Leenhouts K, Hooykaas-Van Slogteren GMS, Hoge JHC, Wullems GJ, Schilperoort RA (1986) Clones from a shooty tobacco crown gall tumor I: deletions rearrangements and amplifications resulting in irregular T-DNA structures and organizations. Plant Mol Biol (in press)

Ream LW (1985) Abstract Int Symp. Plant Mol Biol, Savannah

Rubin RA (1986) Genetic studies on the role of octopine T-DNA border regions in crown gall tumor formation. Mol Gen Genet 202:312-320

Shaw CH, Watson MD, Carter GH, Shaw CH (1984) The right hand copy of the nopaline Ti-plasmid 25 bp repeat is required for tumour formation. Nucl Ac Res 12:6031-6041

Simspon RB, O'Hara PJ, Kwok W, Montoya AM, Lichtenstein C, Gordon MP, Nester EW (1982) DNA from the A6S/2 crown gall tumor contains scrambled Ti-plasmid sequences near its junctions with plant DNA. Cell 29:1005-1014

Slightom JL, Jouanin L, Leach F, Drong RF, Tepfer D (1985) Isolation and identification of T_L-DNA/plant junctions in Convolvulus arvensis transformed by Agrobacterium rhizogenes strain A4. EMBO J 4:3069-3077

Smith EF, Townsend CO (1907) A plant tumor of bacterial origin. Science 25: 671-673

Thomashow LS, Reeves S, Thomashow MF (1984) Crown gall oncogenesis: evidence that a T-DNA gene from the Agrobacterium Ti plasmid pTiA6 encodes an enzyme that catalyzes synthesis of indoleacetic acid. Proc Natl Acad Sci USA 81:5071-5075

Thosmashow MF, Nutter R, Postle K, Chilton MD, Blattner FR, Powell A, Gordon MP, Nester EW (1980) Recombination between higher plant DNA and the Ti plasmid of Agrobacterium tumefaciens. Proc Natl Acad Sci USA 77:6448-6452

Van Lijsebettens M, Inzé D, Schell J, Van Montagu M (1986) Transformed cell clones as a tool to study T-DNA integration mediated by Agrobacterium tumefaciens. J Mol Biol 188:129-145

Wang K, Herrera-Estrella L, Van Montagu M, Zambryski P (1984) Right 25 bp terminus sequence of the nopaline T-DNA is essential for and determines direction of DNA transfer from Agrobacterium to the plant genome. Cell 38: 455-462.

Yadav NS, Vanderleyden J, Bennett DR, Barnes WM, Chilton MD (1982) Short direct repeats flank the T-DNA on a nopaline Ti plasmid. Proc Natl Acad Sci USA 79:6322-6326

Zambryski P, Depicker A, Kruger K, Goodman HM (1982) Tumor induction by Agrobacterium tumefaciens: analysis of the boundaries of T-DNA. J Mol Appl Genet 1:361-370

Zambryski P, Joos H, Genetello C, Leemans J, Van Montagu M, Schell J (1983) Ti plasmid vector for the introduction of DNA into plant cells without alteration of their normal regeneration capacity. EMBO J 2:2143-2150

CLONING OF DNA SEQUENCES FROM AZOSPIRILLUM BRASILENSE, HOMOLOGOUS TO RHIZOBIUM NOD GENES AND AGROBACTERIUM VIR GENES

Jos Vanderleyden[1], Claire Vieille[2], Kris Michiels[1], Giorgio Matassi[2], August Van Gool[1] and Claudine Elmerich[2]

[1]F.A. Janssens Laboratorium Genetica, Katholieke Universiteit Leuven, Kardinaal Mercierlaan 92, B-3030 Heverlee and [2] Unité de Physiologie Cellulaire, Département de Biochimie et Génétique Moléculaire, Institut Pasteur, 75724 Paris Cedex 15, France

INTRODUCTION

Bacteria of the genus Azospirillum (Tarrand et al 1978) are free-living diazotrophs that can readily be isolated from the rhizosphere and roots of grasses. Little is known about the genetics of these bacteria or the molecular biology of their association with plants (for reviews, see Elmerich, 1984; Okon 1985).

No differentiated structures are formed following colonization, but pictures of root hair deformation (Patriquin et al 1983) and changes in root morphology (Kapulnik et al 1985) were reported. We therefore initiated a research program, in order to identify bacterial genes involved in the interaction process. In the initial approach we took advantage of the fact that several genes involved in bacteria plant interaction have been identified in and cloned from Rhizobium (Kondorosi et al 1984) and Agrobacterium (Douglas et al 1985). Hybridization experiments were performed with total DNA of several Azospirillum strains using probes containing the 'common nod genes' and the 'host specific nod genes' of R. meliloti. Homology was found to all probes, although at various degrees. These results have been recently described (Fogher et al 1985). The data presented here show that the chromosomal virulence (chv) region of A. tumefaciens is homologous to DNA sequences in the genomes of two different Azospirillum species. Clones that carry homology either to a probe containing the 'host specific nod genes' or to a probe containing the chv B

NATO ASI Series, Vol. H 4
Recognition in Microbe-Plant Symbiotic and Pathogenic Interactions. Edited by B. Lugtenberg

locus, were isolated from a gene bank of A. brasilense Sp7.

RESULTS AND DISCUSSION

The region of the A. tumefaciens chromosome carrying the chv loci is shown in the upper part of Figure 1. Genome DNA from two Azospirillum species was examined for homology to A. tumefaciens chv by Southern blot hybridization as shown in Figure 2.

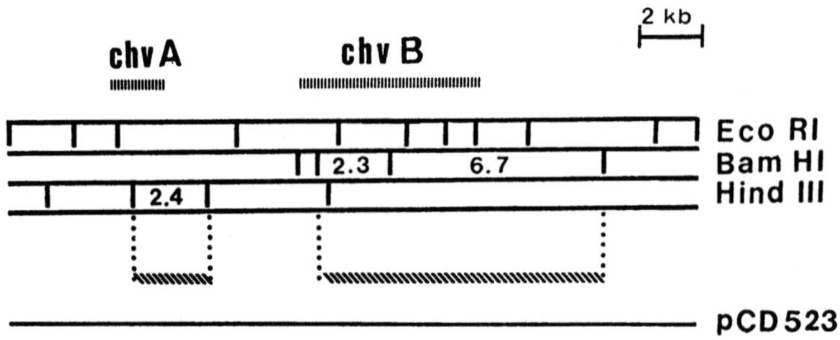

Fig. 1. Physical map of the chromosomal virulence (chv) region of Agrobacterium tumefaciens, according to Douglas et al 1985.

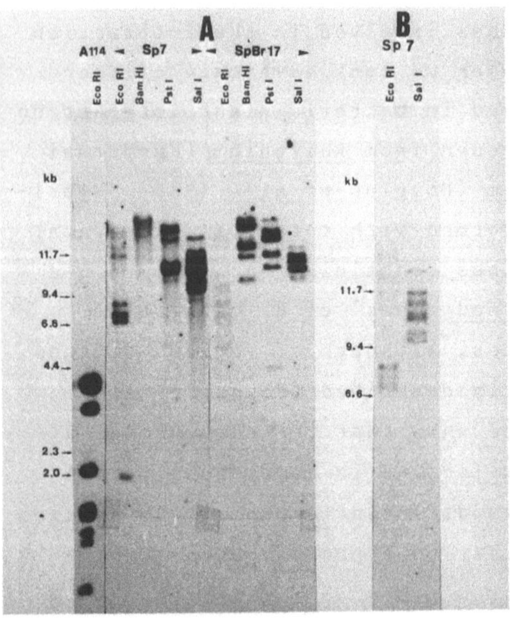

Fig. 2: Southern blot hybridizations between A. tumefaciens chv and digested total DNAs of Azospirillum. A114, A. tumefaciens used as a control; other numbers designate Azospirillum strains.

Table 1. Size of DNA fragments homologous to nod, hsn and chv probes in Azospirillum Sp7[a]

nod		hsn		chvB	
EcoRI	SalI	EcoRI	SalI	EcoRI	SalI
7.2	3.3	12	5.4	7.9	>12
		10	4.0	7.5	11.5
		1.8	2.4	7.2	11(2x)
				6.9	10(2x)
				1.8	9.8
					9.6

[a]Sizes are indicated in kb

In part A of Figure 2, the chv probe was clone pCD523, carrying the chvA and chvB loci in the cosmid vector pLAFR1. In part B of Figure 2, genomic DNA of A. brasilense Sp7 was hybridized with a 2.3 kb BamHI fragment and a 6.7 kb BamHI fragment (pooled), containing most of chvB and some additional DNA. As can be seen in Figure 2, DNA fragments homologous to chvB are present in the genomes of the two Azospirillum species examined. When a 2.4 kb HindIII fragment containing most of chvA was used as a probe, very faint hybridization bands were seen (not shown). In Table 1, the results of the hybridization experiments of total DNA of A. brasilense Sp7 with 'common nod', 'host specific nod' (Fogher et al 1985) and chvB are summarized.

Subsequently, gene banks constituted of Azospirillum Sp7 EcoRI or SalI fragments were constructed in pUC18. Several clones were selected by colony hybridization, and their plasmid DNA was subsequently analyzed by hybridization with a hsn probe or a chv probe. Plasmids pAB501 and pAB502, revealed with the hsn probe, contained a 10 kb EcoRI fragment and pAB503 a 4 kb SalI fragment, which correspond to sizes detected with total DNA of Sp7. Plasmid pAB601, revealed with the chv probe, contained a 6.9 kb fragment, also found in total DNA of Sp7. Although these results do not necessarily imply that the corresponding 'hsn' and 'chv' sequences are functional in Azospirillum, the presence of similar genes can be postulated.

REFERENCES

Douglas CJ, Staneloni RJ, Rubin RA, Nester EW (1985) Identification and genetic analysis of an Agrobacterium tumefaciens chromosomal virulence region. J Bacteriol 161: 850-860

Elmerich Cl (1984) Molecular biology and ecology of diazotrophs associated with non-leguminous plants. Bio/Technology 2: 967-978

Fogher C, Dusha I, Barbot P, Elmerich Cl (1985) Heterologous hybridization of Azospirillum DNA to Rhizobium nod and fix genes. FEMS Microbiol Lett 30: 245-249

Kapulnik Y, Okon Y, Henis Y (1985) Changes in root morphology of wheat caused by Azospirillum inoculation. Can J Microbiol 31: 881-887.

Kondorosi E, Banfalvi Z, Kondorosi A (1984) Physical and genetic analysis of a symbiotic region of Rhizobium meliloti: Identification of nodulation genes. Mol Gen Genet 193: 445-452.

Okon Y (1985) Azospirillum as a potential inoculant for agriculture. Trends in Biotechnology 3: 223-228.

Patriquin DC, Döbereiner J, Jain DK (1983) Sites and processes of association between diazotrophs and grasses. Can J Microbiol 29: 900-915.

Tarrand JJ, Krieg NR, Döbereiner J (1978) A taxonomic study of the Spirillum lipoferum group, with description of a new genus, Azospirillum gen. nov. and two species, Azospirillum lipoferum (Beijerinck) comb. nov. and Azospirillum brasilense sp. nov. Can J Microbiol 24: 967-980.

ATTACHMENT OF AGROBACTERIUM TUMEFACIENS TO PLANT HOST CELLS

Ann G. Matthysse
Department of Biology
University of North Carolina
Chapel Hill, NC 27514
USA

INTRODUCTION

Tumor formation by Agrobacterium tumefaciens involves the transfer of bacterial DNA from the bacteria to the plant host cell. It is thus not surprising that one of the first steps in tumor formation is the attachment of the bacteria to the plant cell surface. This process is more complicated than it was originally thought to be. At least two steps are involved and preliminary evidence suggests that there are additional factors whose role is not yet clearly understood. The first step in bacterial attachment is the

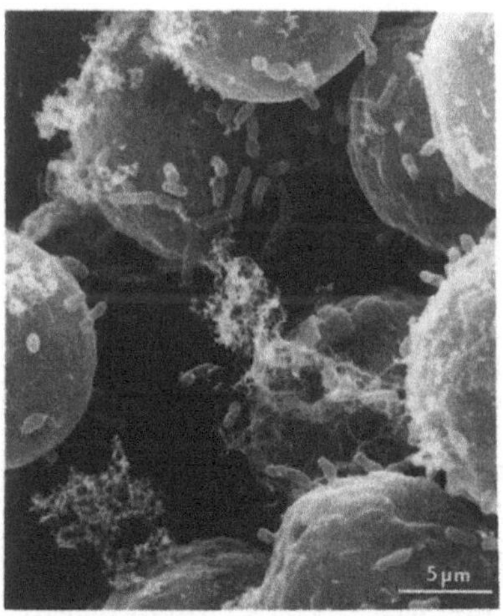

Fig. 1. Scanning electron micrograph showing A. tumefaciens incubated with carrot cells for 20 hours. Note the presence of fibrils and of bacterial aggregates.

NATO ASI Series, Vol. H 4
Recognition in Microbe-Plant Symbiotic and Pathogenic Interactions. Edited by B. Lugtenberg

loose reversible binding of the bacteria to the plant cell surface. At this stage each bacterium is bound directly to a receptor(s) on the host cell. In the second step the bacteria elaborate cellulose fibrils which anchor the bacteria tightly to the plant cell surface. At this stage many bacteria continue to be bound directly to receptors on the host cell, but a much larger number of additional bacteria are trapped and bound indirectly to the plant cell surface by bacterial cellulose fibrils. The result is the formation of large aggregates of bacteria projecting from the host cell surface. A scanning electron micrograph showing Agrobacterium bound both directly and indirectly to carrot cells is shown in Figure 1 and a model of bacterial attachment which is based on these observations is shown in Figure 2(Matthysse et al 1981).

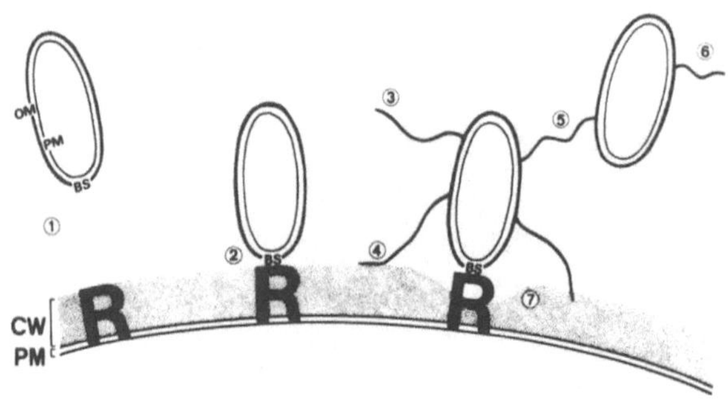

Fig. 2. A model for the attachment of A. tumefaciens to carrot cells. A bacterium attaches to a specific receptor site(R) on the surface of the plant cell(step 1) and is induced to synthesize cellulose fibrils(step 2). The fibrils anchor the bacterium to the surface of the plant cell wall(CW)(step 3) and entrap additional bacteria(step 4). These entrapped bacteria themselves begin to elaborate cellulose fibrils(step 6). This attachment and entrapment of bacteria results in the formation of large aggregates on the surface of the plant cell(Matthysse et al 1981).

THE FIRST STEP IN BACTERIAL ATTACHMENT

In order to test the model for bacterial attachment given in Figure 2 transposon mutants blocked in each of the two proposed steps in bacterial attachment were obtained. Mutants which were unable to bind to plant cells were obtained by screening bacteria into which a transposon had been introduced for their ability to bind to carrot suspension culture cells using a microscopic assay. Five bacterial mutants which were unable to bind to carrot cells as determined by both microscopic and kinetic assays were obtained using this laborious proceedure. All of these mutants were avirulent. Although the isolation of nonbinding mutants was difficult, the nature of the mutation meant that the isolation of revertants of the mutants was not difficult. Revertant bacteria were selected by their ability to bind to carrot cells. Carrot cells were incubated with the bacteria and then collected on a filter which allows the passage of free bacteria; the plant cells which were retained were washed several times and they and any bacteria attached to them and thus retained with them were plated on bacteriological growth media. Revertants of all of the nonbinding mutants were recovered using this procedure which selects for the abillity to bind to carrot cells. The revertants had recovered virulence simultaneously with the ability to bind to carrot cells. This result suggests that the ability to bind to carrot suspension culture cells is required for virulence(Matthysse 1986).

In order to determine which component(s) of the bacterial surface was altered in the nonbinding mutants, the composition of these bacteria was analysed and compared with the parent strain. Unlike the attachment mutants isolated by Douglas et al(1982) our mutants showed no difference in bacterial motility or extracellular polysaccharide production. Our mutants also showed no difference in pili, hydrophobicity, or lipopolysaccharide. All of the mutants were capable of the production of cellulose fibrils. Bacterial spheroplasts do not bind to plant cells. When the polypeptides removed from the bacteria during the production of bacterial spheroplasts were examined using polyacrylamide gel electrophoresis, one or more of three polypeptides of molecular weight between 30 and 40,000 daltons were found to be missing in extracts of nonbinding mutants(Figure 3). These polypeptides reappeared in extracts from the revertants of the mutants(Matthysse 1986).

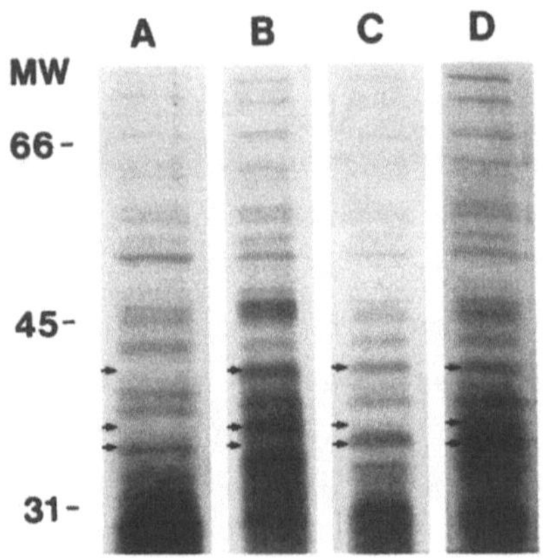

Fig. 3. SDS polyacrylamide gel electrophoresis of surface proteins extracted from three different nonbinding mutants(A, B, and C) and the parent strain(D). Note the absence of one or more of the bands marked by the arrows in the mutants.

Live bacteria are required for binding to plant tissue culture cells, suggesting that the bacteria play an active role in the attachment process. However, the plant cell does not appear to play an active role in this initial phase of bacterial attachment. Bacteria bind to dead carrot cells as well as to live cells. There is little difference in the kinetics of bacterial attachment to living carrot cells or to cells which have been killed with glutaraldehyde or heat treatments(Matthysse et al 1981; Matthysse et al 1982).

THE SECOND STEP IN BACTERIAL ATTACHMENT

Once the bacteria have bound to the plant cell surface they begin to elaborate cellulose fibrils. The synthesis of these fibrils can be induced by the plant cells or by plant cell extracts(Matthysse et al 1981). The chemical nature of the inducer is unknown except that it is not the same as the inducer of the vir region genes of the Ti plasmid.

The role of the cellulose fibrils in bacterial attachment has been examined using transposon mutants of the bacteria. Transposon mutants which fail to elaborate cellulose fibrils were isolated by screening bacteria for lack of fluorescent straining on agar medium containing soytone to induce cellulose synthesis and Calcafluor which stains cellulose resulting in a blue-white fluorescence under ultraviolet light. Seven mutants which did not

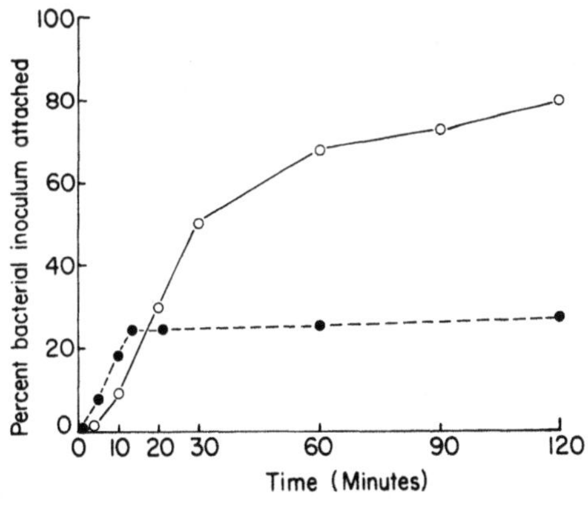

Fig. 4. Time course of attachment of A. tumefaciens to carrot cells. O:wild type bacteria. ●: cellulose-minus mutant bacteria(Matthysse, 1983).

produce cellulose under any conditions tested were isolated. All of these mutants were virulent, although they generally induce smaller tumors than the parent strain. However, when compared with the parent strain they showed alterations in attachment apparently due to the lack of fibrils. The parent strain attaches to plant cells with rather peculiar biphasic kinetics in which the number of bacteria bound never stops increasing with increasing time of incubation(Figure 4). The cellulose-minus mutants do not show this second kinetic phase in their attachment and the number of bacteria bound reaches a saturation plateau after about 20 min. This is the predicted result if the second nonsaturatable phase of attachment is due to the indirect attachment of bacteria via entrapment in cellulose fibrils and the formation of bacterial aggregates(Matthysse 1983).

When the parent bacterial strain is incubated with plant cells for long times(24 hours or more) bacterial attachment and fibril synthesis result in the formation of large aggregates visible to the naked eye composed of bacteria and virtually all of the carrot cells in the culture. Neither cellulose-minus mutant bacteria nor nonbinding mutants are able to form these large aggregates suggesting that they result from the binding of the bacteria to the carrot cells followed by the binding of bacteria to each other via cellulose fibrils(Matthysse 1983).

Although the ability to form cellulose fibrils is not required for bacterial virulence in the laboratory, the fibrils are probably an important virulence factor in nature. All wild type strains of Agrobacterium examined are capable of cellulose fibril formation. In addition when wild type or cellulose-minus bacteria are inoculated into toothpick wounds on Bryophyllum leaves and the wounds washed with water 2 hours after inoculation, tumor formation by the wild type bacteria is unaffected, but no tumors are formed by the mutant bacteria which are washed off the leaf. Thus the ability to make cellulose fibrils may play an important role in the adherance of bacteria to wound sites in naturally occuring situations such as rain(Matthysse 1983).

THE EFFECT OF AUXIN ON BACTERIAL ATTACHMENT

The addition of auxin to inoculated wound sites or the presence or absence of auxin in tissue culture media has no effect on the attachment of wild type Agrobacteria to plant host cells(Gurlitz et al 1986). However, when the wide host range R plasmid pSA is introduced into Agrobacterium the bacteria become avirulent, although they have undergone no irreversible genetic change and recover virulence when cured of the plasmid(Farrand et al 1981). Bacteria containing pSA can be restored to virulence by the addition of auxin to the inoculated wound sites(Chonin et al 1984). Agrobacteria containing pSA fail to bind to carrot suspension culture cells in the absence of auxin although they bind as well as the parent bacteria in the presence of auxin. The effect of auxin is on the bacteria and not on the plant cells since the same results are obtained using either living or heat-killed carrot cells. This result is rather surprising in that if a plant hormone is required for the interaction of the bacteria with the plant cells one would think that the effect of the hormone would be on the plant cells rather than on the bacteria. This observation suggests that there are further complications involved in the recognition and attachment of Agrobacterium to plant host cells which we do not yet understand.

CONCLUSION

When compared to other plant pathogenic bacteria,

Agrobacterium appears to be unique in its initial interactions with plant host cells. For Agrobacteria binding of the bacteria to the plant cell is required for the development of the disease. For most other bacterial plant pathogens binding of the bacteria is not required for the development of disease and indeed may be associated with the initiation of the hypersensitive response and disease resistance.

The initial surface interactions of Agrobacterium with plant host cells appear to resemble more closely the interactions of symbiotic Rhizobia with host cells than those of other pathogens. This is not surprising since the Agrobacteria and Rhizobia are closely related. Both Agrobacteria and Rhizobia are capable of producing bacterial cellulose(Matthysse et al 1981; Napoli et al 1975), but the role of this cellulose in the production of nodules is unknown. For both of these groups of bacteria the initial recognition and interaction with plant host cells is a far from simple event and involves multiple steps and probably cross communication between the bacteria and the plant cells.

REFERENCES

Chernin LS, Lobanok EV, Fomicheva VV, Kartel NA(1984)Crown gall-supressive IncW R plasmids cause a decrease in auxin production by Agrobacterium tumefaciens. Mol Gen Genet 195:195-199

Douglas CJ, Halperin W, Nester EW(1982) Agrobacterium tumefaciens mutants affected in attachment to plant cells. J Bacteriol 152:1265-1271

Farrand SK, Kado CI, Ireland CR(1981)Supression of tumorigenicity by the IncW R plasmid pSA. Mol Gen Genet 181:44-51

Gurlitz RHG, Lamb PW, Matthysse AG(1986)Involvement of carrot cell surface proteins in attachment of Agrobacterium tumefaciens. submitted for publication

Matthysse AG(1983)Role of cellulose fibrils in Agrobacterium tumefaciens infection. J Bacteriol 154:906-915

Matthysse AG(1986)Mutants of Agrobacterium tumefaciens which fail to bind to plant host cells. submitted for publication

Matthysse AG, Holmes KV, Gurlitz RHG(1981)Elaboration of cellulose fibrils by Agrobacterium tumefaciens during attachment to carrot cells. J Bacteriol 145:583-595

Matthysse AG, Holmes KV, Gurlitz RHG(1982)Binding of Agrobacterium tumefaciens to carrot protoplasts. Physiol Plant Pathol 20:27-33

ROLE OF FIMBRIAE AND PILI IN THE ATTACHMENT OF KLEBSIELLA, ENTEROBACTER AND PSEUDOMONAS TO PLANT SURFACES

Timo K. Korhonen[1], Kielo Haahtela[1], Martin Romantschuk[2] and Dennis H. Bamford[2]
Departments of General Microbiology[1] and Genetics[2]
University of Helsinki
SF-00280 Helsinki
Finland

INTRODUCTION

Bacterial fimbriae are filamentous surface proteins, whose only known function is that of adhesion. Fimbrial functions and their biological significance are best known with pathogenic Escherichia coli strains, which possess many fimbrial types that function in different clinical situations. The fimbrial types are differentiated primarily by their binding properties but have chemical and serological differences as well (reviewed by Korhonen et al 1985). Although many of the bacterial species associated with with plants are known to possess fimbriae, their possible role in bacteria-plant interactions has largely remained unexplored. We have been interested in the role of fimbriae and pili in the adhesion of nitrogen-fixing enteric bacteria to the roots of grasses and in the adhesion of a plant pathogen, Pseudomonas syringae, to the leaves of bean and corn. In keeping with the nomenclature suggested by Ottow (1975), we will use the term fimbria for the enterobacterial filaments, which are numerous and not involved in transfer of genetic material, and the term pilus for the filaments of P.syringae, since there are only 1-5 filaments per cell and they function as phage receptors also.

NATO ASI Series, Vol. H 4
Recognition in Microbe-Plant Symbiotic and Pathogenic Interactions. Edited by B. Lugtenberg

ADHESION OF FIMBRIATED ENTERIC BACTERIA TO GRASS ROOTS

Enteric bacteria of the genera Klebsiella and Enterobacter have frequently been isolated from the roots of various plants, mostly grasses, in both tropical and temperate areas (Bally et al 1983; Haahtela and Korhonen 1985). The strains, which fix nitrogen and are classified as associative nitrogen fixers, apparently live attached to root surfaces. All the nitrogen-fixing Klebsiella and Enterobacter strains studied in our laboratory are fimbriated: klebsiellas have type-3 and type-1 fimbriae and the Enterobacter isolates type-1 fimbriae. The latter bind to α-D-mannosides and occur widely in all enterobacterial species and strains. The receptor structure or the binding specificity of the type-3 fimbria are not known, but it is routinely screened by haemagglutination of tannin-treated human erythrocytes (Duguid 1959).

Already in 1959, James Duguid proposed that fimbriae are involved in the adhesion of klebsiellas to plant roots. We have confirmed this by showing that highly purified, ^{125}I-labelled Klebsiella and Enterobacter fimbriae bind specifically to the roots of Poa pratensis, a natural host for these bacteria, and that fimbriae-specific Fab fragments and receptor analoques inhibit bacterial adhesion (Korhonen et al 1983; Haahtela et al 1985; Haahtela and Korhonen 1985). Both fimbrial types are involved in this adhesion process, but tests with model strains and with the nitrogen-fixing isolates showed that the type-3 fimbriae are major adhesins of the Klebsiella strains.

Localization of the adhesion sites for fimbriated klebsiellas on the root surface was determined by two fluorescence microscopic approaches. Firstly, after an in vitro adhesion assay, adherent bacteria on the roots were visualized by indirect immunofluorescence with fimbriae-specific antibodies. Secondly, adhesion assays were performed with fluorochrome-labelled model strains expressing only one fimbrial type or lacking both of

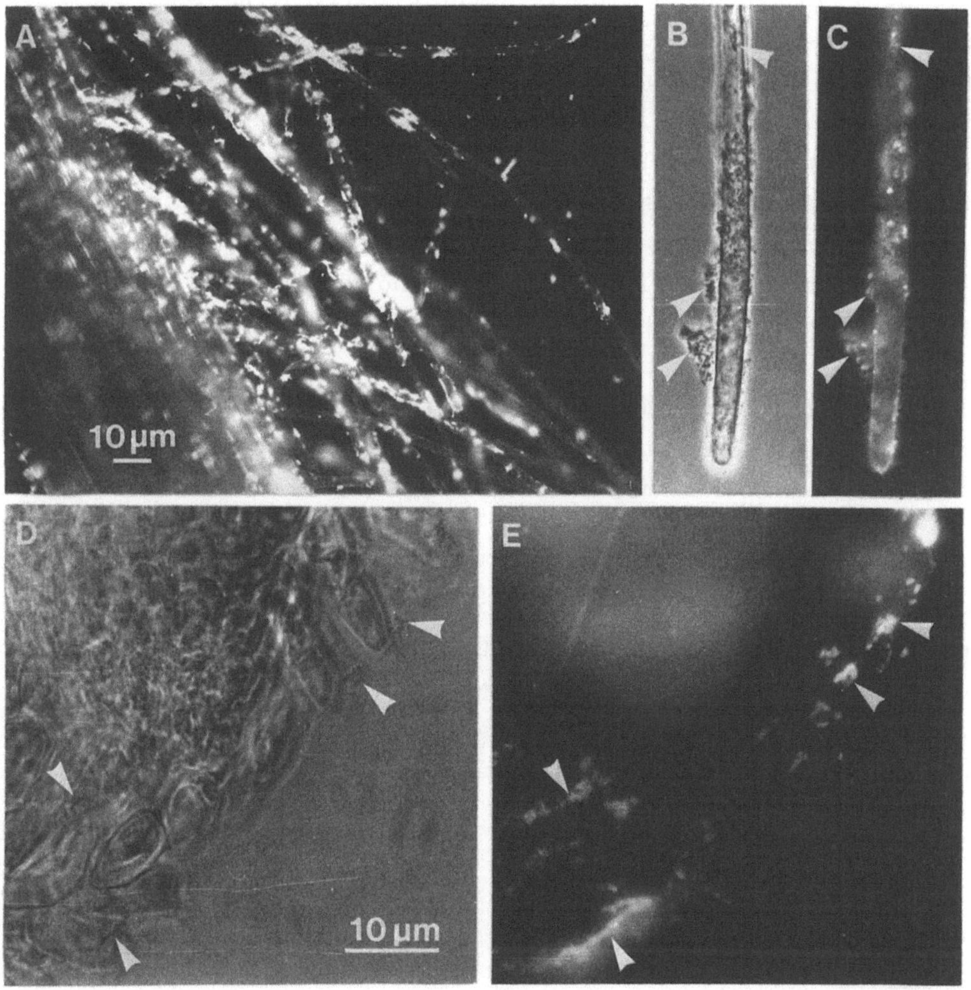

Figure 1. Immunofluorescent study of adhesion of K.pneumoniae strain As to the roots of Festuca rubra (A-C) and of P.pratensis (D and E). Axenic roots were incubated with the bacteria, washed and immunostained with anti-fimbriae antibodies and fluorochrome-labelled second antibodies. In A, adherent bacteria stained with anti-type-3-fimbriae antibodies. Note the strong adhesion to root hairs. In B, a root hair, shown in C for staining with anti-type-1-fimbriae antibodies. Note that only a fraction of the bacterial cells is stained, the other ones most likely expressing type-3 fimbriae. In D, a root cap, shown in E for staining with antitype-3-fimbriae antibodies. Note adhesion to the root cap mucilage and lack of adhesion to the sloughtered-off cells at the root cap.

them. Both methods gave closely similar results. Whereas the non-fimbriated strain adhered anly poorly, the fimbriated bacteria showed an efficient and highly localized adhesion to root hairs (Figs. 1A-1C) and, to a lesser extent, to the root cap mucilage (Figs. 1D and 1E). No differences in the binding sites were observed between the type-1 and the type-3 fimbriae or between the grass species studied. The bacteria also adhered to the root hairs of Trifolium pratense, indicating lack of strict host specificity in fimbriae-mediated adhesion.

We used indirect immunofluorescence also to see whether fimbriae are expressed during an in vitro colonization of grass roots and whether colonization of roots by the klebsiellas localizes to the root sites recognized by the fimbriae. Upon examination of infected P.pratensis roots, we found stained bacteria on root hairs (Fig. 2) and, in very low numbers, also in the root cap mucilage (not shown). This indicates that fimbriae were expressed by the bacteria during colonization and that the colonization took place in the same root sites as did bacterial adhesion. Similar results were obtained for a type-1-fimbriated Enterobacter strain.

We have previously shown by an ^{15}N-dilution method that nitrogen is fixed and transferred to the plant in Klebsiella-P.pratensis association, similar to the one described in Fig. 2 (Haahtela and Kari 1986). Upon examination of the roots, we noticed that the roots of the plants infected with Klebsiella or Enterobacter were larger than those of non-infected plants. The roots of infected plants were densely covered with root hairs and the length of the elongation zone was decreased (Figs. 3A-3D). The morphological changes were less apparent when the plants were supplemented with KNO_3 during growth (compare Figs. 3C and 3D to Figs. 3A and 3B). Also the number of deformed and branched root hairs was increased in infected plants (Figs. 3E-3G).

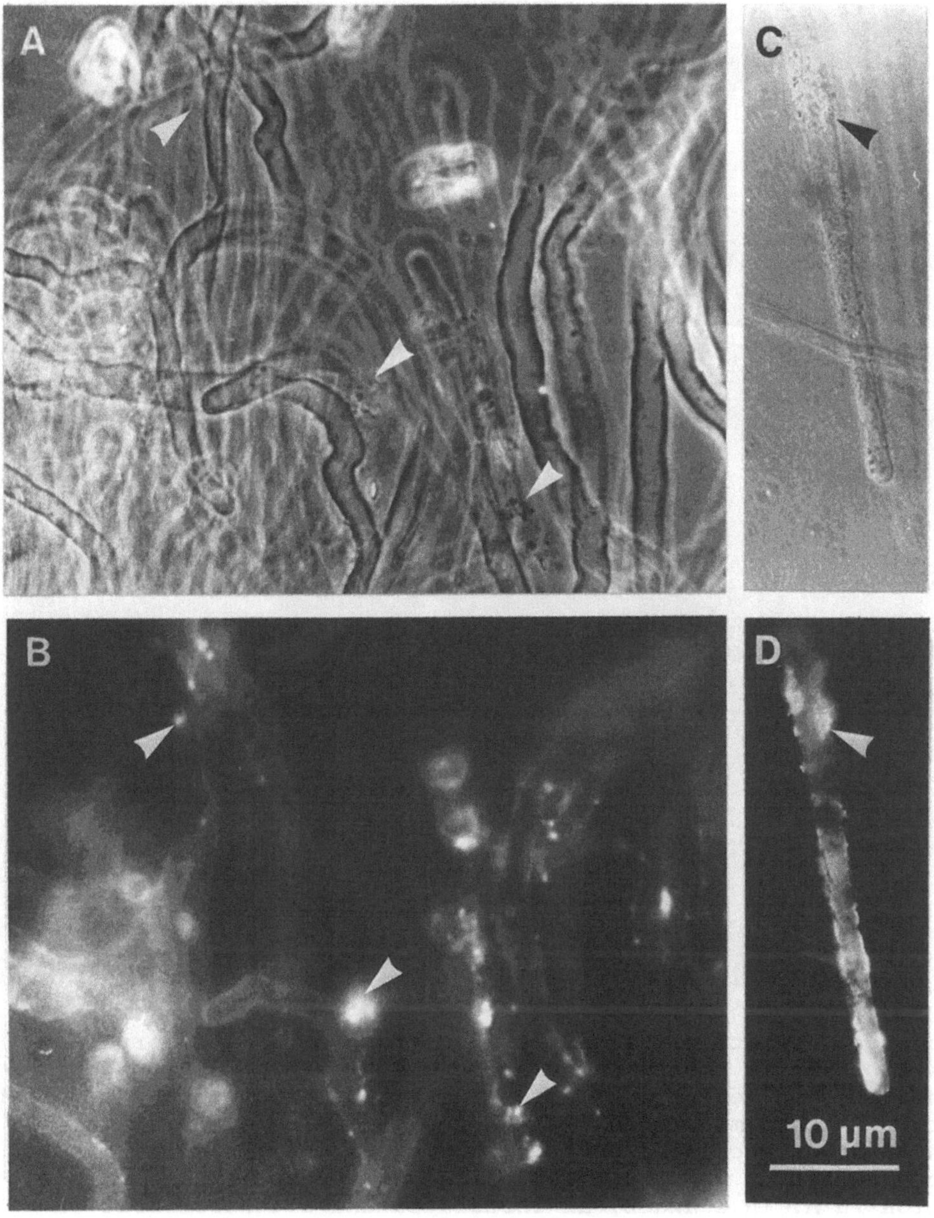

Figure 2. Expression of fimbriae during colonization. Seedlings of P.pratensis were infected with the K.pneumoniae strain As and grown for 5 weeks under greenhouse conditions. After harvesting, the roots were washed, immunostained with anti-fimbriae antibodies and examined by fluorescent microscopy. In A, root hairs of an infected plant, shown in B for the immunostaining. Note bacteria on root hairs (arrowheads). In C, a root hair, shown for staining in D. Arrowhead point bacterial cells reacting with the antiserum.

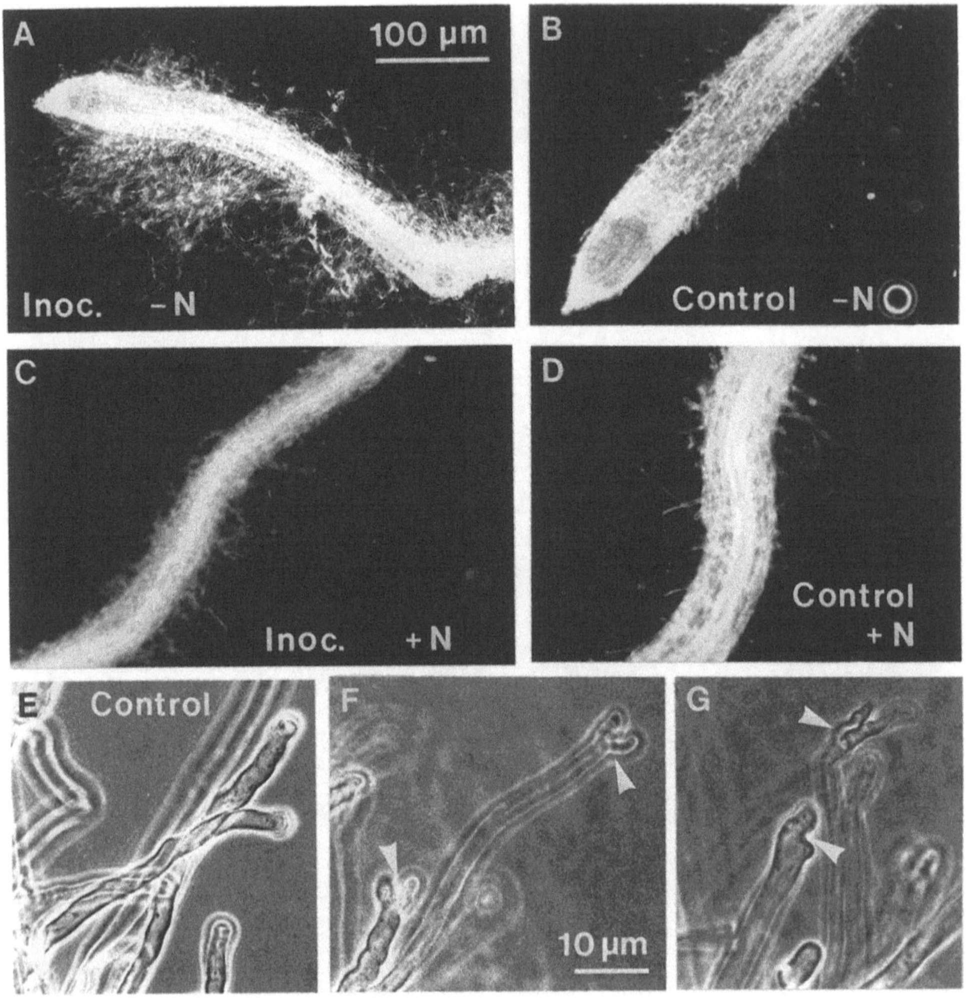

Figure 3. Effect of E.agglomerans strain Am on the morphology of P.pratensis roots. Inoculation and growth of plants were as described in legend to Fig. 2. In A, root of an inoculated plant grown in sand moistened with nitrogen-free nutrient solution. In B, an uninoculated control root grown under similar conditions. Note the numerous root hairs and the short zone of elongation in A. In C and D, an inoculated and a control root of plants grown in nitrogen-containing medium. In E, root hairs of a control plant, and in F-G, root hairs of inoculated plants. Arrowheads point tuned forks and curled tips of root hairs.

ADHESION OF PILIATED P.SYRINGAE TO THE LEAVES OF BEAN AND CORN

P.syringae pv. phaseolicola is the causative agent of Halo blight in bean (Phaseolus vulgaris). The bacterium invades the plant through wounds or through the stomata of the leaves. P.syringae is also an ice nucleation active bacterium that causes frost injury to sensitive plants at temperatures below -2°C (reviewed by Hirano and Upper 1985). Various P.syringae strains function as host for the pilus-specific bacteriophage Ø6 (Bamford et al 1976). In phage adsorption studies (Romantschuk and Bamford 1985), we isolated strains with various levels of pilus expression. We also analyzed these strains for adhesion and pathogenicity to bean leaves (Romantschuk and Bamford 1986). Similar mutants have now been obtained for P.syrinagae pv. syringae, a corn pathogen.

Piliation correlated directly with the adhesive capacity of the pv. phaseolicola (Table 1) and the pv. syringae (Fig. 4) strains. The superpiliated variants (expressing 2-3 pili per cell, the wild type strains had on the average 0.6 pili per cell) adhered clearly more efficiently than the other strains, and the non-piliated variants adhered only poorly. The piliated pv. phaseolicola strains adhered mainly to the stomata of bean leaves, whereas the few non-piliated cells were found evenly distributed on the leaf surface (Fig. 5). Interestingly, the piliated pv. syrinagae strains did not show such a localized adhesion to the leaves of corn; the bacteria were found evenly spread over the leaf surface (Fig. 4).

Upon infecting beans by injecting the various pv. phaseolicola strains directly into leaf veins, no difference in the pathogenicity of the strains was observed (Table 1). However, when the pathogenicity test was performed by the spraying and washing procedure significant differences were found. The non-piliated variants caused no symptoms, whereas the piliated (and adherent) strains diseased the plants (Table 1).

Table 1. Correlation between piliation, adhesion and pathogenicity of P.syringae pv. phaseolicola to P.vulgaris, cultivar Red Kidney. Data modified from Romantschuk and Bamford, 1986.

Strain	Piliation	Adhesion: no. of bacteria		Pathogenicity	
		per mm^2,[a]	per stoma[b]	Injection[c]	Spraying + washing[d]
HB	wild type	2700	80	+	+
MP22.14	superpiliated	5300	200	+	+
MPO.16	non-piliated	550	<10	+	-

[a]Determined by incubating 2.5×10^7 ^{14}C-labelled bacteria for 60 min at room temperature with 15mmx7mm stripes of Red Kidney leaves.
[b]Determined by scanning electron microscopy after incubating 5×10^8 bacteria for 40 min at room temperature on the lower surface of Red Kidney leaves.
[c]Bacteria were injected directly into leaf veins.
[d]Ten ml of bacterial suspension was sprayed onto the plants, followed 20 min later by a second spraying with sterile salt solution.

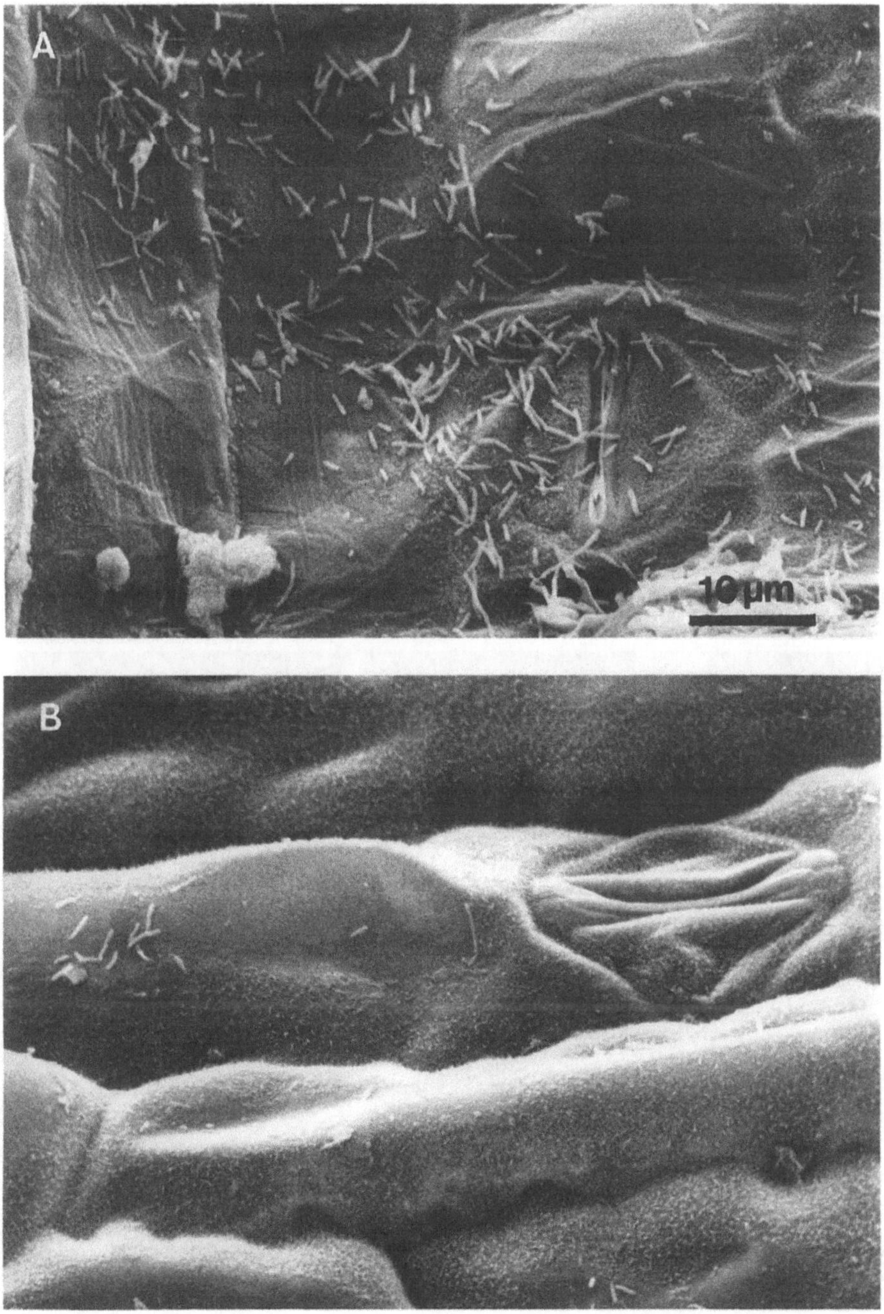

Figure 4. Scanning electron micrographs of Zea mays leaves with adherent P.syringae pv. syringae bacteria. Samples were prepared as described in footnote to Table 1. In A, adhesion of strain HS191, a piliated strain, and in B, adhesion of strain HS191R1, a non-piliated derivative of strain HS191. Note strong and unlocalized adhesion in A and poor adhesion in B.

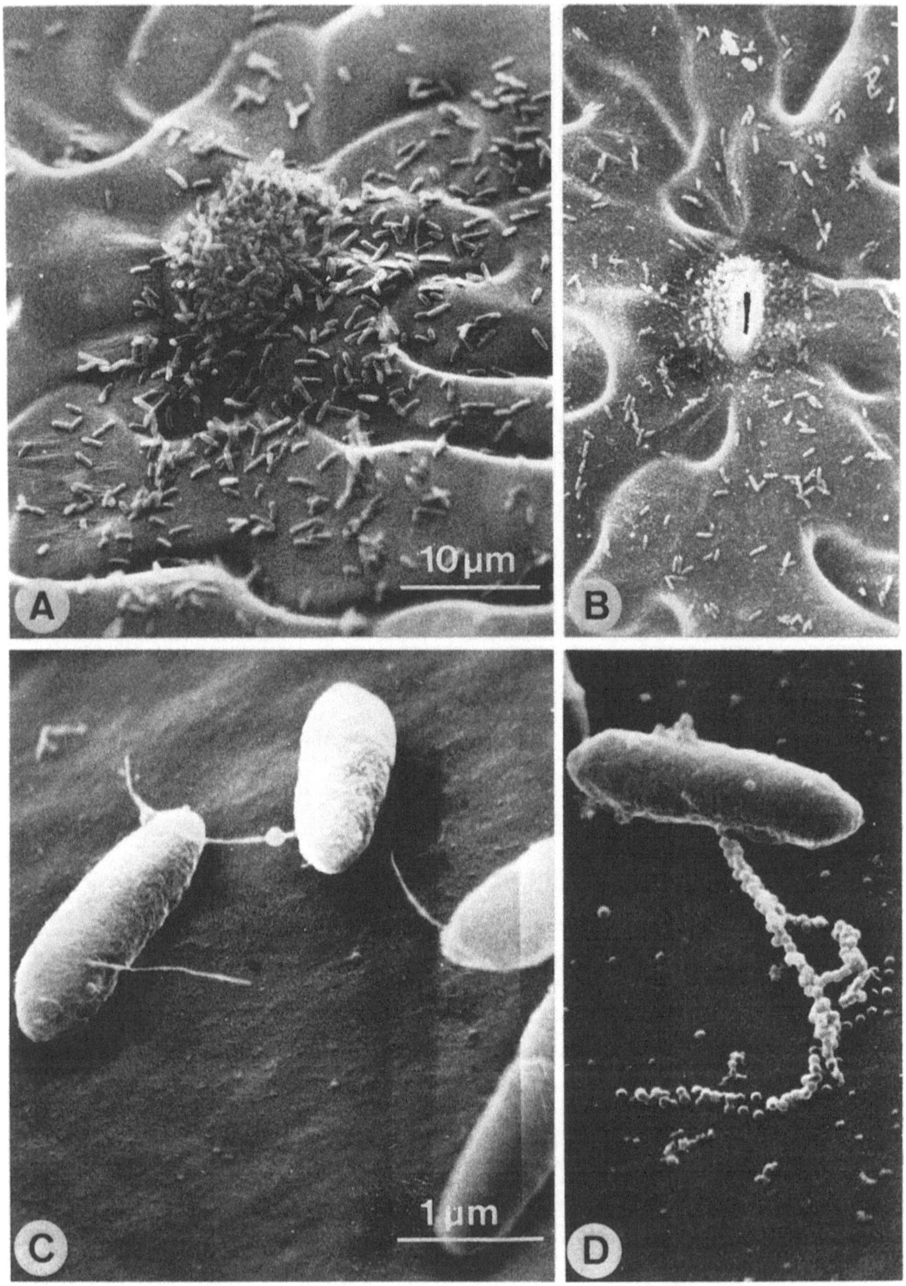

Figure 5. Scanning electron micrographs showing adhesion of P.syringae pv. phaseolicola to the leaves of P.vulgaris. In A, adhesion of piliated strain; in B, of a non-piliated strain. Note the localization of adhesion to the stomata in A. In C, piliated bacteria on the surface of a leaf. In D, a bacterial cell showing pili covered by Ø6 phages.

DISCUSSION

An interesting feature in the fimbria- or pilus-mediated adhesion systems is their localization to only some plant surface domains. The enteric bacteria adhered efficiently to grass root hairs (Figs. 1 and 2) and the P.syringae pv. phaseolicola strains to the stomata on the lower side of bean leaves (Fig. 5). This implies that the receptor structures for these bacteria are unevenly distributed on plant surfaces. This apparently is not the case for all bacteria and all plants, as the piliated P.syringae pv. syringae strains showed an unlocalized adhesion to the leaves of corn (Fig. 4). At present nothing is known about the plant macromolecules acting as receptors for these bacteria. For the enteric bacteria, obvious candidates are the mucilages extruded by the plants. For the pseudomonads, specific adhesion could also result from hydrophobic interactions between the pili and hydrophobic sites on bean and corn leaves.

Both adhesion systems here described are probably highly advantageous for the bacteria. By adhering to root hairs and to the root cap mucilage, the root-associated enteric bacteria gain access to nutrients exudated by the host plant, this would then enhance colonization. Colonization of P.pratensis roots by enteric bacteria results in nitrogen fixation and transfer to the plant (Haahtela and Kari 1986), but more important probably are the changes in root morphology induced by bacterial colonization (Fig. 3). Similarly to the Azospirillum-pearl millet association (Umali-Garcia et al 1980), bacterial colonization increased the number and the deformation of root hairs and decreased the length of the zone of elongation and these effects were suppressed when nitrate was added to the medium. The effects are probably caused by plant growth substances produced by the bacteria.

For P.syringae, adhesion to plant leaves is related to the pathogenic ability of the bacterium (Table 1) and can obviously help the bacteria to invade host plant through stomata. Possibly the adhesion also undirectly increases the probability of frost injuries to sensitive plants. To our knowledge, this is the first example of pili affecting bacterial pathogenicity to a plant. Bacterial adhesion in relation to pathogenicity in plants has often been interpreted as a defence mechanism of the plant, the plant tissue immobilizing incompatible bacteria (Sequeira et al 1977). Stemmer et al (1985) reported that fimbriae, although conferring adhesion capacity to Pseudomonas solanacearum, were not associated with bacterial virulence. However, also the pathogenic variant used by Stemmer et al. was piliated (or fimbriated), possibly expressing pili similar to the ones on P.syringae. The situation with pseudomonads probably is similar to that in enteric bacteria, which possess a number of fimbrial or pilus types, only some of which contribute to bacterial virulence.

REFERENCES

Bamford DH, Palva ET, Lounatmaa K (1976) Ultrastructure and the life cycle of the lipid-containing bacteriophage Ø6. J Gen Virol 32:249-259

Bally R, Thomas-Bauzon D, Heulin T, Balandreau J, Richard C, De Ley J (1983) Determination of the most frequent N_2-fixing bacteria in a rice rhizosphere. Can J Microbiol 29:881-887

Duguid JP (1959) Fimbrae and adhesive properties in Klebsiella strains. J Gen Microbiol 21:271-286

Haahtela K, Kari K (1986) The role of root-associated Klebsiella pneumoniae in the nitrogen nutrition of Poa pratensis and Triticum aestivum as estimated by the method of ^{15}N isotope dilution. Plant and Soil 90:245-254

Haahtela K, Korhonen TK (1985) In vitro adhesion of N_2-fixing enteric bacteria to roots of grasses and cereals. Appl Environ Microbiol 49:1186-1190

Haahtela K, Tarkka E, Korhonen TK (1985) Type-1 fimbria-mediated adhesion of enteric bacteria to grass roots. Appl Environ Microbiol 49:1182-1185

Hirano SS, Upper CD (1985) Ecology and physiology of Pseudomonas syringae. Biotechnology 3:1073-1078

Korhonen TK, Rhen M, Väisänen-Rhen V, Pere A (1985) Antigenic and functional properties of enterobacterial fimbriae. In: Stewart-Tull DES, Davies M (ed) Immunology of the bacterial cell envelope. John Wiley & Sons, New York, p 319

Korhonen TK, Tarkka E, Ranta H, Haahtela K (1983) Type-3 fimbriae of Klebsiella sp: molecular characterization and role in bacterial adhesion to plant roots. J Bacteriol 155:860-865

Ottow JGC (1975) Ecology, physiology and genetics of fimbriae and pili. Annu Rev Microbiol 29:79-108

Romantschuk M, Bamford DH (1985) Function of pili in bacteriophage Ø6 penetration. J Gen Virol 66:2461-2469

Romantschuk M, Bamford DH (1986) The causal agent of halo blight in bean, Pseudomonas syrinage pv. phaseolicola, attaches to stomata via its pili. Microb Pathogen 1:139-148

Sequeira L, Gaard G, De Zoeten GA (1977) Interaction of bacteria and host cell walls: Its relation to mechanisms of induced resistance. Physiol Plant Pathol 10:43-50

Stemmer WPC, Sequeira L (1985) Possible role of fimbriae in attachment of plant-associated bacteria to plant cell walls. In: Szalay AA, Legocky RP (ed) Proceedings of the 2nd international symposium on the molecular genetics of the bacteria-plant interaction. Media Service of Cornell University, Ithaca, New York, p. 199

Umali-Garcia M, Hubbel DH, Gaskins MH, Dazzo FB (1980) Association of Azospirillum with grass roots. Appl Environ Microbiol 39:219-226

RACE SPECIFIC RESISTANCE TO BREMIA LACTUCAE IS EXPRESSED BY LETTUCE CELLS IN SUSPENSION CULTURE

P.F.S. Street, P.M. Rowell, D.N. Crucefix, F. Didehvar & J.W. Mansfield
Biological Sciences Department, Wye College (University of London) Ashford, Kent TN25 5AH, U.K.

INTRODUCTION

The responses of isolated cells and cell suspension cultures to plant pathogens have frequently been reported to differ from those observed in whole plants. In particular, the hypersensitive reaction (HR) characteristic of race specific resistance is usually not expressed in vitro (Fett and Zacharius 1983; Helgeson and Deverall 1983; Ingram 1967; Otsuki et al 1972; Tomiyama et al 1958). During an investigation of the use of cell suspension cultures of lettuce (Lactuca sativa L.) as an assay for elicitors of the HR (Crucefix et al 1984,1986), we found that cultured cells gave differential reactions to avirulent and to virulent isolates of the downy mildew fungus Bremia lactucae Regel. This paper examines the development of the obligate parasite and plant responses in whole cotyledons and cell suspension cultures. Our results indicate that in vitro cultures of B. lactucae with lettuce cells should prove valuable for the assay and isolation of elicitors of the HR, and examination of nutrient transfer to the biotrophic fungus.

Lettuce downy mildew disease is a model system for studies on the recognition processes underlying race specific resistance (Crucefix et al 1984; Maclean et al 1974; Maclean and Tommerup 1979). A gene for gene relationship exists between the pathogen and its host. At least 18 specific resistance factors matched by 18 complementary virulence factors control the relationship (Crute I.R. pers comm. Michelmore and Crute 1982; Norwood and Crute 1984). Here we

NATO ASI Series, Vol. H 4
Recognition in Microbe-Plant Symbiotic and Pathogenic Interactions. Edited by B. Lugtenberg

report interactions between fungal isolates TV (virulence factors 1-10), IM43 (5-10) and NW3 (1, 2, 4, 5, 7-10) and the lettuce cultivars Cobham Green (no known resistance genes) and the near isogenic cv. Mildura (resistance genes Dm-1 and Dm-3).

MATERIALS and METHODS

Fungal cultures. Cotyledons of lettuce cultivar Cobham Green were used to maintain isolates of B. lactucae as described by Crucefix et al (1984).

Cell suspension cultures. Cell suspension cultures were obtained from callus derived from young leaves (6-8wks old). Leaves were surface sterilized in 40% EtOH for 5sec then dipped in 1% (w/v) NaOCl for 10min, washed 3x in SDW, cut into c. $4cm^2$ pieces and plated on a medium of 0.1mg napthylene-acetic acid, 0.1mg kinetin, 0.471g Murashige and Skoog medium without sucrose or hormone (Flow Laboratories), 3g sucrose and 0.8g Difco Bacto agar made up in 100ml distilled water. The medium was dispensed into 9.0cm diam. Petri dishes or 25 x $8cm^3$ multiwell dishes (Sterilin). Leaf callus was grown for 4wks at 24-27°C in the dark. Cell suspension cultures were obtained by placing 0.5-1.0g of friable yellow coloured callus into 20ml of liquid B-5 medium (Alconero 1983; Gamborg and Eveleigh 1968) in a 100ml Erlenmyer flask. Cultures were incubated on a gyrotary shaker at 120r.p.m. (Denley Instruments) in the dark at 24-27°C.

Inoculation with B. lactucae. Cotyledons of cultivars Cobham Green and Mildura were inoculated with suspensions of 10^5 conidiosporangia ml^{-1} in SDW as previously described (Crucefix et al 1984).

Establishing challenges between B. lactucae and cell suspension cultures involved washing 10ml of 2-3 week old cell suspension cultures 2x in SDW and transferring 100μl of resuspended suspension into flat wells of a microtitre plate (Sterilin 96 wells). The same volume (100μl) of spore suspension of B. lactucae was added to the wells and the plate incubated at 14°C in the dark for 12h then light and dark on a 12h cycle (Gallenkamp illuminated incubator). Examination of infection development involved either direct observation of wells on an inverted microscope (Zeiss) or removal of samples to a slide and viewing with a conventional photomicroscope.

Cell and fungal viability. Fungal development and plant cell death in cotyledons of cvs. Cobham Green and Mildura were examined using the staining methods described by Crucefix et al (1984). Viability of cell suspensions and fungal infection structures was determined by staining samples from microtitre wells with fluorescein diacetate (FDA). A stock solution of 10mg FDA in 2ml acetone was prepared from which 20µl was added to 500µl distilled water and approximately 30µl of this added to the same volume taken from the microtitre well. Slides were incubated for 5 min in the dark at 20°C before viewing under incident blue light (FITC excitation and barrier filters). At each time after inoculation two replicate wells were examined and the viability of at least 50 penetrated lettuce cells and associated fungus was determined. Living cells were recognised by their yellow/green fluorescence as demonstrated by Widholm (1972).

RESULTS

Cotyledon infection. Mildura was susceptible to isolate TV, cv. Cobham Green was equally susceptible to all isolates. During the compatible interactions the fungus rapidly developed primary and secondary vesicles within penetrated cells and subsequently formed intercellular hyphae and haustoria as described by Maclean and Tommerup (1979). Resistance in Mildura to both IM43 (conferred by Dm-1 and Dm-3) and NW3 (conferred by Dm-3 alone) was expressed by the HR of epidermal cells following penetration. Growth of the avirulent isolates was usually restricted to the production of a primary vesicle and one or two secondary vesicle initials. Expanded secondary vesicles were not produced but short intra- and occasionally intercellular hyphae sometimes developed from primary vesicles during incompatible interactions. All epidermal cells penetrated by avirulent isolates were necrotic within 12h of inoculation. Necrosis of some penetrated epidermal cells was also observed during compatible interactions 24 and 48h after inoculation but in susceptible tissues, plant cell death followed the rapid production of intercellular hyphae and haustoria by the virulent fungus.

<u>Cell suspension infection</u>. Conidiosporangia germinated rapidly in cell suspension cultures. Germ-tubes which made contact with lettuce cells differentiated appressoria, penetrated and formed primary vesicles (Fig. 1a). In cells from susceptible cultivars, expanded secondary vesicles and hyphae usually developed as shown in Fig. 1b. Many hyphae had grown out of penetrated susceptible cells by 36h after inoculation. In some cases the hyphae penetrated additional cells forming structures resembling the haustoria found in cotyledon mesophyll (Fig. 1c, Table 1). In essence, fungal development was very similar to that observed in the intact cotyledons.

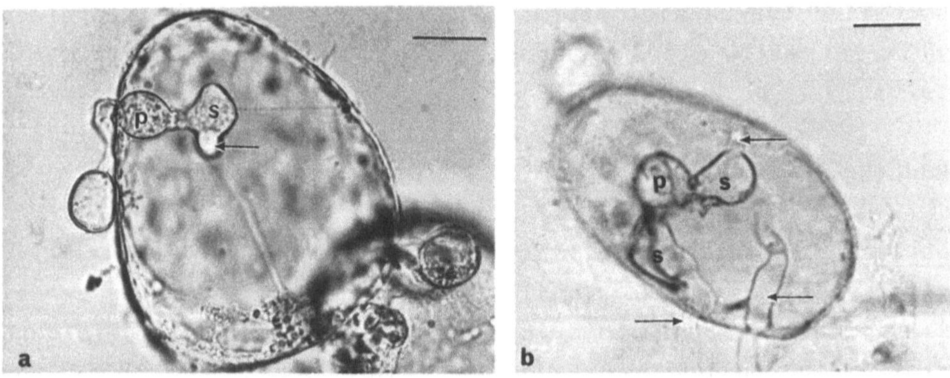

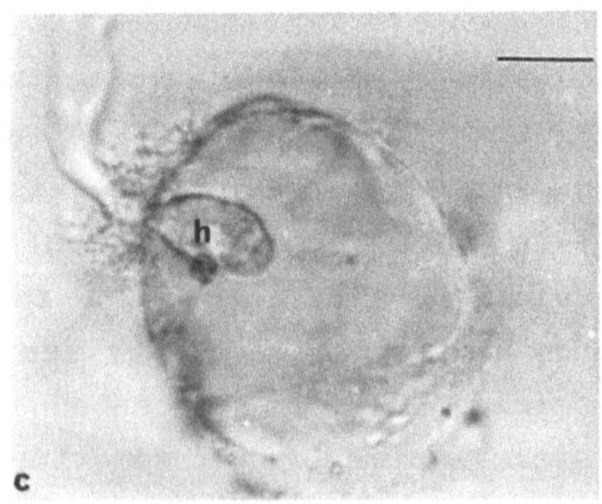

Fig. 1. Cells of cv. Cobham Green in suspension culture infected by the virulent isolate TV. In 1a), note the formation of primary (p) and secondary (s) vesicles and development of a hyphal bud (arrowed), 12h after inoculation. In 1b) and c) taken 36h after inoculation note the development of two secondary vesicles and hyphae (arrowed) which have grown out of the initially invaded cell and in one case (1c) penetrated a second cell forming a structure resembling an haustorium (h). Bars = 20 µm

Quantitative studies on fungal development, summarized for isolate NW3 in Table 1, showed that in cells isolated from the resistant cv. Mildura, clearly defined secondary vesicles were rarely produced by NW3 or IM43. Instead, the fungus developed single narrow hyphae or several small vesicle-like swellings. The distinction between hyphae and poorly formed secondary vesicles was not always clear but fungal growth was usually limited to the production of the small intracellular structures.

Table 1. Development of _B. lactucae_ isolate NW3 in cultured cells of the susceptible cv. Cobham Green and the resistant cv. Mildura.

Cultivar	Time after inoculation (h)	% of infections at each developmental stage[a] I	II	III	IV	V	VI	VII
Cobham Green	6	100	0	0	0	0	0	0
	12	19	0	39	21	21	0	0
	24	0	0	18	12	70	0	0
	36	2	2	7	9	25	42	13
	48	0	0	6	8	47	29	10
Mildura	6	100	0	0	0	0	0	0
	12	38	20	36	6	0	0	0
	24	26	12	22	26	14	0	0
	36	10	53	2	15	20	0	0
	48	8	47	2	33	10	0	0

[a]Stages of development: I, primary vesicle ± bud; II, primary vesicle + narrow hypha; III, primary vesicle + secondary vesicle of at least the same volume; IV, primary vesicle with multiple secondary vesicles; V, primary and secondary vesicles + hyphae growing from secondary; VI, as V but hyphae growing out of penetrated cell; VII as VI + hyphae penetrating a second cell. At each time _c._ 25 first penetrated cells were examined in each of 2 replicates.

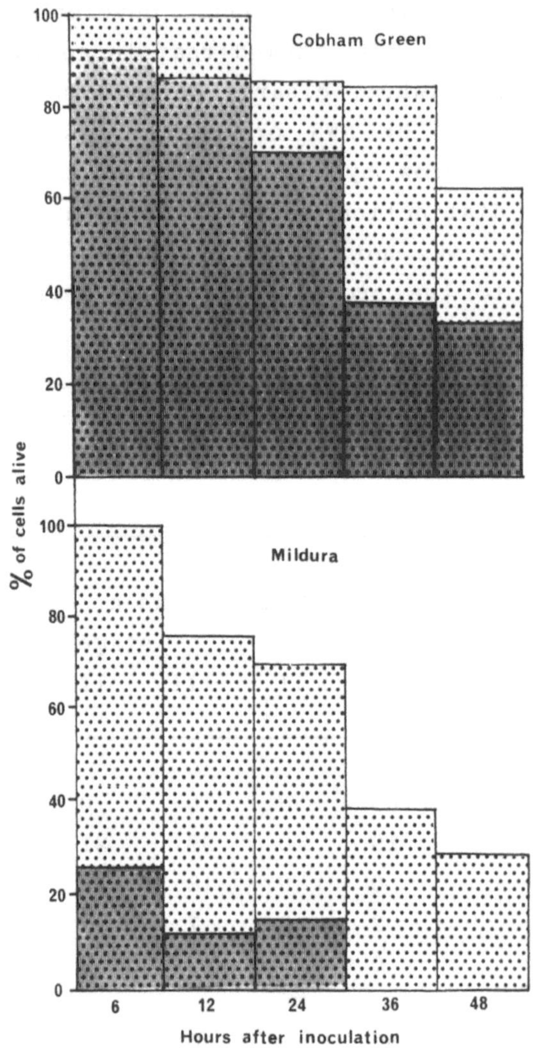

Fig. 2. Viability of intracellular fungal structures (isolate NW3) and penetrated lettuce cells (susceptible cv. Cobham Green and resistant cv. Mildura) in suspension culture. Lightly and darkly shaded columns record fungal and plant cell viability respectively. The % of lettuce cells viable following incubation for 12, 24 and 48h under the experimental conditions but without conidiosporangia were 77, 79 and 49% for cv. Cobham Green and 78, 74 and 51% for cv. Mildura.

Restricted fungal development during all incompatible interactions followed death of penetrated plant cells as indicated by failure to stain with FDA (Fig. 2). Interestingly, FDA staining clearly showed that the fungus did not die for at least 12h following plant cell death. Most penetrated plant cells remained alive in the compatible interactions (all isolates in cv. Cobham Green and TV in cv. Mildura) in spite of the development of large intracellular structures during the first day after inoculation (Fig. 1b, Fig. 2, Table 1). As in the epidermis of susceptible cotyledons, plant cell death usually followed the growth of hyphae out of the initially penetrated cell. Where second cell penetrations occurred (Fig. 1c) the newly penetrated cells were alive 48h after inoculation.

DISCUSSION

Although isolated cultured cells appeared to undergo the HR as expressed in cotyledon epidermal cells, it should be stressed that the differential response of cultured cells to avirulent and virulent isolates was not quite as clear as that observed in the intact plant. Thus, fewer cultured cells were killed during the early stages of incompatible interactions and more cells died in culture than in the intact cotyledon following penetration by virulent isolates. Nevertheless, this is, to our knowledge, the first report of the expression of race specific resistance to a fungal pathogen in isolated cultured plant cells. The HR occurred only in cells penetrated by B. lactucae. No evidence was obtained for the release of diffusible elicitors of the HR by germinating conidiosporangia or by cells undergoing necrosis. These observations suggest that the molecules controlling the HR may be highly localized, perhaps as components of the primary vesicle cell wall as discussed by Crucefix et al (1984).

Phytoalexin accumulation has frequently been implicated as the cause of the inhibition of fungal growth during the HR. Experiments with small groups of dissected cells of French bean and potato have demonstrated that a critical mass of tissue is required for the expression of the HR and inhibition of fungal growth by avirulent isolates of the facultative parasites Colletotrichum lindemuthianum and Phytophthora infestans respectively (Bailey 1982; Skipp and Deverall 1972). Live tissues in contact with the cell undergoing the HR are thought to be the sites of biosynthesis of phytoalexins which restrict fungal development (Bailey 1982; Mansfield et al 1974). The restricted development of B. lactucae in isolated lettuce cells in suspension cultures may not be caused by the accumulation of antifungal compounds but may simply reflect starvation due to the failure of the obligate parasite to absorb nutrients from isolated killed cells.

This work was supported by the Agricultural and Food Research Council, Science and Engineering Research Council and Shell Research Ltd. (C.A.S.E. award to D.N.C.).

REFERENCES

Alconero R (1983) Regeneration of plants from cell suspensions of Lactuca saligna, Lactuca sativa and Lactuca serriola. Hort Science 18:305-307

Bailey JA (1982) Mechanisms of phytoalexin accumulation. In: Bailey JA, Mansfield JW (eds) Phytoalexins. Blackie, Glasgow

Crucefix DN, Mansfield JW, Wade M (1984) Evidence that determinants of race specificity in lettuce downy mildew disease are highly localized. Physiol Plant Pathol 24:93-106

Crucefix DN, Rowell PM, Street PFS, Mansfield JW (1986) A search for elicitors of the hypersensitive reaction in the lettuce downy mildew disease. Physiol Plant Pathol (in press)

Fett WF, Zacharius RM (1983) Bacterial growth and phytoalexin elicitation in soybean cell suspension cultures inoculated with Pseudomonas syringae pathovars. Physiol Plant Pathol 22:151-172

Gamborg OL, Eveleigh DE (1968) Culture methods and detection of glucanases in suspension cultures of wheat and barley. Can J Biochem 46:417-421

Helgeson JP, Deverall BJ (eds) (1983) Use of tissue culture and protoplasts in plant pathology, Academic Press, Sydney

Ingram DS (1967) The expression of R-gene resistance to Phytophthora infestans in tissue cultures of Solanum tuberosum. J Gen Microbiol 49:99-108

Maclean DJ, Sargent JA, Tommerup IC, Ingram DS (1974) Hypersensitivity as the primary event in resistance to fungal parasites. Nature 249:186-187

Maclean DJ, Tommerup IC (1979) Histology and physiology of compatibility and incompatibility between lettuce and the downy mildew fungus, Bremia lactucae Regel. Physiol Plant Pathol 14:291-312

Mansfield JW, Hargreaves JA, Boyle FC (1974) Phytoalexin production by live cells in broad bean leaves infected with Botrytis cinerea Nature 252:316-317

Michelmore RW, Crute IR (1982) A method for determining the virulence phenotype of Bremia lactucae isolates. Trans Br Mycol Soc 79:542-546

Norwood JM, Crute IR (1984) The genetic control and expression of specificity of Bremia lactucae (lettuce downy mildew). Plant Pathology 33:385-400

Otsuki Y, Shimomura T, Takebe I (1972) Tobacco mosaic virus multiplication and expression of the N gene in necrotic responding tobacco varieties. Virology 50:45-50

Skipp RA, Deverall BJ (1972) Relationships between fungal growth and host changes visible by light microscopy during infection of bean hypocotyls (Phaseolus vulgaris) susceptible and resistant to physiologic races of Colletotrichum lindemuthianum. Physiol Plant Pathol 2:357-374

Tomiyama K, Takakuwa M, Takase N (1958) The metabolic activity in healthy tissue neighbouring the infected cells in relation to resistance to Phytophthora infestans (Mont.) De Bary in potatoes. Phytopath Z 31:237-250

Widholm JM (1972) The use of fluorescein diacetate and phenosafranine for determining viability of cultured plant cells. Stain Technol 74:189-194

INTERACTION OF FUNGAL POLYGALACTURONASE WITH PLANT PROTEINS IN RELATION TO SPECIFICITY AND REGULATION OF PLANT DEFENSE RESPONSE

F. Cervone, G. De Lorenzo, L. Degrà and G. Salvi
University of Rome "La Sapienza"
Department of Plant Biology
Piazzale Aldo Moro
00100 Roma Italy

INTRODUCTION

The degradation of plant cell wall by polysaccharidases is a general process by which phytopathogenic microorganisms attack and colonize host plants. Polygalacturonase (PG) is the first cell wall degrading enzyme synthesized by phytopathogenic fungi. It has been suggested that the enzyme is involved in both basic and specific compatibility between fungi and higher plants (Cervone et al 1986). It is well known that PG releases oligosaccharides from the plant cell wall which are capable of eliciting active defense mechanisms in plants (Bruce and West 1982; Walker-Simmons et al 1983) and that proteins which specifically inhibit PG activity (PGIP) are present in the plant cell wall (Albersheim and Anderson 1971). We have therefore explored the possibility that the interaction between PG and PGIP may have relevance in determining whether a plant-fungus system appears compatible or incompatible.

ELICITATION OF PLANT DEFENSE RESPONSE BY HOMOGENEOUS PG

Endopolygalacturonase was purified from a commercial preparation of <u>Aspergillus niger</u> pectinase according to Cervone et al (1977). The enzyme was electrophoretically homogeneous and consisted of a single polypeptide chain with a molecular weight of 33,500 daltons. The galacturonate oligomers produced by the action of PG on polygalacturonic acid, were analyzed by

NATO ASI Series, Vol. H 4
Recognition in Microbe-Plant Symbiotic and Pathogenic Interactions. Edited by B. Lugtenberg

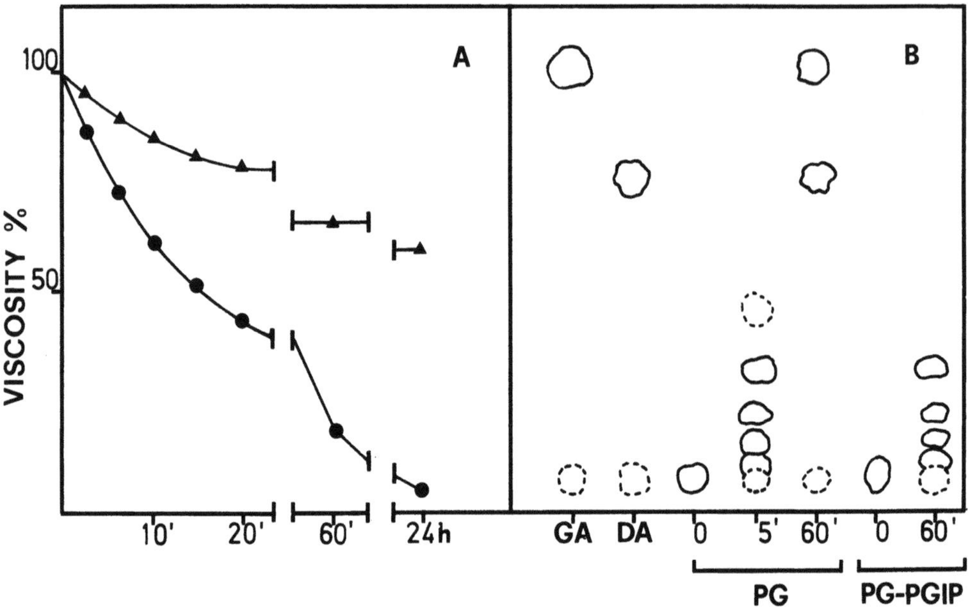

Fig. 1 A) Viscosity changes of polygalacturonic acid incubated with 0.05 units of homogeneous PG alone (●–●) or PG preincubated with 10 μg/ml of purified PGIP (▲–▲). B) Paper chromatogram of the reaction products of PG or PG-PGIP complex incubated with polygalacturonic acid. GA and DA, galacturonic acid and digalacturonic acid, respectively.

descending paper chromatography. Oligomers with a degree of polymerization higher than 4 were released rapidly in the incubation mixture and were later fragmented to a shorter length. The final products of the enzymic hydrolysis were mono- and digalacturonate (Fig. 1B). The action of PG was also tested on cell wall material isolated from hypocotyls of *Vigna unguiculata* Walp. The release of soluble reducing sugars was rapidly detected upon incubation, showing that purified PG is able to attack glycosidic bonds of native wall pectin.

Homogeneous PG (6 μl-drops) was applied to *Vigna* pods from which circular portions (2 mm diameter) of epidermis had been removed with a lancet. After 16 hours of incubation at 25°C in the dark, the appearance of a hypersensitive-like response, characterized by a visible browning at the point where enzyme had been applied, was detected. The response was concomitant with an increase of phenylalanine ammonia lyase activity and lignin-like material. The response intensity was proportional to the quantity of enzyme applied. Active PG was necessary for the eliciting activity since heat-inactivated enzyme or PG fully inactivated with saturating levels of anti-PG IgG did not induce a response. α-D-galacturonate oligomers with a degree of polymerization higher than 4, released by a partial digestion of polygalacturonic acid with PG, elicited a similar response when applied to Vigna pods, whereas D-galacturonic acid and the final PG degradation products of polygalacturonic acid did not. A response, undistinguishable from that induced by α-D-oligogalacturonates or by PG, was also elicited by soluble oligosaccharides released from Vigna cell wall partially digested with PG. The response was similar to the hypersensitive necrosis induced in Vigna pods by artificially inoculated A.niger mycelium. Hypersensitive necrosis restricted the microbial growth and did not let the fungus penetrate deeply into the tissue. Vigna response appeared to be an active mechanism requiring new protein synthesis since it was inhibited by cycloheximide.

PURIFICATION AND CHARACTERIZATION OF PGIP

Six mg of homogeneous A.niger PG were immobilized to 1 g of CNBr-activated Sepharose 4B. The Sepharose-PG conjugate was then used for affinity chromatography purification of PGIP from an extract of Phaseolus vulgaris L. The adsorbed PGIP was fully recovered by eluting with phosphate-buffered saline. The purification factor was higher than 2,800.

Purified PGIP, subjected to isoelectric focusing in the pH range 9-11, focused at pH 9.5 as a single protein peak. A similar pI had been reported for P.vulgaris PGIP by Lafitte et al (1984). The protein described in the latter study is therefore likely to be the same one. This was confirmed by determining its sugar content and molecular weight. Purified PGIP, tested for its ability to inhibit various hydrolases, was specific for PG. Protease inhibitors were ineffective in inhibiting PGIP activity.

The effect of salt concentration and pH on the interaction between PG and PGIP was also investigated. 0.5 M acetate buffer, pH 5.0, dissociated the PG-PGIP complex almost completely. The complex was also dissociated at pH lower than 4.0 and higher than 6.5. Moreover, our results indicated that the inhibitory effect of PGIP depends on its PG-complexing ability. PGIP was active under conditions of pH and salt concentration that favour the formation of the PG-PGIP complex.

The shape of the viscosity decrease curve was comparable to that obtained with PG alone only during the first few minutes of incubation of polygalacturonic acid with the complex PG-PGIP. The curve virtually reached a plateau after 10 min of incubation. Viscosity did not decrease to a significant extent upon longer incubation. After 24 hrs, viscosity of the incubation mixture (PG-PGIP) was still 60% of the initial value, whereas that of the control (PG alone) was minimal after 1 hr (Fig. 1A). The degradation products of polygalacturonate were analyzed by paper chromatography. Oligogalacturonates with a degree of polymerization higher than 4 were still present, while PG alone had completely degraded the substrate to the final products (Fig. 1B). Galacturonate oligomers released by the complex PG-PGIP, were able to elicit the hypersensitive response in Vigna.

In conclusion, these data show that a specific recognition occurs between fungal PG and plant PGIP and that the pattern of oligosaccharides with eliciting activity released from the cell wall is strongly influenced by the formation of the PG-PGIP complex.

DISCUSSION

The role of PGIP in vivo is somewhat complicated. The protein may be involved in a general mechanism of resistance against all invading pathogens (Albersheim and Anderson 1971; Fisher et al 1973), or may prevent the plant defense response by reducing the production of eliciting oligosaccharides released by PG of specific pathogens (Hoffman and Turner 1984). Our data show that, by complexing PG, PGIP delays the full degradation of polygalacturonic acid and leads to the formation of stable (24 hrs or longer) oligomers which exhibit a hypersensitive response eliciting activity. We suggest that the action of PGIP in vivo is that of counteracting the fungal invasion by helping exogenous polygalacturonases in eliciting the plant defense response (Fig. 2). In the light of these results, the differential specificity of PGIP towards PG from different plant pathogens should be investigated in greater detail. The possibility exists that a different degree of recognition (different affinity) between the various PG's and PGIP's may lead to different kinetics of formation of cell wall degradation products. It is worth mentioning that a fine control of cell wall degradation could regulate the rate or timing of plant reaction to an invading pathogen, and that resistant and susceptible reactions are, in some cases, similar phenomena (J. Mansfield, personal communication). The resistant response, in these cases, is characterized by a rapid hypersensitive necrosis, restricting the microbial growth soon after the penetration. The susceptible response is, on the other hand, characterized by a delayed host cell death.

In conclusion, we would like to emphasize that evidence is presently being gathered on the following points. (i) The fate of a plant-fungus interaction may depend on the oligosaccharides formed during the plant cell wall degradation. (ii) The rate or the timing of the oligosaccharide production may be

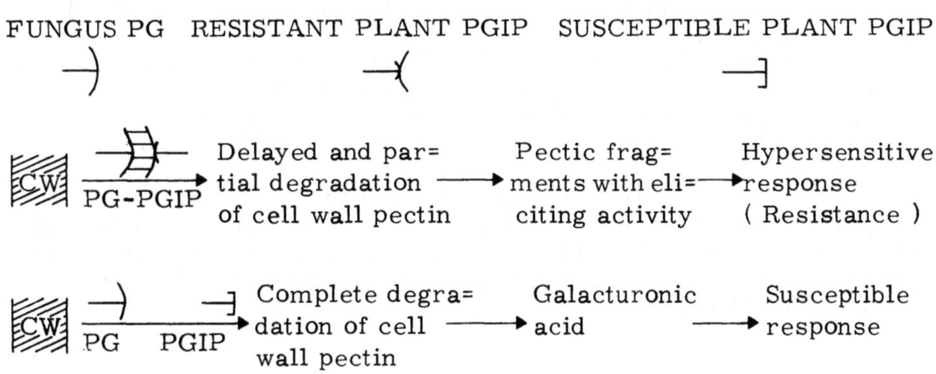

Fig. 2. Schematic representation of the possible regulation mechanism of plant defense response by the complex PG-PGIP.

regulated by PG activity. (iii) PG activity may be regulated by recognition proteins (PGIP) localized in the cell wall.

ACKNOWLEDGEMENT

This work was supported by "V.V. Landi" Foundation and by C.N.R. - Special Grant I.P.R.A. (subproject 1, publ. n. 922).

REFERENCES

Albersheim P, Anderson AJ (1971) Proteins from plant cell walls inhibit polygalacturonase secreted by plant pathogens. Proc Natl Acad Sci U.S.A. 68: 1815-1819

Bruce RJ, West CA (1982) Elicitation of casbene synthetase activity in Castor Bean. The role of pectic fragments of the plant cell wall in elicitation by fungal polygalacturonase. Plant Physiol 69: 1181-1188

Cervone F, Scala A, Foresti M, Cacace MG, Noviello C (1977) Endopolygalacturonase from Rhizoctonia fragariae: purification and characterization of two isoenzymes. Biochem biophys Acta 482: 379-385

Cervone F, De Lorenzo G, Salvi G, Camardella L (1986) Molecular evolution of fungal polygalacturonase. In: Bailey J (ed) Biology and molecular biology of plant-pathogen interactions. Plenum Press, New York (in press)

Fisher ML, Anderson AJ, Albersheim P (1973) Host-pathogen interactions VI. A single plant protein efficiently inhibits endopolygalacturonases secreted by Colletotrichum lindemuthianum and Aspergillus niger. Plant Physiol 51: 489-491

Hoffman RM, Turner JG (1984) Occurrence and specificity of an endopolygalacturonase inhibitor in Pisum sativum. Physiol Plant Pathol 24: 49-59

Lafitte C, Barthè JP, Montillet JL, Touzè A (1984) Glycoprotein inhibitors of Colletotrichum lindemuthianum endopolygalacturonase in near isogenic lines of Phaseolus vulgaris resistant and susceptible to anthrachnose. Physiol Plant Pathol 25: 39-53

Walker-Simmons M, Hadwiger C, Ryan CA (1983) Chitosans and pectic polysaccharides both induce the accumulation of the antifungal phytoalexin pisatin in pea pods and antinutrient proteinase inhibitors in tomato leaves. Biochem Biophys Res Commun 110: 194-199

A MODEL SYSTEM FOR THE STUDY OF FUNGUS - HOST SURFACE INTERACTIONS: ADHESION OF PHYTOPHTHORA MEGASPERMA TO PROTOPLASTS AND MESOPHYLL CELLS OF SOYBEAN.

Hans R. Hohl and Sylvia Balsiger
University of Zürich
Institute of Plant Biology
Zollikerstr. 107
CH-8008 Zürich
Switzerland

INTRODUCTION

Cell surface phenomena are important but little understood aspects of fungal pathogen - host plant interactions (Bracker and Littlefield 1973; Albersheim and Prouty 1975; Keen 1982; Yoshikawa et al 1983; Hermanns and Ziegler 1984; Furuichi et al 1980). Using plasmolysis techniques Nozue et al (1979) demonstrated tight adherence of host plasma membranes (extrahaustorial membranes) to haustoria of Phytophthora infestans in potato and Coffey (1983) has made similar observations with Albugo candida on cabbage. Physical contact and adhesion seem to be important early steps of infection yet are very little studied in fungal as compared to bacterial systems (e.g. Rao et al 1982; Dazzo and Gardiol 1984; Matthysse 1984; Érsek et al 1985). In all likelihood these processes involve not only surface molecules preformed on the host and parasite surfaces but also induced interactions (e.g. biosynthetic activities) preceding or following the first physical contact.

One obvious reason for this lack of information in fungal parasitism (and symbiosis) is the difficulty encountered in studying contact phenomena in situ with filamentous fungi

NATO ASI Series, Vol. H 4
Recognition in Microbe-Plant Symbiotic and Pathogenic Interactions. Edited by B. Lugtenberg

penetrating deeply into host tissue and making contact with the host cells at unpredictable sites. Thus, only few reports deal with the properties of these interfaces. They are mostly ultrastructural and histochemical in nature (e.g. Hächler and Hohl 1982; Coffey 1983; Mendgen et al 1985) and provide some information, e.g. on the presence of polysaccharides and/or glycoproteins at these interfaces.

This paper describes some results obtained with a novel method developed to study and experimentally manipulate these early, dynamic host - parasite interactions, in particuclar adhesion processes between fungal hyphae and host cell surfaces.

MATERIAL AND METHODS

Growth of the pathogen and production of cysts. - Race 6 of Phytophthora megasperma f.sp. glycinea Kuan and Erwin (Pmg) was used throughout. In one experiment Phytophthora infestans (Mont.) de Bary, strains 134 and 193, was also included. Stock cultures of Pmg were grown on Borlotti bean agar (Kuhn and Hohl 1985), of P. infestans on rye-glucose medium (Bodenmann et al 1985) and transferred at 4-weekly intervals. Zoospores and cysts of Pmg were obtained according to Eye et al (1978). For high yields of zoospores it was important to grow the fungus on lima bean agar. Encystment was induced by shaking the zoospore suspension for 1 min on a Vortex mixer.

For experimental work 10,000 cysts in 0.9 ml of distilled water were pipetted into wells of FALCON 3047 24-well, flat-bottom tissue culture plates containing round, acid-cleaned glass coverslips. If no microscopy was involved coverslips were omitted so that the cysts attached directly to the plastic bottom of the well. 0.1 ml of pea broth was added to each cyst-containing well. For preparing the broth 75 g of commercial, freshly frozen garden peas were cooked in 250 ml of water and filtered. To this were added 10 g of sucrose, 1 g of asparagine, 0.25 g of $MgSO_4 . 7 H_2O$, and 0.5 g of KH_2PO_4 and the medium filled up to make a total of 1 L. The fungal cultures were incubated overnight (16 h) at room temperature. During this period a regular network of hyphae had developed which was rinsed twice with sterile water and was then ready for the addition of the protoplast suspension.

Host plants. - Soybean seedlings of cv. Harosoy and Harosoy 63 (obtained from Dr. J. Paxton, Urbana, Ill., USA) were grown for 5 to 6 d at 22 - 25 C in the dark in vermiculite-containing plastic boxes.

Isolating protoplasts. - Hypocotyls were finely sliced (slices appr. 1 mm thick) and the tissue fragments transferred to charcoal-agar plates (agar 200 mg, gelatine 300 mg, activated charcoal 100 mg, in 10 ml of distilled water) containing 10 ml of the following isolation medium: 2 % cellulase Onozuka R-10 (SERVA, Heidelberg, FRG) and 0.1 % pectolyase Y-23 (SEISHIN Pharmacytical Comp., Noda, Chiba, Japan) in a near-isotonic buffer composed of mannitol 0.4 M, $CaCl_2 . 2 H_2O$ 4 mM, MES 10 mM, and adjusted to pH 5.8 with NaOH. The material was incubated for 3 h at 28 C on a shaker at 50 rpm. Following this the suspension was filtered through a 150 µm nylon mesh and the protoplasts further cleaned on a Ficoll step gradient (15 and 5 % Ficoll, topped with the mannitol-buffer where the intact protoplasts collected after centrifugation at 180 g for 15 min). The purified protoplasts were washed once with the mannitol-buffer and then counted. Cell viability was checked with 0.02 % methylene blue (Berliner et al 1978) and fluorescein diacetate (FDA) according to Widholm (1972). About 90 % of freshly isolated protoplasts were viable.

Isolation of mesophyll cells. - Leaf mesophyll cells from 12 d old Harosoy plantlets were isolated according to Cosio et al (1983), except that no methyl cellulose was added to the isolation medium.

Adhesion of protoplasts and mesophyll cells to fungal hyphae. - 50'000 protoplasts or cells respectively in 0.5 ml of buffered mannitol solution were pipetted into the wells containing the hyphal network. The cultures were incubated at room temperature (22-23 C). To measure adhesion the supernatant was removed with a pipet, 1 ml of buffered mannitol solution was added, removed, replaced and finally removed again. This washing procedure has to be standardized and is critical for reproducibility of the test. Finally the well was kept moist by the addition of a small amount of buffered mannitol solution until the number of adhering protoplasts was determined. With this procedure only very few protoplasts adhered directly to the plastic or glass.

For microscopical determinations the mannitol solution was removed from the wells and the coverslip placed upside down on a microscope slide together with a drop of glycerol - buffered mannitol solution (1:1). The edge of the coverslip was sealed with nail polish and the protoplasts were counted from 10 different fields. The number of protoplasts was then transformed into numbers per mm^2 of the hyphal network.

For enzymatic determination of protoplasts the solution in the wells was replaced with 200 μl of para-nitrophenyl-α- D - mannopyranoside (2 mg/ml in 0.2 M sodium acetate buffer of pH 5.0). Incubation at room temperature for at least one h was followed by the addition of 1 ml of 0.2 M sodium carbonate solution and the extinction measured at 400 nm. The value obtained was subtracted from that of a control with only the hyphal network present and the final activity expressed as mU/ml.

Test compounds used. - In some experiments the following compounds were added to the wells harbouring hyphae 1 h prior to the addition of the protoplasts: snail digestive enzyme (BOEHRINGER, Mannheim, GFR) 5 μl/ml, cellulase (Onozuka R-10, SERVA, Heidelberg, FRG) 4 mg/ml, proteinase K (MERCK, Darmstadt, FRG) 4 mg/ml, cycloheximide (SIGMA, St. Louis, USA) 2 μg/ml, actinomycin D (SIGMA) 50 μg/ml; 2-deoxy-D-glucose (dglc) from SIGMA, N-Acetyl-D-glucosamine (glcNAc), methyl-α -D-mannopyranoside (met-man), D-galactose (gal), N-Acetyl - D galactosamine (galNAc), and L(+)-rhamnose monohydrate (rha), the last 5 from FLUKA, Buchs, Switzerland, and all 6 at 0.1 M.

Chemical fixation of protoplasts and mycelium. - Protoplasts were fixed for 15 min at room temperature with 3 % paraformaldehyde in mannitol-buffer of pH 5.8, washed and resuspended in the mannitol-buffer. The mycelium adhering to the coverslips was fixed similarly, washed and then exposed to the protoplast suspension.

RESULTS

Formation of the hyphal network. - Although we did not explore this point systematically it was evident that adhesion of germinating cysts and development of the mycelial network

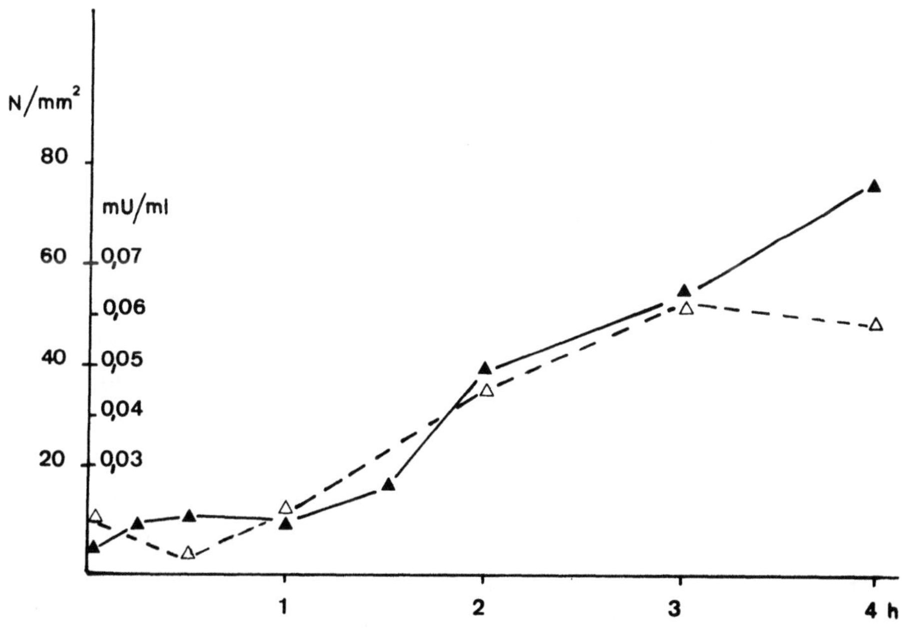

Fig. 1. Adhesion of soybean protoplasts to fungal hyphae. (h) hours after deposit of protoplast suspension. Protoplasts expressed as number N / mm^2 of mycelium (solid line), and as α-mannosidase activity (dotted line).

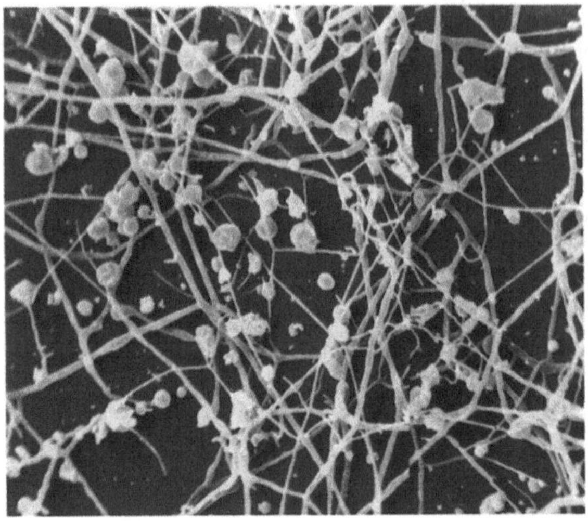

Fig. 2. Hyphal network with adhering soybean protoplasts. The small globules are fungal cysts, x 250.

varied with the fungal species, the nutrient broth and the properties of the substrate used. With the method finally adopted Pmg attached better than Phytophthora infestans. The nutrient medium had to be adjusted to give (1) solid attachment, and (2) minimal number of hyphae growing away from the substrate and into the liquid nutrient broth. Too many loose hyphae might trap protoplasts mechanically and thus interfere with the test. The pea broth used proved to be best among several other media or concentrations tested. Finally, germ tubes and hyphae generally attach more firmly to the plastic material of the wells than to the glass coverslips. In case of difficulties one might use plastic coverslips although we did not explore this possibility.

Adhesion of soybean protoplasts to hyphae of Pmg. - Fig. 1 shows the kinetics of adhesion of protoplasts to hyphal surfaces averaged from 5 separate experiments. After a lag phase of approximately 90 min adhesion of protoplasts increased with time. After 4 h it reached from 80 to 100 protoplasts / mm^2 of mycelial area. For the next about 14 h an additional increase of approximately 60 % was noted. Results obtained with the enzyme test were generally comparable to

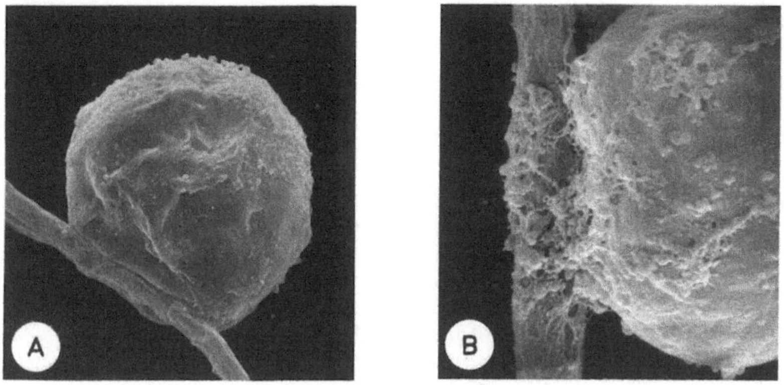

Fig. 3. Adhesion of soybean protoplasts to fungal hyphae. (A) Often no conspicuous material is visible at the site of contact, x 2500. (B) Sometimes the site of adhesion is covered by partly fibrillar material, x 3900.

those from microscopic counts. However, direct counting is the safer method since (1) rupture of some of the adhering protoplasts may lead to a loss in enzyme activity, and (2) the enzyme itself may adhere tightly to the fungal network (unpublished results) which leads to an overestimate of the number of adhering protoplasts. Fig. 2 is a SEM picture of protoplasts adhering to the hyphal network. Many contact sites did not show elaborations of extracellular material (Fig. 3a) while some did reveal the presence of fibrillar or coarse material (Fig. 3b).

Adhesion of isolated, walled soybean mesophyll cells to hyphae of Pmg. - Fig. 4 demonstrates that walled cells adhered to hyphae with similar kinetics although the lag period was less well pronounced.

Effect of various treatments or compounds on adhesion of protoplasts to hyphae of Pmg. - The results of these experiments are summarized in fig. 5. Addition of cycloheximide or actinomycin D prevented attachment as did chemical fixation of the hyphae. Killing the protoplasts, however gave no inhibition. Adhesion was also prevented in the presence of some polysaccharide and protein degrading enzymes pointing to carbohydrates and/or proteins as essential

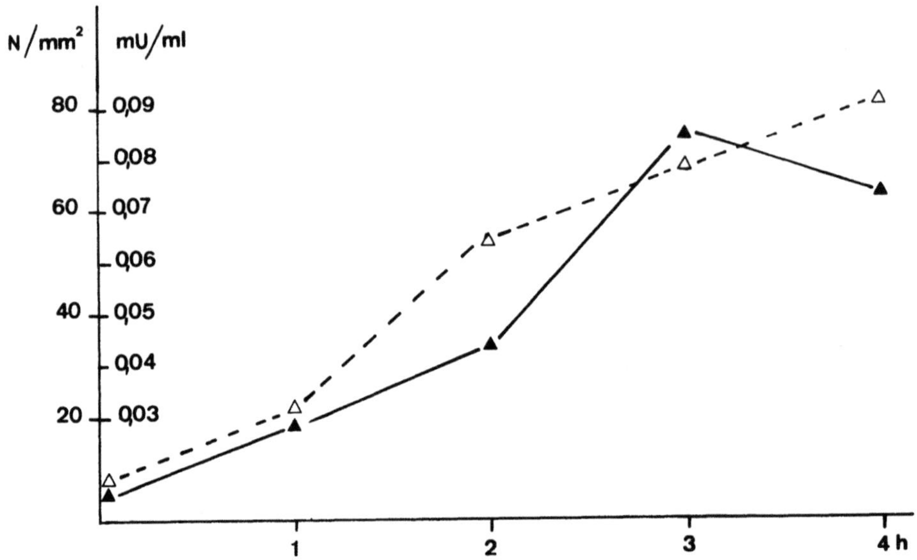

Fig. 4. Adhesion of isolated soybean cells to fungal mycelium. (h) hours after deposit of cell suspension. Solid line: Cells/mm^2 of mycelium; dotted line: α-mannosidase activity of adhering cells.

components of this process. Varying the calcium ion concentration in the incubation medium from 0 - 10 mM had no influence. Finally of six sugar haptens studied three (dglc, glcNAc, met-man) prevented adhesion while gal, galNAc, and rha did not.

DISCUSSION

With the method described in this paper it is possible to routinely and quantitatively measure and observe the adhesion of protoplasts or isolated walled cells to fungal hyphae. It allows extensive manipulation of these interactions by e.g. addition of enzymes, potential inhibitors of biosynthetic processes, lectins, antibodies, host or fungal products, by treatments such as cold, heat, and pH shock, or by changing environmental parameters such as types of ions present, ionic strength, pH, and nutrition of the fungus. The multiwell system is flexible and allows a multitude of experiments and controls with a minimum of cell material and chemicals.

Protoplasts as well as isolated walled cells adhere to hyphae within 1 to 2 h. Attachment of Agrobacterium tumefaciens takes place on walled cells and on protoplasts and in both instances shows the same bacterial strain specificity (Matthysse et al 1982). This indicates a basic similarity of receptor molecules on both plasma membrane and outer wall surface.

Particularly with protoplasts the lag phase for attachment to hyphae of about 90 min is well pronounced. It appears to be somewhat shorter for walled cells. Using a different technique (time-dependent removal of isolated mesophyll cells from a suspension by freely floating hyphal colonies) Hargreaves and Keon (1983) noticed binding within about 15 min of isolated

Fig. 5. Effect of various compounds or treatments on adhesion between soybean protoplasts and fungal hyphae

Compound/treatment	Adhesion
None (control)	+
Cycloheximide	-
Actinomycin D	-
Protoplasts fixed with paraformaldehyde	+
Fungal Hyphae fixed with paraformaldehyde	-
Snail digestive juice	-
Proteinase K	-
Cellulase	-
Ca-ions 0 - 10 mM	+
Galactose	+
N-Acetyl-galactosamine	+
Rhamnose	+
Deoxyglucose	-
N-Acetyl-glucosamine	-
Methyl-mannoside	-

for details see "Material and Methods"

mesophyll cells from barley leaves to hyphae of the barley pathogen Pyrenophora teres and of non-pathogenic fungi. Binding of these mesophyll cells appeared to be mediated through an affinity of the plant cell wall (possibly cellulose) and a component of the extracellular polymeric material (possibly glycoproteins) produced by the fungal hyphae. The difference for the minimum time necessary for adhesion in their and our system (15 vs. 90 min) could possibly be explained on a nutritional basis: These authors grew their fungi in a rich medium where the fungi could produce copious amounts of extracellular material immediately available for binding of the cells, whereas in our system the fungus is grown in a dilute medium where extracellular material possibly forms only upon a contact signal.

We did not observe differences in the adhesion of protoplasts from the susceptible and resistant cultivar. Thus adhesion per se is certainly not a specific response. Nozue et al (1979) also described an unspecific adherence of host plasma membranes to infecting hyphae of compatible and incompatible races of Phytophthora infestans. However, the possibility should not be disregarded that the tight adhesion monitored with our test system is possibly accompanied, pre- or antedated by mechanically more subtle specific and hapten-reversible interactions as evident in bacterial adhesion to plant surfaces (Dazzo and Gardiol 1984) and by our experiment with hapten sugars discussed in the end. Furthermore the work of Érsek et al (1985) on attachment of Pseudomonas syringae pv. glycinea to isolated single leaf cells of resistant soybean cultivars also revealed an early adhesion process inhibited by the addition of rhamnose or glucose.

Adhesion in our system occurred equally well in the absence of added calcium ions. Mathysse (1984) also reported no need for divalent cations for attachment of Agrobacterium to plant cell surfaces. The results from chemical fixation, and from the inhibitor studies are evidence that the fungus and not the protoplast produced the adhesive material and that the latter is not preformed. The lag phase of about 90 minutes also indicates that the adhesive material is not preformed and appears to be produced only upon contact or some other signal from the protoplasts. The microbial partner also plays the active part in the attachment of Pseudomonas syringae pv. glycinea (Érsek et al 1985), of Rhizobium (Dazzo and Gardiol 1984), and of Agrobacterium tumefaciens (Matthysse 1984; Matthysse et al 1982) to their respective host cells.

Occasionally we have observed fibrillar extracellular material at the contact sites of both susceptible and a resistant cv. reminiscent of the fibrillar material around bacteria attaching to host cell walls (e.g. Dazzo and Gardiol 1984; Mathysse 1984). While in the latter case fibrils are produced by the microbial cells, in our study the presently available evidence (unpublished results) points to the protoplast, i.e. the host, as producer of this material. Since adhesion normally occurs without the presence of this material it obviously is not essential for attachment but probably represents a response of the host cell to its contact with the pathogen.

The inhibitory action of glcNAc and met-man but not of galactose, galNAc and rhamnose indicates a possible involvement of lectin-mediated binding processes in adhesion. The enzyme studies are consistent with this view as are several published reports: Keen and Legrand (1980) observed that the fungus binds concanavalin A and produces cell wall bound glycoproteins with glucose and mannose residues. Gibson et al (1982) found twice as much soybean agglutinin in resistant compared to susceptible cultivars. Furuichi et al (1980) implicate the potato lectin in the binding of germ tubes of Phytophthora infestans to potato cell membranes. Nozue et al (1980) describe that chitobiose (the hapten of wheat germ agglutinin and potato lectin) added to potato tuber cells infected by Phytophthora infestans inhibited the hypersensitive response as well as binding between host plasmalemma and infecting hyphae. Finally, Kogel et al (1985) have shown that reagents with affinity for wheat plasma galactoconjugates suppress the hypersensitive response in the wheat stem rust interaction.

The system described in the present paper should allow us an approach to address even more directly the question of lectins and other compounds and mechanisms involved in adhesion and recognition in host - pathogen interactions.

Acknowledgement. This research was supported by Swiss National Science Foundation grants 3.331-0.82 and 3.240-0.85 to H.R.H. We would like to thank Dr. Christa Werner for many helpful hints and Mr. U. Jauch for preparing the SEM pictures.

REFERENCES

Albersheim P, Anderson-Prouty AJ (1975) Carbohydrates, proteins, cell surfaces, and the biochemistry of pathogenesis. Annu Rev Plant Physiol 26: 31-52

Berliner MD, Wood NL, Damico J (1978) Vital and calcofluor staining of Cosmarium and its protoplasts. Protoplasma 96: 39-46

Bodenmann J, Heiniger U, Hohl HR (1985) Extracellular enzymes of Phytophthora infestans: endo-cellulase, β-glucosidases, and 1,3- β-glucanases. Can J Microbiol 31: 75-82

Bracker CE, Littlefield LJ (1973) Structural concepts of host-pathogen interfaces, 159-318. In: Byrde RJW, Cutting CV (eds) Fungal pathogenicity and the plant's response. Academic Press, London New York, p 499

Coffey MD (1983) Cytochemical specialization at the haustorial interface of a biotrophic fungal parasite, Albugo candida. Can J Bot 61: 2004-2014

Cosio EG, Servaites JC, McClure JW (1983) Isolation and photosynthetic characteristics of mesophyll cells from developing leaves of soybean. Physiol Plant 59: 595-600

Dazzo FB, Gardiol AE (1984) Host specificity in Rhizobium - legume interactions, 3 - 31. In : Verma DPS, Hohn Th (eds) Genes involved in microbe - plant interactions. Springer - Verlag, Wien and New York, p 393

Érsek T, Gáborjányi R, Höltzl P, Király Z (1985) Sugar-specific attachment of Pseudomonas syringae pv. glycinea to isolated single leaf cells of resistant soybean cultivars. Phytopath Z 113: 260-270

Eye LL, Sneh B, Lockwood JL (1978) Inoculation of soybean

seedlings with zoospores of Phytophthora megasperma var. sojae for pathogenicity and race determination. Phytopathol 68: 1769-1773

Furuichi N, Tomiyama K, Doke N (1980) The role of potato lectin in the binding of germ tubes of Phytophthora infestans to potato cell membrane. Physiol Plant Pathol 16: 249-256

Gibson DM, Stack S, Krell K, House J (1982) A comparison of soybean agglutinin in cultivars resistant and susceptible to Phytophthora megasperma var. sojae (Race 1). Plant Physiol 70: 560-566

Hargreaves JA, Keon JPR (1983) The binding of isolated mesophyll cells from barley leaves to hyphae of Pyrenophora teres. Plant Cell Reports 2: 240-243

Hermanns R, Ziegler E (1984) Localization of α-mannan in the hyphal wall of Phytophthora megasperma f.sp. glycinea and its possible relevance to the host-pathogen interaction of the fungus with soybeans (Glycine max). Phytopath Z 109: 363-366

Hächler H, Hohl HR (1982) Histochemistry of papillae in potato tuber tissue infected with Phytophthora infestans. Bot Helv 92: 23-31

Keen NT (1982) Specific recognition in gene-for-gene host-parasite systems. Adv Plant Pathol 1: 35-82

Keen NT, Legrand M (1980) Surface glycoproteins: evidence that they may function as the race specific phytoalexin elicitors of Phytophthora megasperma f.sp. glycinea. Physiol Plant Pathol 17: 175-192

Kogel KH, Schrenk F, Sharon N, Reisener HJ (1985) Suppression of the hypersensitive response in wheat stem rust interaction by reagents with affinity for wheat plasma membrane conjugates. J Plant Physiol 118: 343-352

Kuhn H, Hohl HR (1985) Stimulatory and inhibitory effects of glyceollin and indole-3-acetic acid (IAA) on sporulation and growth of Phytophthora megasperma f.sp. glycinea. Bot Helv 95: 301-310

Matthysse AG (1984) Interaction of Agrobacterium tumefaciens with the plant cell surface, 33 - 54. In: Verma DPS, Hohn Th (eds) Genes involved in microbe - plant interactions. Springer, Wien and New York, p 393

Matthysse AG, Holmes KV, Gurlitz RHG (1982) Binding of Agrobacterium tumefaciens to carrot protoplasts. Physiol Plant Pathol 20: 27-33

Mendgen K, Lange M, Bretschneider K (1985) Quantitative estimation of the surface carbohydrates on the infection structures of rust fungi with enzymes and lectins. Arch Microbiol 140: 307-311

Nozue M, Tomiyama K, Doke N (1979) Evidence for adherence of host plasmalemma to infection hyphae of both compatible and incompatible races of Phytophthora infestans. Physiol Plant Pathol 15: 111-115

Nozue M, Tomiyama K, Doke N (1980) Effect of N,N'-di-acetyl-D-chitobiose, the potato lectin hapten and other sugars on hypersensitive reaction of potato tuber cells infected by incompatible and compatible races of Phytophthora infestans. Physiol Plant Pathol 17: 221-228

Rao SS, Lippincott BB, Lippincott JA (1982) Agrobacterium adherence involves the pectic portion of the host cell walls and is sensitive to the degree of pectin methylation. Physiol Plant 56: 374-380

Widholm JM (1972) The use of fluorescein diacetate and phenosaffranine for determining viability of cultured plant cells. Stain Technol 47: 189-194

Yoshikawa M, Keen NT, Wang M-C (1983) A receptor on soybean membranes for a fungal elicitor of phytoalexin accumulation. Plant Physiol 73: 497-506

ULTRASTRUCTURAL STUDIES OF SURFACE INTERACTIONS DURING ADHESION AND INFECTION BY ERICOID ENDOMYCORRHIZAL FUNGI.

V. Gianinazzi-Pearson (1), P. Bonfante-Fasolo (2) and J. Dexheimer (3)
(1) Station d'Amélioration des Plantes, INRA, BV 1540, 21034 Dijon, France, (2) Centro di studio sulla Micologia del Terreno del CNR, Dipartimento di Biologia Vegetale dell'Università, 10125 Turin, Italy and (3) Laboratoire de Biologie des Ligneux, Université de Nancy I, 54506 Vandoeuvre-Les-Nancy, France

INTRODUCTION

Although mycorrhizae, symbiotic associations between roots and fungi, are widespread, they differ considerably in their specificity depending on the host plant and fungus involved (Gianinazzi-Pearson 1984). Ericoid endomycorrhizal associations show the greatest degree of specificity and for this reason are of particular interest for research into symbiotic fungal-plant interactions. The fungal symbionts are ascomycetes, some of which have been identified as *Pezizella ericae* Read (Read 1974; Vegh et al 1979) ; they only develop mycorrhizal infections with a small number of plant genera in the Ericaceae (e.g. *Vaccinium*, *Erica*, *Gaultheria*, *Calluna*, *Rhododendron*,...).

The fine structure of ericoid mycorrhiza has been well described in both natural and axenically synthesized associations (Bonfante-Fasolo and Gianinazzi-Pearson 1979, 1981 ; Bonfante-Fasolo et al 1981 ; Duclos 1981 ; Duddridge and Read 1982), but nothing is known about the cellular basis of specificity and the processes determining recognition or compatibility between the symbionts. Recent observations on ultrastructural events occurring at cell surfaces during infection establishment have, however, begun to provide some clues.

FUNGAL WALL MODIFICATIONS AND CELL CONTACT PRIOR TO INFECTION.

In pure culture, hyphae of ericoid mycorrhizal fungi are often surrounded by a loose weft of ill-defined material (Figure 1). When fungi are inoculated into axenic cultures of host plants, there is an important increase in the amount of exocellular material and its becomes organized into an extensive fibrillar sheath which radiates out from the hyphal wall (Figure 2). As mycelium grows over the root surface, this fibrillar sheath comes into close contact with the outer wall layer of host cells to which it seems to attach the hyphae. Once cell to cell contact has been established between the symbionts, external hyphae swell to form an appressorium from which a thinner septate hypha develops and penetrates the host cell wall (Figure 3).

Examination of a large number of fungal isolates has shown that they can vary in their ability to produce exocellular fibrils (Gianinazzi-Pearson and Bonfante-Fasolo 1986). The amount of fibrillar material produced, quantified using computer aided image analysis (CAIA), is closely correlated ($r = 0.91$; $P = 0.01$) to the ability of the fungi to infect a host plant, highly infective isolates having the most developed fibrillar sheaths (Table 1). It therefore appears likely that the first cellular recognition event in the establishment of ericoid mycorrhiza is the attachment or adhesion of the fungal symbiont to the root surface, and that this process is mediated by the exocellular fibrils produced at the hyphal surface. This is very similar to the role attributed to fibrillar sheaths of pathogenic fungi (Murray and Maxwell 1975 ; Evans et al 1981) and fibrillar capsules of symbiotic or pathogenic bacteria (Dazzo and Gardiol 1984 ; Matthysse 1984) during infection of host tissues.

It has been possible to identify some components of the symbiont surfaces in ericoid mycorrhizae using probes such as cytochemical stains and fluorescent lectins.PATAg-positive 1-4 polysaccharides (Thiéry 1967), Alcian blue mucopolysaccharides (Behnke and Zelander 1970) and proteins revealed by the Swift reaction for cystine-containing proteins (Swift 1968) or by

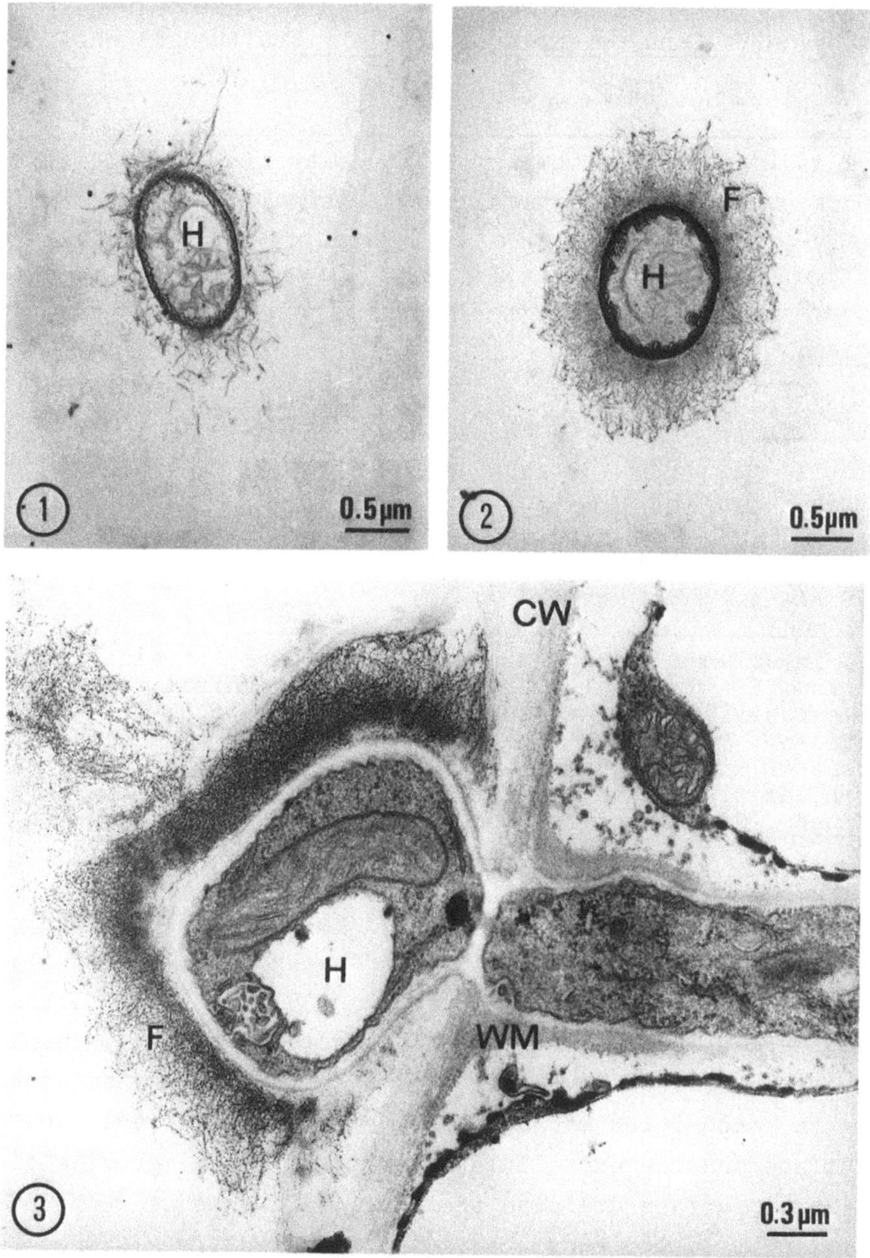

Figures 1 to 3 : Hyphae of Pezizella ericae strain A (1) growing in water agar, (2) developing around roots of Vaccinium corymbosum in axenic culture, (3) penetrating a host cell after attachment to the root surface of Calluna vulgaris. F, exocellular fibrillar sheath ; H, hypha ; CW, plant cell wall ; WM, host wall material. (1) and (2) PATAg reaction, (3) lead citrate / uranyl acetate staining.

Table 1 : Mycorrhizal establishment by different ericoid fungal isolates on _Calluna vulgaris_ plants _in vitro_ (unpublished data)

Fungal isolate	% number of mycorrhizal roots	Intensity of mycorrhizal infection(1)	Amount of extracellular fibrillar material(2)
P. ericae			
strain A	75.3 a	2.7	0.47 a
Strain VIII	77.4 a	2.0	0.40 b
Strain IX	73.4 ab	2.3	0.30 c
Strain XV	52.3 bc	1.7	0.25 d
Strain VI	36.8 c	1.3	0.25 d
Strain XXII (3)	0 d	0	0.08 e

Values in columns showing a common letter are not significantly different (P = 0.05). (1) Frequency of mycorrhizal cells along roots classified as 0, absent ; 1, occasional ; 2, frequent ; 3, abundant. (2) Values obtained from computer aided image analysis of micrographs of hyphal sections. (3) Isolate weakly mycorrhizal with _Erica carnea_ (Duclos 1981) and _Gaultheria procumbens_ (Gianinazzi-Pearson and Bonfante-Fasolo 1986).

weakly acid phosphotungstic acid said to reveal glycoproteins (Huet et al 1974), are localized in the exocellular fibrils of ericoid mycorrhizal fungi (Figures 4 and 5). Although further analyses are necessary, it cannot be excluded that glycoproteins may be present in the fibrillar sheath and may be contributing to recognition mechanisms at the hyphal surface. The outer surface and the exocellular fibrils of hyphae growing close to the root surface are also the site of intense acid phosphatase activity ; such enzyme activity is absent from hyphae developing at some distance from the root (Figure 6). This suggests that physiological changes are occurring at the fungal surface as hyphae approach the surface of roots ; the possibility that these are involved in adhesion processes and are specifically induced by the host tissues is being investigated. In fact, acid phosphatase activity has been proposed to play a

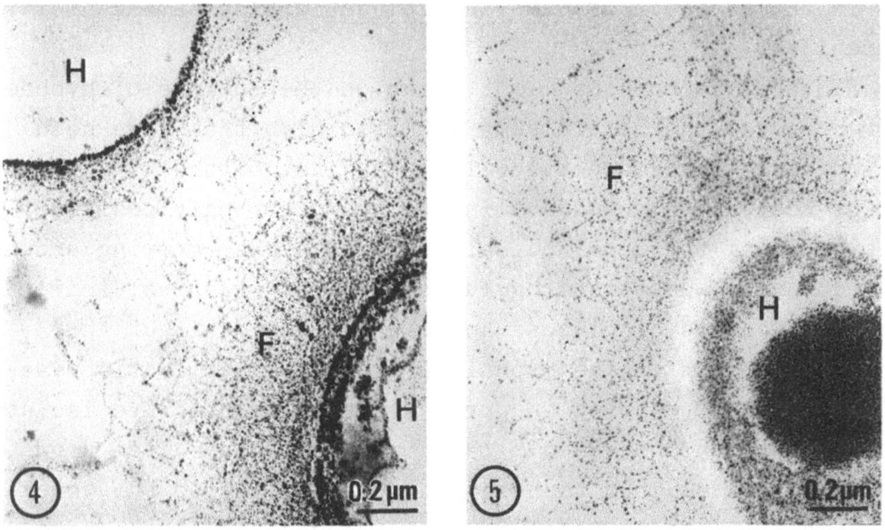

Figures 4 and 5 : Hyphae of P. ericae strain A developing around roots of V. corymbosum and stained for (4) polysaccharides (PATAg reaction) and (5) proteins (Swift reaction). F, exocellular fibrillar sheath ; H, hypha.

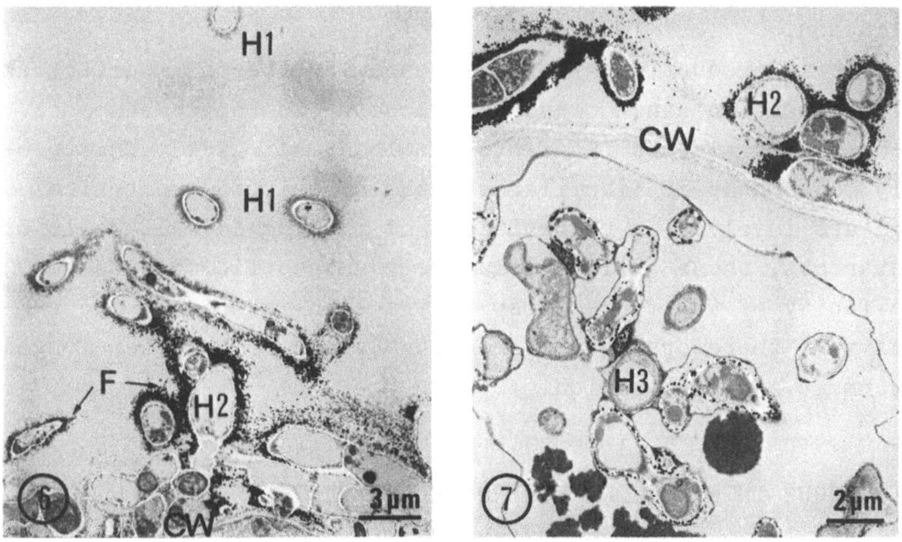

Figures 6 and 7 : Localization of acid (pH 5.0) phosphatase activity of the fungal symbiont (isolate IX) developing in association with Erica cinerea. Enzyme activity is indicated by electron-dense lead phosphate deposits. (6) Phosphatase activity, absent from hyphae at some distance from the root (H1), appears in the exocellular fibrillar sheath (F) of hyphae at the root surface (H2). (7) After cell infection only weak enzyme activity is associated with the wall of some intracellular hyphae (H3). CW, plant cell wall.

role in the adhesion of *Candida albicans* during pathogenicity (Tronchin et al 1980).

Since glycoproteins might occur on the surface of hyphae, sugar residues that could be associated with them have been studied using fluorescent labelled lectins. Of the different lectins tested, only Concavalin A gives a strongly positive reaction along the longitudinal walls of hyphae growing around host roots and wheat germ agglutinin is not reactive, indicating the presence of glucose and mannose residues but the absence of chitin on the hyphal surface. Similar lectin reactivity has been observed in pure cultures of infective isolates (Perotto and Bonfante-Fasolo 1986) and chitin has been shown, using a colloidal gold technique, to be limited to the inner hyphal wall (Bonfante-Fasolo, unpublished data). It appears from these observations that the distribution of surface sugar residues of hyphae does not change with the passage of the fungal endophyte from a saprophytic to a symbiotic phase, even though the hyphal surface shows an important morphological modification.

Fibril production, fungal adhesion to the cell surface and cell penetration can, however, also occur in the presence of non-host root tissues (Bonfante-Fasolo et al 1984). This probably means that if fibrillar sheaths of ericoid mycorrhizal fungi are involved in recognition and binding to cell walls of plant roots, these will be relatively unspecific processes. Specificity of the fungus-host interaction in ericoid mycorrhiza must therefore result from events occurring after fungal attachment and penetration.

HOST-FUNGUS INTERACTIONS DURING CELL INFECTION

In ericoid endomycorrhiza, intracellular development of the fungal symbiont is limited to cells surrounding the central cylinder of the host root. Hyphae penetrate host cell walls by a combination of enzymic processes and physical pressure, and a compatible relationship develops with both symbionts showing intense cellular activity (Bonfante-Fasolo and

Gianinazzi-Pearson 1979, 1981 ; Bonfante-Fasolo et al 1981 ; Duddridge and Read 1982). The host plasmalemma closely extends around the fungus as it enters and develops within the host cell. Material resembling inner host wall accumulates in the interfacial matrix between the host plasmalemma and the wall of the invading hypha. This process is, however, somehow affected by the actively growing hyphae since the quantity of wall material deposited in the host-fungus interface greatly diminishes with development of the intracellular hyphae. In spite of this, the host plasmalemma does not appear to be adversely modified ; it retains a normal three-layered structure, ATPase and neutral phosphatase activities (Gianinazzi-Pearson et al 1984).

At the same time the host cell appears to exert some control over the activity of the fungal symbiont. Fungal development is contained within each infected cell and lateral spread of hyphae from cell to cell does not occur in host tissues, so that each cell encloses an individual infection unit composed of hyphal loops or coils. Once hyphae have entered living host cells, the exocellular fibrillar sheath is no longer produced and the wall-associated phosphatase activity strongly decreases (Figure 7). These changes in fungal behaviour are linked to the continual presence of the host plasmalemma and host protoplast around the developing hyphae. They have not been observed when hyphae occasionally infect moribund host cells or non-host tissues. In the latter, the plant plasmalemma does not enclose the fungal symbiont, there is rapid disorganisation of cell contents and no apparent control over fungal activity. Hyphal growth is not limited to individual cells but lateral spread from cell to cell occurs and exocellular fibrils are always present around the intracellular hyphae (Bonfante-Fasolo et al 1984 ; Gianinazzi-Pearson 1986). This change in symbiont behaviour can be compared to that of certain ineffective Rhizobium mutants during nodulation, where early loss of the peribacteroid membrane changes the symbiotic compartmentation in the nodule to a more parasitic type of interaction (Vance 1983). It would therefore appear that, as previously suggested (Smith 1979), in symbiotic associations sequestration of the endosymbionts by host membrane offers some sort of control.

CONCLUSION

It is now possible to recognize two distinct stages in the formation of ericoid endomycorrhizal associations. The initial step in the infection process prior to cell penetration involves unspecific attachment or adhesion of the fungal symbiont to the root surface, which appears to be determined by the nature of the hyphal surface. This is followed by cell infection and the establishment of a specific cellular compatibility between the symbionts. Host-fungus recognition apparently requires contact between the fungal wall and the host plasmalemma but further research is necessary to determine whether specificity in ericoid endomycorrhiza results from interactions between complementary molecules at the surface of both partners.

ACKNOWLEDGEMENTS

The authors are grateful to M. Giraud and M. Malval for technical assistance and to L. Gribaudo for CAIA programming.

REFERENCES

Behnke O, Zelander T (1970) Preservation of intercellular substances by the cationic dye Alcian blue. In preparative procedures for electron microscopy. J Ultrastruct Res 31 : 424-438

Bonfante-Fasolo P, Gianinazzi-Pearson V (1979) Ultrastructural aspects of endomycorrhiza in the Ericaceae. I. Naturally infected hair roots of Calluna vulgaris L. Hull. New Phytol 83 : 739-744

Bonfante-Fasolo P, Berta G, Gianinazzi-Pearson V (1981) Ultrastructural aspects of endomycorrhiza in the Ericaceae. II. Host-endophyte relationships in Vaccinium myrtillus. New Phytol 89 : 219-244

Bonfante-Fasolo P, Gianinazzi-Pearson V (1982) Ultrastructural aspects of endomycorrhiza in the Ericaceae. III. Morphology of the dissociated symbionts and modifications occurring during their reassociation in axenic culture. New Phytol 91 : 691-704

Bonfante-Fasolo P, Gianinazzi-Pearson V, Martinengo L (1984) Ultrastructural aspects of endomycorrhiza in the Ericaceae. IV. Comparison of infection by Pezizella ericae in host and non-host plants. New Phytol 98 : 329-333

Dazzo FB, Gardiol AE (1984) Host specificity in Rhizobium-legume interactions. In : Verma DPS, Hohn Th (eds) Genes involved in microbe-plant interactions. Springer-Verlag, Vienna New York, p 3

Duclos JL (1981) Contribution à l'étude expérimentale des endomycorhizes des Ericacées. Etude sur Erica carnea L. et quelques variétés horticoles : aspects nutritionnels et ultrastructuraux 3ème cycle doctorat thesis. Lyons University

Duddridge JA, Read DJ (1982) An ultrastructural analysis of the development of mycorrhizas in Rhododendron ponticum. Can J Bot 60 : 2345-2356

Evans RC, Stempen H, Stewart SJ (1981) Development of hyphal sheaths in Bipolaris maydis race r. Can J Bot 59 : 453-459

Gianinazzi-Pearson V (1984) Host-fungus specificity, recognition and compatibility in mycorrhizae. In : Verma DPS, Hohn Th (eds) Genes involved in microbe-plant interactions. Springer-Verlag, Vienna New York, p 225

Gianinazzi-Pearson V (1986) Cellular modifications during host-fungus interactions in endomycorrhizae. In : Bailey J (ed) Biology and molecular biology of plant pathogen interactions. Plenum Publ Corp, New York (in press)

Gianinazzi-Pearson V, Bonfante-Fasolo P (1986) Variability in wall structure and mycorrhizal behaviour of ericoid fungal isolates. In : Gianinazzi-Pearson V, Gianinazzi S (eds) Physiological and genetical aspects of mycorrhizae. INRA-Presse, Paris, p 575

Gianinazzi-Pearson V, Dexheimer J, Gianinazzi S, Jeanmaire C (1984) Plasmalemma structure and function in endomycorrhizal symbioses. Z Pflanzenphysiol 114 : 201-205

Huet M, Benchimol S, Berlinguet J-C, Castonguay C, Cantin M (1974) Cytochimie ultrastructurale des cardiocytes de l'oreillette humaine IV. Digestion des granules spécifiques par les protéases. J Microscopie 21 : 147-158

Matthysse AG (1984) Interaction of Agrobacterium tumefaciens with the plant cell surface. In : Verma DPS, Hohn Th (eds) Genes involved in microbe-plant interactions. Springer-Verlag, Vienna New York, p 34

Murray GM, Maxwell DP (1975) Penetration of Zea mays by Helminthosporium carbonum. Can J Bot 53 : 2872-2883

Perotto S, Bonfante-Fasolo P (1986) Lectin binding on surfaces of ericoid mycorrhizal fungi. In : Gianinazzi-Pearson V, Gianinazzi S (eds) Physiological and genetical aspects of mycorrhizae. INRA-Presse, Paris, p 581

Read DJ (1974) Pezizella ericae sp. nov. the perfect state of a typical mycorrhizal endophyte of Ericaceae. Trans Brit mycol Soc 63 : 381-383

Smith D (1979) From extracellular to intracellular : the establishment of a symbiosis. Proc Royal Soc London B204 : 115-130

Swift JA (1968) The electron histochemistry of cystine containing proteins in thin transverse sections of human hair. J Royal Microscop Soc 88 : 449-460

Thiéry JP (1967) Mise en évidence des polysaccharides sur coupes fines en microscopie électronique. J Microscopie 6 : 978-1018

Tronchi G, Poulain D, Biguet J (1980) Localisation ultrastructurale de l'activité phosphatasique acide chez Candida albicans. Biol Cell 38 : 147-152

Vegh I, Fabre E, Gianinazzi-Pearson V (1979) Présence en France de Pezizella ericae Read, champignon endomycorhizogène des Ericacées horticoles. Phytopath Z 96 : 231-243

Vance CP (1983) Rhizobium infection and nodulation : a beneficial plant disease ? Ann Rev Microbiol 37 : 399-424

STRUCTURAL MODIFICATIONS OF THE FUNGAL WALL BEFORE AND DURING VAM SYMBIOSIS

P. Bonfante-Fasolo, C. Marzachi' and B. Testa
Centro di Studio sulla Micologia del Terreno del CNR
Dipartimento di Biologia Vegetale dell' Universita'
Viale Mattioli 25, 10125 Torino - Italy

INTRODUCTION

VA mycorrhizae provide a good example of two highly integrated eucaryotes which exchange information and energy. The cellular compatibility, typical of the established symbiosis, is the result of recognition events, the molecular basis of which is still unknown. At the ultrastructural level, however, we can identify some important events as short range interactions, that is host/fungus contacts mediated by cell walls or similar material (Bonfante-Fasolo and Gianinazzi Pearson, 1986).

It is now clear that the cell wall is not an inert and passive envelope for cells. In contrast, it represents an active device, modulable and sensitive to the environment (Roland and Vian 1979). Moreover, it has now been accepted that some specific cell wall components, oligosaccharins, play an active and regulatory role in pathogen/plant systems (Albersheim and Darvill 1985). This stresses the importance of knowing much more about the symbiont walls in mycorrhizal systems. The aim of the present paper was to describe the wall structure of the fungus during its life cycle, giving particular attention to the pre-infective phase, which has never been analyzed so far.

OBSERVATIONS ON THE FUNGAL WALLS

The following materials were used. Spores of Glomus versiforme Berch (Daniels and Trappe), G.caledonium (Nicol. & Gerd.) Trappe & Gerd., G.fasciculatum (Thaxter sensu Gerd.) Gerd. & Trappe, mycelia obtained from the germination of G.caledonium spores on Petri dishes containing water-agar and mycorrhizae collected from plants of Vitis vinifera L. inoculated with the fungi listed above. In addition to the conventional ultrastructural techniques, all samples were studied with (i) fluorescent lectins to identify surface sugare residues (Bonfante-Fasolo and Perotto, unpublished) and (ii) cytochemical probes based on the molecular affinity, such as WGA/CG (wheat germ agglutinin/colloidal gold) and chitinase/CG, in order to localize chitin distribution on the thin sections (Vian et al 1983; Roberts et al 1983).

The spores of VAM fungi usually show a complex and double wall. Whereas fibrils in the primary wall are localized in parallel, the secondary wall can show a typical helicoidal structure due to the presence of layers in which regularly ordered fibrils are rotated through a small angle with respect to the previous layers (Bonfante-Fasolo and Vian, 1984) (Fig. 1). By using cytochemical probes such as WGA/CG and chitinase/CG, we can demonstrate that chitin fibrils occur both in primary and secondary walls, irrespective of their three-dimensional architecture (Fig. 2). Labeling does not occur in the electron-dense line separating the two walls (Fig. 2). A fibrillar texture is evident in the hyphae originating from germinating spores, even when the helicoidal texture is not so evident. Among the different FITC-labeled lectins, only WGA, specific for N-acetylglucosamine residues, specifically binds to the germinated mycelium.

As we have previously reported (Bonfante-Fasolo and Grippiolo 1982), after root penetration, the fungal wall gradually thins out and becomes amorphous in structure. At the ultrastructural level, chitin fibrils are no longer visible. However, WGA/FITC strongly binds to the intracellular fungus on fresh sections of the mycorrhizal root and N-acetylglucosamine residues can be localized on the intraradical hyphal walls

using the WGA/CG probe (Fig. 3). Labeling does not occur, on the contrary, using the chitinase/CG probe.

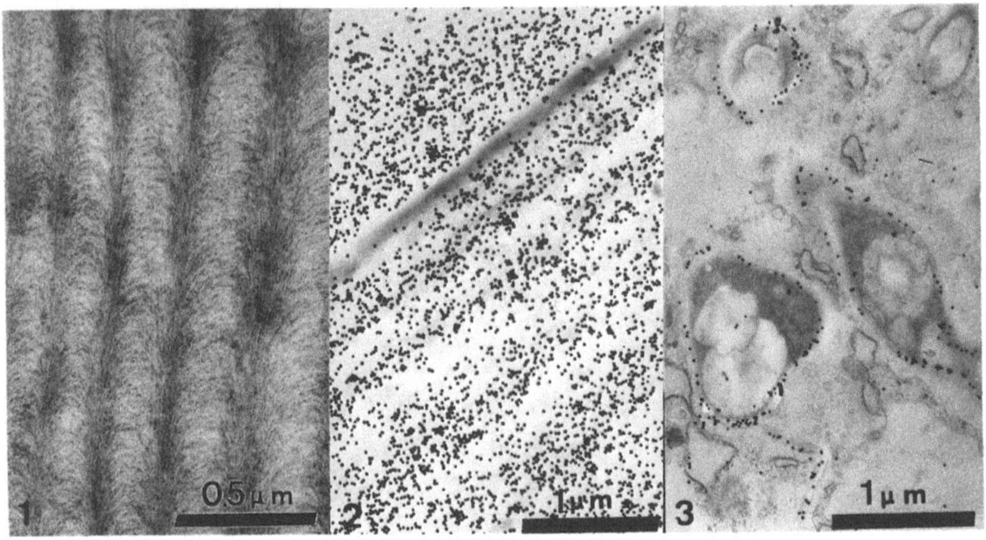

Figures 1, 2 and 3. For explanation, see text.

CONCLUSIONS

The comparison between ultrastructural observations and sugar-specific labeling at the light and electron microscopic level, shows that there are important modifications in the fungal wall during VAM symbiosis. When passing from the preinfective to the colonization phase, the fungal wall undergoes striking changes which extend beyond the already known merely morphological features, and which include additional modifications in chitin structure and localization. In fact during the life cycle of VAM fungi, chitin may be present as highly polymerized and ordered fibrils with a helicoidal pattern, as ordered parallel fibrils and lastly as amorphous short chains. All these macromolecular forms are also mirrored by the different responses to WGA/CG and chitinase/CG treatments. While chitinase/CG only binds to chitin fibrils, WGA/CG requires only two N-acetylglucosamine residues for its specific binding, which

enables differentiation of the macromolecular states of the chitin. The results show that the fungal symbiont wall represents a modulable structure: in the soil where the fungus must survive - perhaps for a long time - the wall is thick and complex. The same is true for the wall of the hypha that contacts the host plant before the infection. The wall completely changes and becomes thin and amorphous at the moment of the establishment of a functional mycorrhiza. We suggest that the complex wall morphogenesis of the fungus might be host-controlled.

ACKNOWLEDGEMENT

Research work was supported by CNR, Italy. Special grant I.P.R.A. - Sub-project 1: Paper N.865.

REFERENCES

Albersheim P, Darvill AG (1985) Oligosaccharins. Scientific American 250: 44-50

Bonfante-Fasolo P, Gianinazzi-Pearson V (1986) Wall and Plasmalemma modifications in mycorrhizal symbiosis. In: Gianinazzi-Pearson V, Gianinazzi S (eds) Physiological and genetical aspects of mycorrhizae. INRA-Presse, Paris, p. 65

Bonfante-Fasolo P, Grippiolo R (1982) Ultrastructural and cytochemical changes in the wall of a vesicular-arbuscular mycorrhizal fungus during symbiosis. Can J Bot 60: 2303-2312

Bonfante-Fasolo P, Vian B (1984) Wall texture in the spore of a vesicular-arbuscular mycorrhizal fungus. Protoplasma 120: 51-60

Roberts RL, Bowers B, Slater ML, Cabib E (1983) Chitin synthesis and localization in cell division cycle mutants of Saccharomyces cerevisiae. Mol Cell Biol 3: 922-930

Roland JC, Vian B (1979) The wall of the growing plant cell: its three dimensional organization. Int Rev Cytol 61: 129-166

Vian B, Brillouet JM, Satiat-JeuneMaitre B (1983) Ultrastructural visualization of xylans in cell walls of hardwood by means of xilanase-gold complex. Biol Cell 49: 179-182

OCCURRENCE OF RACE-SPECIFIC ELICITORS IN THE HOST PATHOGEN INTERACTION TOMATO-*CLADOSPORIUM FULVUM*

Pierre J.G.M. de Wit and Ingrid M.J. Toma
Agricultural University
Department of Phytopathology
Binnenhaven 9,
6709 PD Wageningen
The Netherlands

INTRODUCTION

Cladosporium fulvum Cooke (syn. *Fulvia fulva*), the causal agent of tomato leaf mould, is an economically important disease in glasshouse tomato crops and also in field grown tomatoes, where humid and high temperature conditions prevail. The disease was first described in 1883 in England, but presumably originates from South America (Butler and Jones 1949). *C. fulvum* is a biotroph entering the host through natural openings like stomata. After penetration the fungus colonizes the intercellular space around the leaf mesophyll cells. Under natural conditions the infection cycle takes 2 to 3 weeks. Ten to twelve days after inoculation conidiophores emerge through stomata and produce conidia. Until 1930 no genes for resistance were available in commercially grown tomatoes. Since a number of resistance genes have been introduced in a number of tomato cultivars. Resistance factors were obtained from several *Lycopersicon* species such as *Lycopersicon pimpinellifolium, L. hirsutum, L. peruvianum* var. *glabatrum* and *L. glandulosum*. In Canada Bailey and later Kerr have done much of the breeding for resistance to *C. fulvum* in tomato (Bailey and Langford 1947; Kerr and Bailey 1964). However, soon after introduction of new resistant varieties, resistance was overcome due to appearance of new races of the fungus. At present many races exist and many genes for resistance are known.

The phenomenon of race-specific resistance is frequently observed within many biotrophic and obligate parasites. Flor (1942) who studied the relationship between flax and flax rust introduced the gene-for-gene hypothesis, which now is thought to hold for many other host pathogen interactions including the interaction between *C. fulvum* and tomato (Table I). The gene-for-gene concept implies that for each gene conditioning avirulence in the parasite, there is a corresponding gene conditioning resistance in the host plant.

NATO ASI Series, Vol. H 4
Recognition in Microbe-Plant Symbiotic and Pathogenic Interactions. Edited by B. Lugtenberg

Table I. Differential interactions of races 0, 2, 4, 5, 2.4, 2.4.5 and 2.4.5.9 of *Cladosporium fulvum* with tomato cultivars carrying different genes for resistance to *Cladosporium fulvum*.

Tomato cultivars	Genes for resistance	Races of *Cladosporium fulvum*[1]						
		0	2	4	5	2.4	2.4.5	2.4.5.9
		Genotypes of races[2]						
		$A_2A_4A_5A_9$	$a_2A_4A_5A_9$	$A_2a_4A_5A_9$	$A_2A_4a_5A_9$	$a_2a_4A_5A_9$	$a_2a_4a_5A_9$	$a_2a_4a_5a_9$
Moneymaker	none	C[3]	C	C	C	C	C	C
Near-isogenic line Cf2 of Moneymaker	Cf2	I[4]	C	I	I	C	C	C
Near-isogenic line Cf4 of Moneymaker	Cf4	I	I	C	I	C	C	C
Near-isogenic line Cf5 of Moneymaker	Cf5	I	I	I	C	I	C	C
Near-isogenic line Cf9 of Moneymaker	Cf9	I	I	I	I	I	I	C
Sonato	Cf2Cf4	I	I	I	I	C	C	C
Sonatine	Cf2Cf4Cf9	I	I	I	I	I	I	C

[1] Races are named by their genes for virulence.

[2] Virulence (a) is supposed to be a recessive trait and avirulence (A) is supposed to be a dominant trait.

[3] C = compatible interaction (the cultivar is susceptible).

[4] I = incompatible interaction (the cultivar is resistant).

THE COMPATIBLE AND INCOMPATIBLE INTERACTION

A number of light and electron microscopal studies have shown that growth of *C. fulvum* in susceptible plants under ideal environmental conditions can be considered a compatible interaction comparable to that of other biotrophic parasites (Lazarovits and Higgins 1976a, b; De Wit 1977). Initially in compatible interactions host cells do not respond to the presence of the pathogen. No signs of presence of cell wall degrading enzymes or cell breakdown are observed. The fungus grows in the intercellular space around mesophyll cells and no haustorium-like structures are produced. Most probably the fungus lives on sugars, amino acids, etc. occurring in the intercellular space. Also mannitol, which increases significantly in the intercellular space of compatible interactions could function as a carbon source. Mannitol can only be used by the fungus and not by the host.

The incompatible interaction is characterized by a typical hypersensitive response (HR) visualized by formation of wall apposition (callose containing) and necrosis. Also leakage of electrolytes, accumulation of phytoalexins and pathogenesis-related (PR) proteins can be observed in incompatible interactions (Dow and Callow 1979; De Wit and Flach 1979; De Wit and Kodde 1981a; De Wit and Van der Meer 1986; De Wit et al 1986). The phytoalexins occurring in tomato leaves are not sesquiterpenes such as rishitin occurring in tomato fruits but polyacetylenes (De Wit and Flach 1979; De Wit and Kodde 1981a). Of the PR-proteins occurring in tomato leaves, a protein with a molecular weight of 14.000 D (P14) is most distinct. In incompatible interactions this protein accumulated very soon after fungal penetration and its accumulation is an early determinant of incompatibility comparable with the accumulation of phytoalexins. Most probably a number of charged isomers of P14 accumulate as can be visualized by polyacrylamide gel electrophoresis under low pH non-denaturing conditions and Western blotting with antibodies raized against P14 (De Wit et al 1986). The resistance response is usually restricted to the guard cells and a few epidermal and spongy mesophyll cells at the point of stomatal penetration. Fungal growth is restricted to the few responding cells.

In conclusion there are many advantages to use the interaction between *C. fulvum* and tomato as a model system to study gene-for-gene relationships. (i) Many genes for resistance to various races of *C. fulvum* are available in near-isogenic lines of tomato. (ii) A number of resistance genes (Cf2, Cf4, Cf5 and Cf9) give clear-cut protection suggesting that blockage of fungal development occurs only at one or a few points (compare this with the many genes involved in the *Rhizobium*-legume relationship; see other chapters in this volume). (iii) The fungus is a biotroph growing intercellularly between mesophyll cells in tomato leaves, which makes the interaction an ideal model system to study recognition events between plant and pathogen, as the inter-surface exchange of molecules is confined to the apoplast, a region which is readily explorable.

SPECIFIC INDUCTION OF THE RESISTANCE RESPONSE

Two arrangements of gene-for-gene systems have frequently been studied, the so-called quadratic check (Fig. 1) and the reciprocal check (Fig. 2). Examination of the quadratic check leads to the conclusion that three different genotypical combinations (A_1/r_1, a_1/r_1 and a_1/R_1) all lead to a compatible

Genotypes of parasite isolates	Genotypes of host lines	
	R_1R_1	r_1r_1
A_1	I	C
a_1	C	C

Fig. 1. The 'quadratic check'. Possible combinations between two host lines and two haploid parasite isolates differing in resistance and parasitic capability, respectively. R_1 and A_1 are alleles for resistance and avirulence, respectively, and r_1 and a_1 are the opposite alleles for susceptibility and virulence. C = compatible or susceptible; I = incompatible or resistant.

Genotypes of parasite isolates	Genotypes of host lines	
	$R_1R_1r_2r_2$	$r_1r_1R_2R_2$
A_1a_2	I	C
a_1A_2	C	I

Fig. 2. The 'reciprocal check'. Possible combinations between two host lines and two haploid parasite isolates differing in resistance and parasitic capability, respectively. A_1 and A_2 are different alleles for specific avirulence on R_1 and R_2, respectively; a_1 and a_2 are the opposite alleles for virulence. C and I as in Fig. 1.

genotype while only one unique genotypical combination (A_1/R_1) leads to incompatibility. This can be interpreted to mean that specificity is controlled by the interaction between the gene products of the dominant A and R alleles. In all other cases it is envisaged that there is no recognition or interaction between gene products and as a consequence compatibility results. The reciprocal check is an extension of the quadratic check. This arrangement can be interpreted to mean that A_1 interacts specifically with R_1 and not with R_2 and that A_2 interacts specifically with R_2 and not with R_1. There is a strict one-gene-for-one-gene relationship. In addition, this arrangement implies that incompatibility is epistatic to compatibility.

In the *C. fulvum*-tomato system there has been an extensive search for products of avirulence genes, the so-called race-specific elicitors, which would specifically induce incompatibility. When different races of *C. fulvum* were grown on synthetic medium elicitors could be obtained from culture filtrates as well as from mycelial cell walls. These elicitors appeared to be peptidogalactoglucomannans, which were potent inducers of necrosis and callose (Lazarovits and Higgins 1979; Lazarovits et al 1979), electrolyte leakage (Dow and Callow 1979) and the accumulation of phytoalexins (De Wit and Roseboom 1980; De Wit and Kodde 1981b). However, their activity appeared neither race- nor cultivar-specific. Irrespective of the genes for resistance present in all cultivars, the above-mentioned responses were induced in similar quantities. The elicitor activity of the peptidogalactoglucomannan could be destroyed by α-mannosidase or alkali. The conclusion from this part of the research was that *in vitro* no race-specific elicitors could be found. This means that these elicitors are not produced or the isolation conditions of cell wall elicitors in particular have been so harsh that specificity could have been lost during the isolation. Probably growth in aqueous shake nutrient solutions induces the fungus to secrete the nonspecific extracellular glycoproteins of which the composition is strongly dependent on the composition of the medium (De Wit and Kodde 1981b). Growth of a biotroph on synthetic medium is probably not the best way to search for race-specific elicitors, which only have a function *in vivo* in the host plant. The inability to find race-specific elicitors *in vitro* prompted us to go back to the interaction *in vivo*. This decision was also inspired by the idea to be able to discriminate between two catching on hypotheses: the 'specific elicitor-receptor model' and the 'nonspecific elicitor-specific suppressor model' (Ellingboe 1982). The last model would imply that all races of the pathogen would in addition to nonspecific elicitors produce specific suppressors, which would work in concert to give the race-specific outcome. These suppressors, it would be envisaged, would indeed be produced in compatible interactions *in vivo* where there is no visible resistance response. The *C. fulvum*-tomato interaction would be an excellent model system to discriminate between presence of race-specific suppressors or race-specific elicitors. As the fungus grows intercellularly between leaf mesophyll cells intercellular fluids obtained from compatible interaction by *in vacuo* infiltration and centrifugation would contain the above mentioned putative products. Studies with intercellular fluids from many race-cultivar interactions revealed that there was only support for the presence of race-specific elicitors but not for presence of race-specific suppressors (De Wit and Spikman 1982; Higgins and De Wit 1985). The race-specific elicitors induce chlorosis and necrosis in resistant but not in susceptible cultivars. The specificity of the elicitors was not determined by the genes for resistance present in the susceptible cultivar from which the intercellular fluids were obtained, but by the race of the invading fungus, i.e. by the avirulence genes present in the race (De Wit et al 1984). Why race-specific elicitors are only produced by the fungus growing in the intercellular space of susceptible tomato leaves, but not on synthetic medium is still an open question. The host, however, is not acting simply as a substrate, as most likely there is a host-mediated effect on the release, production or stability of specific elicitors by the fungus.

PURIFICATION AND IDENTIFICATION OF A RACE-SPECIFIC ELICITOR

Once the occurrence of race-specific elicitors in intercellular fluids of compatible interactions of *C. fulvum* had been established, purification and identification of one of the race-specific elicitors could be started. For this purpose near-isogenic line of cv. Moneymaker carrying the Cf9-gene for resistance or cv. Sonatine, a commercially grown cultivar carrying the same gene, were used as assay-plants to search for the elicitor produced by avirulence gene A9. Plants carrying the Cf9-gene for resistance are very useful for this purpose as they respond very intensively to elicitors produced by races 0, 2, 4, 5, 2.4 and 2.4.5 but not to elicitors produced by race 2.4.5.9. Elicitor activity was significantly destroyed after treatment with the proteinase Pronase (Calbiochem), indicating that the elicitor was proteinaceous in origin. Gel chromatography (Biogel, BIO-RAD) and ion exchange chromatography (DEAE-Sephadex; CM-Sephadex) revealed that the elicitor was a rather small and basic protein or glycoprotein. Polyacrylamide gel electrophoresis under low pH non-denaturing conditions revealed that necrosis-inducing activity always coincided with a peptide of which the molecular weight was around 4000 D (De Wit et al 1985). This peptide appears to be present in intercellular fluids of all race-cultivar interactions except the interaction with race 2.4.5.9, which is the only race that can grow on cultivars carrying resistance gene Cf9. As all races which induce necrosis on cultivars carrying the Cf9-gene for resistance contain avirulence gene A9, while race 2.4.5.9 does not contain this avirulence gene, it is tempting to assume that the purified elicitor is the product of avirulence gene A9. At present the elicitor can be purified in sufficient amounts by preparative gel electrophoresis and consecutive reversed phase HPLC to determine its amino acid composition and sequence.

FUTURE STRATEGIES TO ISOLATE AVIRULENCE GENE A9 OF *CLADOSPORIUM FULVUM* AND RESISTANCE GENE Cf9 OF TOMATO

Once the complete amino acid sequence of the elicitor is known it will be possible to isolate the gene producing the elicitor. With help of the amino acid sequence a synthetic DNA-probe of at least 18 bases long can be synthesized which can be used to screen genomic DNA, a gene library or a cDNA library of *C. fulvum*. Hybridizing DNA fragments can be selected and cloned. The cloned gene can be used initially to transform a bacterium pathogenic to tomato. If the gene is expressed in such a bacterium its effects on tomato cultivars carrying Cf9-genes could be studied. Also races of *C. fulvum* which do not carry the A9-gene could be transformed into A9-races.

The elicitor can also be very useful to isolate the Cf9 resistance gene of tomato. The elicitor would bind to a receptor (the putative product of the Cf9 gene) and the elicitor-receptor complex can be isolated if the elicitor carries a label. After removal of the elicitor from the receptor, the receptor can be purified and its amino acid sequence determined. Here again a synthetic DNA probe can be produced for screening a genomic, gene or cDNA library. The Cf9 gene could be cloned and used to transform tomato plants not carrying this gene.

The strategies described above may sound too optimistic but time will tell us whether this optimism was justified.

REFERENCES

Bailey D L, Langford A N (1947) The development through breeding of greenhouse tomato varieties resistant to leaf mould. Ontario Hort Expt Sta Rept 1945-46: 56-60

Butler E J, Jones S G (1949) Tomato leaf mould, *Cladosporium fulvum* Cooke. In: Butler E J, Jones S G (eds) Plant Pathology, MacMillen & Co. LTD. London, p 671

De Wit P J G M (1977) A light and scanning-electron microscopic study of infection of tomato plants by virulent and avirulent races of *Cladosporium fulvum*. Neth J Plant Pathol 83: 109-122

De Wit P J G M, Flach W (1979) Differential accumulation of phytoalexins in tomato leaves, but not in fruits after inoculation with virulent and avirulent races of *Cladosporium fulvum*. Physiol Plant Pathol 15: 257-267

De Wit P J G M, Kodde E (1981a) Induction of polyacetylenic phytoalexins in *Lycopersicum esculentum* after inoculation with *Cladosporium fulvum*. Physiol Plant Pathol 18: 143-148

De Wit P J G M, Kodde E (1981b) Futher characterization and cultivar-specificity of glycoprotein elicitors from culture filtrates and cell walls of *Cladosporium fulvum* (syn. *Fulvia fulva)*. Physiol Plant Pathol 18: 297-314

De Wit P J G M, Roseboom P H M (1980) Isolation, partial characterization and specificity of glycoprotein elicitors from culture filtrates, mycelium and cell walls of *Cladosporium fulvum* (syn. *Fulvia fulva)*. Physiol Plant Pathol 16: 391-408

De Wit P J G M, Spikman G (1982) Evidence for the occurrence of race and cultivar-specific elicitors of necrosis in intercellular fluids of compatible interactions of *Cladosporium fulvum* and tomato. Physiol Plant Pathol 21: 1-11

De Wit P J G M, Van der Meer F E (1986) Accumulation of the pathogenesis-related tomato leaf protein P14 as an early indicator of incompatibility in the interaction between *Cladosporium fulvum* (syn. *Fulvia Fulva)* and tomato. Physiol Mol Plant Pathol 28: 203-214

De Wit P J G M, Hofman J E, Aarts J M M J G (1984) Origin of specific elicitors of chlorosis and necrosis occurring in intercellular fluids of compatible interactions of *Cladosporium fulvum* (syn. *Fulvia fulva)* and tomato. Physiol Plant Pathol 24: 17-23

De Wit P J G M, Hofman J E, Velthuis G C M, Kuć J A (1985) Isolation and characterization of an elicitor of necrosis isolated from intercellular fluids of compatible interactions of *Cladosporium fulvum* (syn. *Fulvia fulva)* and tomato. Plant Physiol 77: 642-647

De Wit P J G M, Buurlage M B, Hammond K E (1986) The occurrence of host, pathogen and interaction-specific proteins in the apoplast of *Cladosporium fulvum* (syn. *Fulvia fulva)*-infected tomato leaves. Physiol Mol Plant Pathol 28: 000-000

Dow J M, Callow J A (1979) Leakage of electrolytes from isolated leaf mesophyll cells of tomato induced by glycopeptides from culture filtrates of *Fulvia fulva* (Cooke) *Ciferri* (syn. *Cladosporium fulvum*) Physiol Plant Pathol 15: 27-34

Ellingboe A H (1982) Genetic aspects of active defense. In: Wood R K S (ed) Active defense mechanisms in plants, vol. 37 Plenum Press, New York, London, p 179

Flor H H (1942) Inheritance of pathogenicity in *Melampsora lini*. Phytopathology 32: 653-669

Higgins V J, De Wit P J G M (1985) Use of race and cultivar specific elicitors from intercellular fluids for characterizing races of *Cladosporium fulvum* and resistant tomato cultivars. Phytopathology 75: 645-690

Kerr E A, Bailey D L (1964) Resistance to *Cladosporium fulvum* Cooke obtained from wild species of tomato. Can J Bot 42: 1541-1554

Lazarovits G, Higgins V J (1976a) Histological comparison of *Cladosporium fulvum* race 1 in immune resistant and susceptible tomato varieties. Can J Bot 54: 224-234

Lazarovits G, Higgins V J (1976b) Ultrastructure of susceptible, resistant and immune reactions of tomato to races of *Cladosporium fulvum*. Can J Bot 54: 235-249

Lazarovits G, Higgins V J (1979) Biological activity and specificity of a toxin produced by *Cladosporium fulvum*. Phytopathology 69: 1056-1061

Lazarovits G, Bhullar B S, Sugiyama H J, Higgins V J (1979) Purification and partial characterization of a glycoprotein toxin produced by *Cladosporium fulvum*. Phytopathology 69: 1062-1068

STUDIES ON THE ROLE OF CARBOHYDRATES IN HOST-MICROBE INTERACTIONS

Peter Albersheim, Alan G. Darvill, Janice K. Sharp, Keith R. Davis and Steven H. Doares
Complex Carbohydrate Research Center and Dept. of Biochemistry
University of Georgia and USDA
Russell Research Center, P.O. Box 5677
Athens, Georgia U.S.A.

Our laboratory team has long pursued two seemingly unrelated lines of research--one, elucidating the structure and function of the matrix polysaccharides of the walls of growing plant cells, and the other, studying the manner by which plants defend themselves against disease. Results from each project led us to discover that structurally defined fragments of cell-wall polysaccharides are chemical messages with highly specific regulatory properties, many of which are involved in host-microbe interactions. These regulatory molecules (oligosaccharins) are released from the cell wall by means of enzymes. This paper will present evidence to show that the cell walls are the repository of numerous, different oligosaccharins that can regulate triggering of plant defenses against pathogens and other types of stress.

THE PURIFICATION AND STRUCTURAL CHARACTERIZATION OF A β-GLUCAN OLIGOSACCHARIDE ELICITOR OF PHYTOALEXINS

Plants synthesize phytoalexins as a general defense mechanism against potential pathogens (Bailey and Mansfield 1982; Brooks and Watson 1985; Darvill and Albersheim 1984; Smith and Banks 1986). Over a decade ago our research team learned that oligoglucoside fragments of a major structural polysaccharide of the walls of a variety of fungi elicit plant cells to synthesize *de novo* the enzymes responsible for catalyzing the synthesis of phytoalexins (antibiotics) (Albersheim and Valent 1978; Ayers et al 1976b; Ayers et al 1976c; Darvill and Albersheim 1984; Ebel et al 1976). The β-glucan elicitors of phytoalexin accumulation were first extracted from isolated mycelial walls of fungi by autoclaving the walls (Ayers et al 1976a). The crude solubilized elicitor preparation thus obtained has been used by a wide variety of investigators on numerous plant systems (Chappell and Hahlbrock 1984; Cramer et al 1985; Dixon et al 1983a; Dixon et al 1983b;

NATO ASI Series, Vol. H 4
Recognition in Microbe-Plant Symbiotic and
Pathogenic Interactions. Edited by B. Lugtenberg

Ebel et al 1976; Hauffe et al 1986; Kuhn et al 1984; Robertsen 1986; Roby et al 1985; Schmelzer et al 1985; Showalter et al 1985).

A mixture of oligoglucosides of various sizes was obtained by partial acid hydrolysis of the cell-wall polysaccharides of the fungus *Phytophthora megasperma* var. *glycinea*. The smallest elicitor-active glucan fragment was shown, by gel-permeation chromatography, to be about the size of a heptaglucoside. It was later discovered that about three hundred heptaglucosides were present in the mixture of variously sized oligoglucosides (Sharp et al 1984b). The heptaglucoside elicitor was so important that this laboratory spent nearly 10 years attempting to purify it to homogeneity.

The goal of this study was to isolate one or more homogeneous, elicitor-active heptaglucosides for structural characterization. The discovery that only a small portion of the mixture of heptasaccharide fragments was elicitor-active meant that the amount of homogeneous elicitor that could be isolated for structural characterization would be extremely limited. The available methods for separation and structural analysis of complex carbohydrates were inadequate for such small quantities of sample. Therefore, the h.p.l.c. separation of underivatized oligosaccharides was improved (Sharp et al 1984b), and a microscale glycosyl-sequence analysis technique was developed (Waeghe et al 1983).

The improved methods proved worthwhile for in 1984 we were able to report the purification to homogeneity of one elicitor-active and seven elicitor-inactive hexa(β-D-glucopyranosyl)-D-glucitols isolated from the mycelial walls of *Phytophthora megasperma* (Sharp et al 1984c). The elicitor-active and six of the elicitor-inactive hexa(β-D-glucopyranosyl)-D-glucitols were constructed of terminal, 3-, 6-, and 3,6-linked glucopyranosyl residues (Fig. 1).

Structural analysis of derivatized oligosaccharide fragments obtained from small amounts (30 to 50 μg) of the hexa(β-D-glucopyranosyl)-D-glucitols also required improved understanding of the fragmentation of such molecules in electron-impact mass spectrometry. A new fragment ion was identified that was diagnostic of 3-linked D-glucopyranosyl residues in per-*O*-alkylated oligosaccharide-alditols (Sharp and Albersheim 1984). Identifying the origin of the diagnostic fragment ion and developing improved analytical methods allowed characterization of the complete primary structures of the one elicitor-active and seven elicitor-inactive hexa(β-D-glucopyranosyl)-D-glucitols (Sharp et al 1984c).

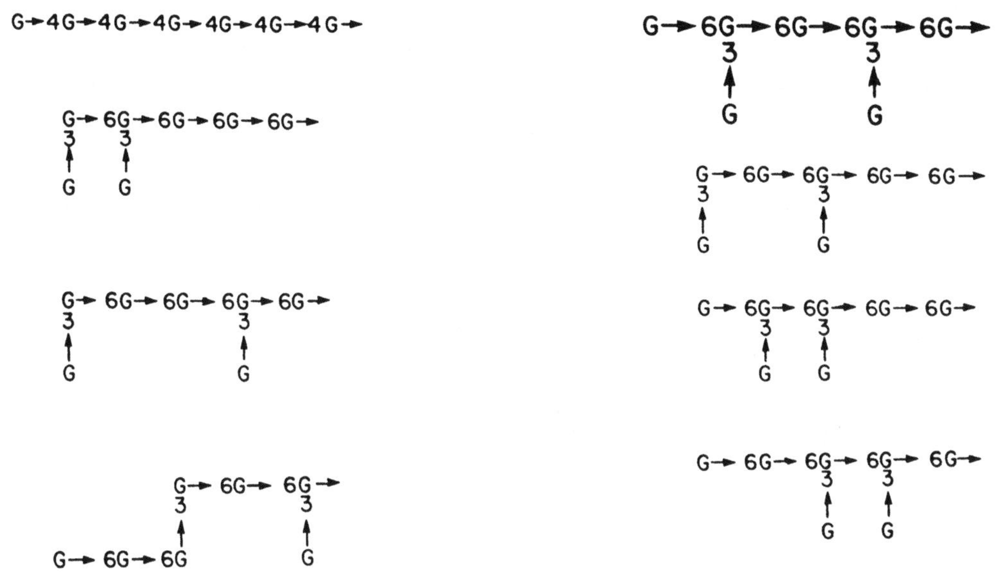

Figure 1. Structures of eight hepta-β-glucoside-alditols. The structure presented in bold print in the top right-hand corner was active as an elicitor of phytoalexin accumulation. The seven other hepta-β-glucoside-alditols were inactive as elicitors at 25 times the concentration of the active elicitor and did not interfere with the active hepta-β-glucoside-alditol. All of the glucosyl residues have been shown to be in the β-anomeric configuration, to have the D absolute configuration, and to be in the pyranoid ring-form. The glucose residue at the reducing end of each hepta-β-glucoside was converted to glucitol to aid in purification by preventing equilibration of this residue between its α and β forms, although the unreduced hepta-β-glucoside was equally active as an elicitor.

The determined structure of the active heptaglucoside was used by Ossowski *et al.* (1984) to synthesize the molecule by organic chemical means. The structures and the elicitor activities of the synthetic and the mycelial-wall-derived hexa(β-D-glucopyranosyl)-D-glucitols were compared (Sharp et al 1984a). The specific elicitor activities, retention times on reversed-phase liquid chromatography columns, glycosyl-linkage compositions, ^{1}H-n.m.r. spectra, and glycosyl-sequence analyses of the synthetic and mycelial-wall-derived molecules were indistinguishable. Studies using the synthesized molecule provided conclusive proof that the elicitor activity was associated with the structure proposed for the elicitor-active mycelial-wall-derived hexa(β-D-glucopyranosyl)-D-glucitol (Sharp et al

1984a). Furthermore, chemical synthesis of an elicitor-inactive heptaglucoside with a structure extremely similar to that of the elicitor-active heptaglucoside confirmed the structural specificity of the active elicitor.

The purification and characterization of the hexa(β-D-glucopyranosyl)-D-glucitol elicitor was important to the efforts to determine whether complex carbohydrates are biologically active regulatory molecules. The active heptaglucoside elicitor was the only elicitor present in the mixture of about three hundred heptaglucosides released by the partial hydrolysis of the *Phytophthora* mycelial walls (Sharp et al 1984b). This observation, in conjunction with the isolation of two elicitor-inactive glycosyl-linkage isomers and four other isomers from the same polymer, established that the structural requirements of the elicitor-active hexa(β-D-glucopyranosyl)-D-glucitol were extremely precise.

The hexa(β-D-glucopyranosyl)-D-glucitol elicitor of phytoalexins was the first fully characterized oligosaccharin. Realizing that the elicitor-active molecule was a highly defined carbohydrate structure prompted speculation that other oligosaccharins might be components of the cell-wall polysaccharides of plant cells and the extracellular matrix polysaccharides of animal cells. Gaining the knowledge that only nanomoles of a carbohydrate elicitor could elicit production of specific mRNAs (Chappell and Hahlbrock 1984; Cramer et al 1985; Dixon et al 1983a; Dixon et al 1983b; Kuhn et al 1984; Schmelzer et al 1985; Showalter et al 1985) and translation of these messengers into enzymes involved in plant defense provided fundamental support for our hypothesis that complex carbohydrates participate in the chemical message systems of organisms.

PURIFICATION AND STRUCTURAL CHARACTERIZATION OF THE ENDOGENOUS ELICITOR OF PHYTOALEXINS

Injury of plant cells by bacteria often elicits the accumulation of phytoalexins. We hypothesized and proved that injuring plant cells causes the release of an oligosaccharide fragment of a plant cell-wall polysaccharide that, like the heptaglucoside from fungal cell walls, elicits the synthesis of phytoalexins (Hahn et al 1981). Initial experiments with this "endogenous elicitor," the first recognized biologically active carbohydrate from plant cell walls, were performed with oligosaccharides released from plant cell walls by partial acid hydrolysis. The elicitor was first shown to be a fragment of a pectic polysaccharide by the predominance of galactosyluronic acid in the elicitor-active fractions, and by the ability of an homogeneous endo-α-1,4-galacturonase to destroy the elicitor activity of the fragments (Hahn et al 1981). We then established that this endogenous elicitor was a dodeca-α-1,4-D-galacturonide (Nothnagel et al 1983). Lee and West (1981) demonstrated that an endo-α-1,4-polygalacturonase secreted by *Rhizopus stolinifer*, a fungal pathogen of castor beans, was itself an elicitor of phytoalexins. Bruce and West (1982) then showed that this endopolygalacturonase functioned by releasing oligogalacturonides from the walls of plant cells, confirming the observation of Hahn et al (1981) that oligogalacturonide-rich wall fragments elicited phytoalexin accumulation.

During the same time period, we demonstrated that bacteria also secrete enzymes that release oligogalacturonide fragments from plant cell walls and that the oligogalacturonides initiate phytoalexin accumulation in plants (Davis et al 1986a; Davis et al 1986b). Again, linear α-1,4-linked oligogalacturonides were found to be the endogenous eliciting molecules. Our laboratory also established that bacteria and fungi can cause the release of the endogenous phytoalexin elicitors from plant cell walls by a second mechanism--the microbes can injure plant cells, releasing a plant enzyme that solubilizes elicitor-active oligosaccharide fragments from plant cell walls (Lyon and Albersheim 1982).

Thus, there are apparently several mechanisms by which phytoalexin accumulation, one of the defense responses of plants, can be activated. Plants can recognize oligosaccharide fragments of fungal cell walls that are released by enzymes present constitutively in the cell walls of plants. Additionally, fungi and bacteria can secrete enzymes capable of releasing oligosaccharide elicitors from the walls of plant cells. And finally, injury of plants by microbes can cause the release of a plant enzyme that solubilizes fragments of plant cell walls that elicit phytoalexin accumulation. The latter mechanism may explain how abiotic elicitors such as heavy metals, UV light, organic solvents, and partial freezing can result in accumulation of phytoalexins in plant tissues. Abiotic elicitors are likely to work by activating plant enzymes that release oligogalacturonide elicitors from plant cell walls.

The studies on elicitors prompted an investigation of the biochemical basis for the ability of phytoalexins to inhibit bacterial growth (Weinstein and Albersheim 1983). The evidence obtained indicated that the antibacterial activity of several phytoalexins was due to a non-specific interaction with the bacterial membrane, causing the membrane to be leaky. These observations agree with earlier studies on the mechanism of action of phytoalexins and support the hypothesis that phytoalexins exert their antibiotic effect by a general disruption of membrane function (Bailey and Deverall 1971; Lyon 1980; Oku et al 1976; Skipp and Bailey 1976; VanEtten and Bateman 1971). Our results provided biochemical evidence for the proposal that phytoalexins are multiple-site toxicants that affect a variety of membrane-associated processes (Keen 1981; VanEtten and Bateman 1971), and contradicted the hypothesis of a single site of action for phytoalexins (Kaplan et al 1980; Koeppe and Paxton 1981). A mechanism of action in which phytoalexins alter the structural integrity of the membrane would account for the various physiological functions that many of these antibiotics have been observed to affect (Keen 1981; Koeppe and Paxton 1981; Lyon 1980; Lyon 1978; Oku et al 1976; Slayman and VanEtten 1974; VanEtten and Bateman 1971). An ability of phytoalexins to work through a non-specific process (*i.e.*, disruption of membrane integrity), would make it difficult for microbes to develop (evolve) insensitivity to phytoalexin binding or attack. Furthermore, an ability to affect cells by altering the structural integrity of their membranes would account for the widely observed, broad-spectrum antibiotic activity of phytoalexins.

SYNERGISM BETWEEN THE HEPTA-β-GLUCOSIDE AND OLIGOGALACTURONIDE ELICITORS

In 1985, we observed that the hexa-β-D-glucosyl-glucitol elicitor from fungal cell walls can act synergistically with: i) an oligogalacturonide elicitor from plant cell walls; ii) with enzymes that release the oligogalacturonide elicitor: and iii) with abiotic treatments that are believed to release the oligogalacturonide elicitor (Davis et al 1986c). When the hexa-β-glucosyl-glucitol and oligogalacturonide elicitors are applied simultaneously to plant tissues, their combined elicitor activity is as much as 35-fold higher than the sum of the responses of these elicitors assayed separately. The results suggest that oligogalacturonides act as signals of tissue damage, and as such, can enhance the response of plant tissues to other elicitor-active molecules during initiation of phytoalexin accumulation. This hypothesis is supported by reports that oligogalacturonide and glucan elicitors can activate other defense responses of plants, such as accumulation of lignin (Hammerschmidt et al 1984; Robertsen 1986), hydroxyproline-rich glycoproteins (Clarke et al 1981; Hammerschmidt et al 1984; Roby et al 1985; Showalter et al 1985), and protease inhibitors (Bishop et al 1984). In other words, these molecules appear to be general activators of plant defense responses.

AN OLIGOSACCHARIN THAT MAY ELICIT THE HYPERSENSITIVE RESISTANCE RESPONSE IN PLANTS

Hypersensitive resistance to microbial infection is characterized by rapid death of plant cells at the point of attempted infection. An oligosaccharide fragment prepared from the walls of either suspension-cultured maize or sycamore cells by partial acid hydrolysis, when applied to either of the two types of cells in liquid culture, causes cell death. Cell death was demonstrated with vital dyes (Yamazaki et al 1983), by failure of the cells to form colonies on agar medium (Fry et al 1983), and by the inhibition of incorporation of [^{14}C]leucine into the proteins of these cells (Yamazaki et al 1983). Since the heptaglucoside and oligogalacturonide elicitors of phytoalexins do not inhibit protein synthesis in the maize and sycamore cells (unpublished results), the elicitor of the hypersensitive resistance response (as shown by inhibition of incorporation of [^{14}C]leucine) is almost certainly a different oligosaccharin than the elicitors of phytoalexin accumulation. We hypothesize that i) the putative hypersensitive-inducing oligosaccharin is released by enzymes from the walls of plant cells during microbial infection of plants, and ii) the initiation of the hypersensitive resistance response by the released oligosaccharin is the key step in eliciting those responses of plants capable of defending them against potential pathogens.

We have evidence that the plant pathogen *Pyricularia oryzae* secretes an enzyme capable of releasing the putative hypersensitive-inducing oligosaccharin that kills plant cells. The enzyme, which by itself kills plant cells, has been purified. This enzyme appears to be a pectin lyase, as the killing activity co-purifies through several columns with the pectin lyase. The purified enzyme appears to kill plant cells by releasing, from the walls of plant cells, an oligosaccharin that initiates cell death. This observation agrees with the long demonstrated ability of pectic-degrading enzymes to kill plant cells (Yamazaki et al 1983). Elucidating the mechanisms leading to hypersensitive resistance could lead to an understanding of the mechanisms underlying the race/cultivar specificity exhibited by the most destructive plant pathogens.

A DETERMINATION OF WHETHER THE POLYSACCHARIDES SECRETED BY RHIZOBIUM ARE INVOLVED IN THE SPECIFICITY OF HOST-SYMBIONT SELECTION

Various laboratories have suggested that species-specific polysaccharides are factors involved in determining the specificity of host-symbiont interaction (Bauer 1981; Graham 1980; Verma and Long 1983; Vincent 1977). We have structurally characterized the acidic polysaccharides and neutral cyclic β-glucans secreted by various *Rhizobium* species in order to determine whether these molecules are species-specific (Åman et al 1982; Dell et al 1983; Dudman et al 1983a; Dudman et al 1983b; Franzén et al 1983; McNeil et al 1986). These studies demanded improved techniques for determining the structures of complex carbohydrates, and led to the structural characterization of the largest repeating units of any repeating-unit polysaccharides yet identified in nature (respectively, 8, 9, 10, and 11 glycosyl residues per unit).

Our most definitive study of repeating-unit polysaccharides (McNeil et al 1986) involved isolating the octasaccharide repeating units from the acidic polysaccharides secreted by *Rhizobium trifolii* strain NA30, *R. trifolii* strain LPR5, *R. leguminosarum* strain LPR1, and *R. phaseoli* strain LPR49. (*R. trifolii* is the symbiont of clover, *R. leguminosarum*, of peas, and *R. phaseoli*, of beans). The repeating units were obtained by treating the polysaccharides with an endolyase obtained from a bacteriophage capable of lysing the *Rhizobium* under investigation. The glycosyl sequence and the structures and locations of the non-glycosyl substituents were shown to be identical for repeating units derived from all of these polysaccharides, except for that derived from the polysaccharide produced by *R. trifolii* NA30. Therefore, the discernible structural features of the acidic polysaccharides secreted by *Rhizobium* species cannot be the determinant of host specificity. In support of this conclusion is the observation that *R. trifolii* LPR5045, generated by curing *R. trifolii* LPR5 of its Sym plasmid (the Sym plasmid is required for symbiosis and host specificity, therefore, *R. trifolii* LPR5045 was not capable of nodulating clover), secreted a polysaccharide having the same structure (including identities and locations of non-glycosyl substituents) as that of the polysaccharide secreted by its plasmid-containing parent. Thus, the structural genes that encode for synthesis of the acidic polysaccharide secreted by *R. trifolii* LPR5045 are not located on the Sym plasmid, and neither are the genes that encode for synthesis and attachment of non-glycosyl substituents of the polysaccharide.

The possibility remains that a quantitatively minor component of the acidic polysaccharides could be a host-specific determinant. For example, if a single repeating unit in 50 had an altered sequence of glycosyl residues, and if this specific sequence were to be recognized by host plants, the analytical procedures used in our studies would not detect the critical sequence. A bioassay that is at least semi-quantitative (for example, induced mitosis in cortical cells [Calvert et al 1984; Turgeon and Bauer 1982]) would have to be used to detect and purify rare, bioactive repeating units of the acidic polysaccharides secreted by *Rhizobium* species if such units exist.

OLIGOSACCHARIDES AND PLANT-MICROBE INTERACTIONS: FUTURE DIRECTIONS

The role of plant cell-wall polysaccharides in host-microbe interactions is now known to be more than an inert barrier function. Complex carbohydrates, and specifically, oligosaccharide fragments of plant and fungal cell walls, have been demonstrated to display biological regulatory activities and are critical active components in plant defense responses against microbes. The regulatory activities are specific and reproducible; oligosaccharides from fungi and plants have been shown to individually and synergistically elicit phytoalexins (antibiotics) in several plant species, as well as to elicit other defense responses and a plant hypersensitive-like response.

Future research directions lean toward understanding how oligosaccharins control gene expression in plants. Do oligosaccharins work through a receptor? Where is the receptor located? Are secondary messages, such as phosphoinositols and protein kinases, involved? Eventually, we will have to answer how oligosaccharins can result in the activation of genes, that is, in the production of transcripts from genes that were not being transcribed before the release of the oligosaccharins. We also must learn how two oligosaccharides can synergistically interact to induce phytoalexin synthesis. Understanding the mechanisms by which two regulatory molecules can control gene expression could help to explain the general phenomenon of regulation in plants.

The earliest physiologically observed defense response and the one most widely seen in nature is the hypersensitive-resistance response, a localized region of plant cell death around the site of attempted invasion by a pathogen. Through this self-sacrifice of host cells, the plant is capable of stopping the spread of many pathogens. The hypersensitive response has been studied for many years as part of the genetics of race specificity (Azad and Kado 1984; Klement et al 1984). This response is evoked only by those races of a pathogen against which a plant can defend itself. Thus, whether or not the hypersensitive response is elicited is likely to be a key to understanding race/cultivar specificity which, in turn, is a characteristic of the most devastating pathogens of plants. The hypersensitive response is not only important in host/pathogen interactions, it is also widely seen as a resistance response to microbes that are not pathogenic on that plant (Darvill and Albersheim 1984; Yamazaki et al 1983).

The mechanism by which hypersensitive resistance is induced is not known. We hypothesize that the hypersensitive-resistance response is the result of the release of oligosaccharides from the walls of the host cells by enzymes secreted by attacking pathogens. The question of whether the wall fragments that induce cell death are involved in the *in vivo* elicitation of hypersensitive resistance can only be answered by studying a host/pathogen combination at the biochemical, genetic, and molecular genetic levels. Success in these studies could lead to understanding of race/cultivar-specific pathogenesis as well as the more general hypersensitive-resistance response.

REFERENCES

Albersheim P, Valent BS (1978) Host-pathogen interactions in plants. Plants, when exposed to oligosaccharides of fungal origin, defend themselves by accumulating antibiotics. J Cell Biol 78:627-643

Åman P, Franzén L-E, Darvill JE, McNeil M, Darvill AG, Albersheim P (1982) Host-symbiont interactions X. The structure of the acidic polysaccharide secreted by *Rhizobium phaseoli* strain 127 K38. Carbohydr Res 103:77-100

Ayers A, Ebel J, Valent B, Albersheim P (1976a) Host-pathogen interactions IX. Quantitative assays of elicitor activity and characterization of the elicitor present in the extracellular medium of cultures of *Phytophthora megasperma* var. *sojae*. Plant Physiol 57:751-759

Ayers A, Ebel J, Valent BS, Albersheim P (1976b) Host-pathogen interactions X. The fractionation and biological activity of an elicitor isolated from the mycelial walls of *Phytophthora megasperma* var. *sojae*. Plant Physiol 57:760-765

Ayers A, Valent BS, Ebel J, Albersheim P (1976c) Host-pathogen interactions XI. Composition and structure of wall-released elicitor fractions. Plant Physiol 57:766-774

Azad HR, Kado CI (1984) Relation of tobacco hypersensitivity to pathogenicity of *Erwinia rubrifaciens*. Phytopathology 74:61-64

Bailey JA, Deverall BJ (1971) Formation and activity of phaseollin in the interaction between bean hypocotyls (*Phaseolus vulgaris*) and physiological races of *Colletotrichum lindemuthianum*. Physiol Plant Pathol 1:435-449

Bailey JA, Mansfield JW (eds) (1982) Phytoalexins. Halstead Press, John Wiley, New York

Bauer WD (1981) Infection of legumes by *Rhizobia*. Annu Rev Plant Physiol 32:407-409

Bishop PD, Pearce G, Bryant JE, Ryan CA (1984) Isolation and characterization of the proteinase inhibitor-inducing factor from tomato leaves. Identity and activity of poly- and oligogalacturonide fragments. J Biol Chem 259:13172-13177

Brooks CJW, Watson DG (1985) Phytoalexins. Natural Product Reports 2:427-459

Bruce RJ, West CA (1982) Elicitation of casbene synthetase activity in castor bean. The role of pectic fragments of the plant cell wall in elicitation by a fungal endopolygalacturonase. Plant Physiol 69:1181-1188

Calvert HE, Pence MK, Pierce M, Malik NSA, Bauer WD (1984) Anatomical analysis of the development and distribution of *Rhizobium* infections in soybean roots. Can J Bot 62:2375-2384

Chappell J, Hahlbrock K (1984) Transcription of plant defence genes in response to UV light or fungal elicitor. Nature 311:76-78

Clarke JA, Lisker N, Lamport DTA, Ellingboe AH (1981) Hydroxyproline enhancement as a primary event in the successful development of *Erysiphe graminis* in wheat. Plant Physiol 67:188-189

Cramer CL, Ryder TB, Bell JN, Lamb CJ (1985) Rapid switching of plant gene expression induced by fungal elicitor. Science 227:1240-1243

Darvill AG, Albersheim, P (1984) Phytoalexins and their elicitors - a defense against microbial infection in plants. Ann Rev Plant Physiol 35:243-275

Davis KD, Darvill AG, Albersheim P, Dell A (1986a) Host-pathogen interactions XXIX. Oligogalacturonides released from sodium polypectate by endopolygalacturonic acid lyase are elicitors of phytoalexins in soybean. Plant Physiol 80:568-577

Davis KD, Darvill AG, Albersheim P, Dell A (1986b) Host-pathogen interactions XXX. Characterization of elicitors of phytoalexin accumulation in soybean released from soybean cell walls by endopolygalacturonic acid lyase. Z Naturforsch 41:39-48

Davis KD, Darvill AG, Albersheim P (1986c) Host-pathogen interactions XXXI. Several biotic and abiotic elicitors act synergistically in the induction of phytoalexin accumulation in soybean. Plant Mol Biol 6:23-32.

Dell A, York WS, McNeil M, Darvill AG, Albersheim P (1983) Host-symbiont interactions XIV. The cyclic structure of β-D-(1→2)-linked D-glucans secreted by *Rhizobia* and *Agrobacteria*. Carbohydr Res 117:185-200

Dixon RA, Dey PM, Lamb CJ (1983a) Phytoalexins: enzymology and molecular biology. In: Meister A (ed) Advances in Enzymology. John Wiley and Sons, New York, pp. 1-136.

Dixon RA, Dey PM, Lawton MA, Lamb CJ (1983b) Phytoalexin induction in French bean. Intercellular transmission of elicitation in cell suspension cultures and hypocotyl sections of *Phaseolus vulgaris*. Plant Physiol 71:251-256

Dudman WF, Franzén L-E, Darvill JE, McNeil M, Darvill AG, Albersheim P (1983a) Host-symbiont interactions XI. The structure of the acidic polysaccharide secreted by *Rhizobium phaseoli* strain 127 K36. Carbohydr Res 117:141-156

Dudman WF, Franzén L-E, McNeil M, Darvill AG, Albersheim P (1983b) Host-symbiont interactions XIII. The structure of the acidic polysaccharide secreted by *Rhizobium phaseoli* strain 127 K87. Carbohydr Res 117:169-183

Ebel J, Ayers A, Albersheim P (1976) Host-pathogen interactions XII. The response of suspension-cultured soybean cells to the elicitor isolated from *Phytophthora megasperma* var. *sojae*, a fungal pathogen of soybeans. Plant Physiol 57:775-779

Franzén L-E, Dudman WF, McNeil M, Darvill AG, Albersheim P (1983) Host-symbiont interactions XII. The structure of the acidic polysaccharide secreted by *Rhizobium phaseoli* strain 127 K44. Carbohydr Res 117:157-167

Fry SC, Darvill AG, Albersheim P (1983) Amino acid transport and protein synthesis, possible primary targets of biologically active cell wall fragments. In: Interactions Between Nitrogen and Growth Regulators in the Control of Plant Development. Brit Plant Growth Reg Group, Monograph 9, 33-43

Graham TL (1980) Recognition in *Rhizobium*-legume symbioses. In: Atherly A, Giles K (eds) Biology of the Rhizobiaceae. Academic Press, New York, pp 127-248

Hahn MG, Darvill AG, Albersheim P (1981) Host-pathogen interactions XIX. The endogenous elicitor, a fragment of a plant cell wall polysaccharide that elicits phytoalexin accumulation in soybeans. Plant Physiol 68:1161-1169

Hammerschmidt R, Lamport DTA, Muldoon EP (1984) Cell wall hydroxyproline enhancement and lignin deposition as an early event in the resistance of cucumber to *Cladosporium cucumerinum*. Physiol Plant Pathol 24:43-47

Hauffe KD, Hahlbrock K, Scheel D (1986) Elicitor-stimulated furanocoumarin biosynthesis in cultured parsley cells: S-adenosyl-L-methionine:bergaptol and S-adenosyl-L-methionine:xanthotoxol O-methyltransferases. Z Naturforsch 41c:228-239

Kaplan DT, Keen NT, Thomason IJ (1980) Studies on the mode of action of glyceollin in soybean. Incompatibility of the root knot nematode *Meloidogyne incognita*. Physiol Plant Pathol 16:319-325

Keen NT (1981) Evaluation of the role of phytoalexins. In: Staples RC (ed) Plant Disease Control. J Wiley and Sons, New York, pp. 155-177

Klement Z, Farkas GL, Lovrekovich L (1984) Hypersensitive reaction induced by phytopathogenic bacteria in the tobacco leaf. Phytopathology 54:474-477

Koeppe DE, Paxton JD (1981) Glyceollin: phytotoxicity at complex I of soybean mitochondrial electron transport. Plant Physiol 67:S-108

Kuhn DN, Chappell J, Boudet A, Hahlbrock K (1984) Induction of phenylalanine ammonia-lyase and 4-coumarate:CoA ligase mRNAs in cultured plant cells by UV light or fungal elicitor. Proc Natl Acad Sci USA 81:1102-1106

Lee S-C, West CA (1981) Polygalacturonase from *Rhizopus stolonifer*, an elicitor of casbene synthetase activity in castor bean (*Ricinus communis* L.) seedlings. Plant Physiol 67:633-639

Lyon GD (1980) Evidence that the toxic effect of rishitin may be due to membrane damage. J Exp Bot 31:957-966

Lyon GD (1978) Attenuation by divalent cations of the effect of the phytoalexin rishitin on *Erwinia carotovora* var. *atroseptica*. J Gen Microbiol 109:5-10

Lyon GD, Albersheim P (1982) Host-pathogen interactions XXI. Extraction of a heat-labile elicitor of phytoalexin accumulation from frozen soybean stems. Plant Physiol 70:406-409

McNeil M, Darvill J, Darvill AG, Albersheim P, van Veen R, Hooykaas P, Schilperoort R, Dell A (1986) The discernible structural features of the acidic polysaccharides secreted by different *Rhizobium* species are the same. Carbohydr Res 146:307-326

Nothnagel EA, McNeil M, Albersheim P, Dell A (1983) Host-pathogen interactions XXII. A galacturonic acid oligosaccharide from plant cell walls elicits phytoalexins. Plant Physiol 71:916-926

Oku H, Ouchi S, Shiraishi T, Utsumi K, Seno S (1976) Toxicity of a phytoalexin, pisatin, to mammalian cells. Proc Jpn Acad 52:33-36

Ossowski PA, Pilotti Å, Garegg PJ, Lindberg B (1984) Synthesis of a glucoheptose and a glucooctaose that elicit phytoalexin accumulation in soybean. J Biol Chem 259:11337-11340

Robertsen B (1986) Elicitors of the production of lignin-like compounds in cucumber hypocotyls. Physiol Mol Plant Path 28:137-148

Roby D, Toppan A, Esquerré-Tugayé M-T (1985) Cell surfaces in plant-microorganism interactions V. Elicitors of fungal and of plant origin trigger the synthesis of ethylene and of cell wall hydroxyproline-rich glycoprotein in plants. Plant Physiol 77:700-704

Schmelzer E, Somssich I, Hahlbrock K (1985) Coordinated changes in transcription and translation rates of phenylalanine ammonia-lyase and 4-coumarate:CoA ligase mRNAs in elicitor-treated *Petroselinum crispum* cells. Plant Cell Reports 4:293-296

Sharp JK, Albersheim P (1984) An electron-impact mass-spectrometric fragment-ion that identifies 3-linked glucopyranosyl residues in per-*O*-alkylated, linear β-D-glucopyrano-oligosaccharide-alditols. Carbohydr Res 128:193-202

Sharp JK, Albersheim P, Ossowski P, Pilotti P, Garegg P, Lindberg B (1984a) Comparison of the structures and elicitor activities of a synthetic and a mycelial-wall-derived hexa(β-D-glucopyranosyl)-D-glucitol. J Biol Chem 259:11341-11345

Sharp JK, Valent B, Albersheim P (1984b) Purification and partial characterization of a β-glucan fragment that elicits phytoalexin accumulation in soybean. J Biol Chem 259:11312-11320

Sharp JK, McNeil M, Albersheim P (1984c) The primary structures of one elicitor-active and seven elicitor-inactive hexa(β-D-glucopyranosyl)-D-glucitols isolated from the mycelial walls of *Phytophthora megasperma* f.sp. *glycinea*. J Biol Chem 259: 11321-11336

Showalter AM, Bell JN, Cramer CL, Bailey JA, Varner JE, Lamb CJ (1985) Accumulation of hydroxyproline-rich glycoprotein mRNAs in response to fungal elicitor and infection. Proc Natl Acad Sci USA 82:6551-6555

Skipp R, Bailey J (1976) The effect of phaseollin on the growth of *Colletotrichum lindemuthianum* in bioassays designed to measure fungi toxicity. Physiol Plant Pathol 9:253-263

Slayman CL, VanEtten HD (1974) Are certain pterocarpanoid phytoalexins and steroid hormones inhibitors of membrane ATPase. Plant Physiol 54:S-24

Smith DA, (1982) Toxicity of phytoalexins. In: Bailey JA, Mansfield JW (eds) Phytoalexins. J Wiley and Sons, New York, pp. 218-252

Smith DA, Banks SW (1986) Biosynthesis, elicitation and biological activity of isoflavonoid phytoalexins. Phytochemistry 25:979-995

Turgeon BG, Bauer WD (1982) Early events in the infection of soybean by *Rhizobium japonicum*. Time course and cytology of the initial infection process. Can J Bot 60:152-161

VanEtten H, Bateman D (1971) Studies on the mode of action of the phytoalexin phaseollin. Phytopathology 61:1363-1372

Verma DPS, Long S (1983) The molecular biology of *Rhizobium*-legume symbiosis. Int Rev Cytol, Suppl 14:211-245

Vincent JM (1977) *Rhizobium*: general microbiology. In: Hardy RWF, Silver WS (eds) A Treatise on Dinitrogen Fixation, Section III: Biology. Wiley, New York, pp 277-366

Waeghe TJ, Darvill AG, McNeil M, Albersheim P (1983) Determination, by methylation analysis, of the glycosyl-linkage compositions of microgram quantities of complex carbohydrates. Carbohydr Res 123:281-304

Weinstein LI, Albersheim P (1983) Host-pathogen interactions XXIII. The mechanism of the antibacterial action of glucinol, a pterocarpan phytoalexin synthesized by soybeans. Plant Physiol 72:557-563

Yamazaki N, Fry SC, Darvill AG, Albersheim P (1983) Host-pathogen interactions XXIV. Fragments isolated from suspension-cultured sycamore cell walls inhibit the ability of the cells to incorporate [^{14}C]leucine into proteins. Plant Physiol 72:864-869

BIOCHEMICAL INTERACTIONS OF PLANTS WITH POTENTIALLY PATHOGENIC FUNGI

K. Hahlbrock, B. Cuypers, C. Douglas, K. H. Fritzemeier,
H. Hoffmann, F. Rohwer, D. Scheel, W. Schulz
Max-Planck-Institut für Züchtungsforschung, Abteilung
Biochemie, D-5000 Köln 30, FRG

INTRODUCTION

As in two previous written versions of progress reports on current research on this topic (Hahlbrock et al 1985, Kombrink et al 1986), the present knowledge of the complex biochemical interactions of plants with potentially pathogenic microorganisms (Hahlbrock and Scheel 1986) is summarized as an abbreviated introduction in an up-dated scheme (Fig. 1). It is important to note that this scheme represents the average of many typical plant-microorganism interactions, but not necessarily each individual case. Moreover, not all parts of the multi-component defense response have been studied in a sufficiently large number of plants to draw conclusions as to their significance in disease resistance. Two existing case studies involving fungal genetics have revealed that cutin (Kolattukudy 1985, Dickman and Patil 1986) or phytoalexins (Kistler and VanEtten 1984) can be decisive components of the plant's defense response.

Besides phytoalexins, the most extensively investigated component is the highly branched phenylpropanoid metabolism, which comprises a large number of distinct pathways leading to various low-molecular-weight as well as polymeric or wall-bound compounds (Hahlbrock and Scheel 1986).

Our own work has concentrated primarily on two parts of the plant's defense response, phytoalexin synthesis and various involvements of phenylpropanoid pathways. In the non-host resistance response of parsley (*Petroselinum crispum*) leaves to the soybean

NATO ASI Series, Vol. H 4
Recognition in Microbe-Plant Symbiotic and Pathogenic Interactions. Edited by B. Lugtenberg

pathogen, *Phytophthora megasperma* f. sp. *glycinea* (Pmg), phenylpropanoid-derived furanocoumarin phytoalexins are rapidly accumulated (see Chapter by Scheel et al), indicating an overlap of this and other defense-related phenylpropanoid pathways. By contrast, the typical sesquiterpenoid phytoalexins of potato (*Solanum tuberosum*) tubers are of different biosynthetic origin and allow the function of phenylpropanoid derivatives in disease resistance to be investigated independently of the occurrence and synthesis of this class of phytoalexins.

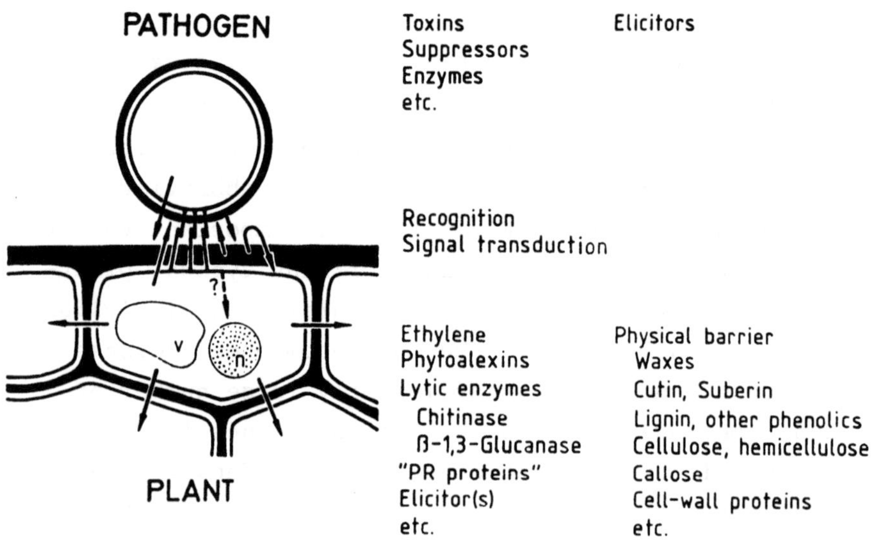

Fig. 1. Up-dated version of a scheme (Hahlbrock et al 1985, Kombrink et al 1986) illustrating the most frequently occurring biochemical mechanisms and structural components (Hahlbrock and Scheel 1986) involved in plant-pathogen interactions.

This report will focus on two current projects on phenylpropanoid metabolism in the two systems, followed by a brief discussion of a working hypothesis for further research in this area. In the parsley system, our studies of gene activation in elicitor-treated cell suspension cultures have advanced far enough to estimate the number of genes encoding enzymes of phenylpropanoid metabolism and study their involvement in defense-related pathways. In potato, cultivars are available with operationally defined genes conferring resistance to certain pathotypes of *Phytophthora infestans* (Pi). Thus the importance and relative timing of activation of phenylpropanoid pathways in compatible and incompatible interactions of the two organisms can be investigated. The working hypothesis is primarily concerned with the design of experiments aimed at the elucidation of gene activation mechanisms in plants, including the possible function of 'resistance genes'.

GENES ENCODING ENZYMES OF PHENYLPROPANOID METABOLISM IN PARSLEY

Treatment of cultured parsley cells with a crude elicitor preparation from hyphal cell walls of Pmg has previously been shown to result in cell browning and the accumulation of furanocoumarin phytoalexins in the culture medium. Extent and timing of these reactions are dependent on culture age as well as elicitor concentration and mimic in many respects the rapid formation of necrotic spots and the accumulation of furanocoumarins at Pmg-infection sites in parsley leaves (see chapter by Scheel et al for further details). Both the browning reaction, whose biochemistry is poorly understood, and the biosynthesis of furanocoumarins are likely to require the activation of the enzymes of general phenylpropanoid metabolism together with those of the respective specific pathways. Phenylalanine ammonia-lyase (PAL) and 4-coumarate:CoA ligase (4CL), the first and last enzymes of general phenylpropanoid metabolism (Fig. 2), catalyze the branch-point reactions from primary metabolism (PAL) and to various specific pathways (4CL). Therefore, they are particularly interesting from a regulatory point of view. The involvement of gene activation in the coordinated induction of these two enzymes in elicitor-treated cells has been demonstrated (Chappell and Hahlbrock 1984). Both enzymes occur in the form of isoenzymes, e. g. in bean (PAL) (Bolwell et al 1985) and soybean (4CL) (Knobloch and Hahlbrock 1975), and the existence of more than one gene encoding each of these enzymes in parsley seemed possible.

We have used a combination of cDNA and genomic DNA clones to determine the copy number of PAL and 4CL genes in cultured parsley cells. Analysis of restriction fragmentation patterns

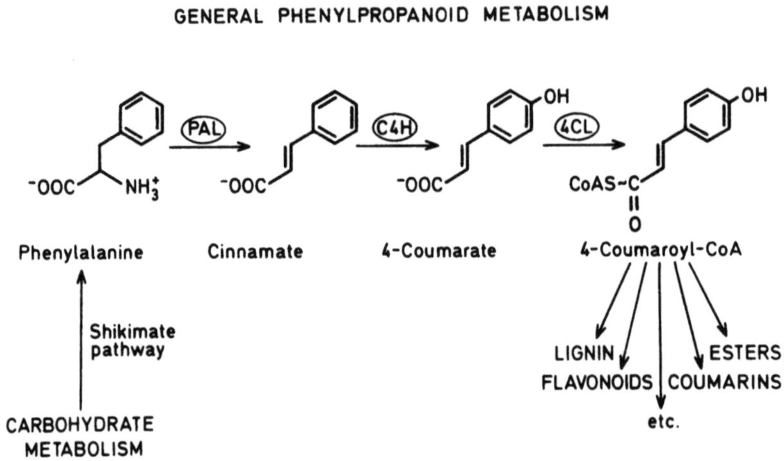

Fig. 2. Reactions catalyzed by the enzymes of general phenylpropanoid metabolism (PAL = phenylalanine ammonia-lyase, C4H = cinnamate 4-hydroxylase, 4CL = 4-coumarate:CoA ligase).

of cloned genomic DNA, hybridization studies ('genomic Southerns') with various subcloned genomic or cDNA fragments, as well as partial sequence analysis of some cloned materials have revealed the presence of two, possibly three, different PAL genes and two different 4CL genes in this system.

Under all physiological and environmental conditions tested thus far, PAL and 4CL were coordinately regulated in parsley and many other plants, and a variety of different types of endogenous and exogenous stimuli have been found to affect the regulation of their activities. The small numbers of genes, relative to the large number of possible effectors, indicated that the same genes can be activated by different stimuli. Recent results have demonstrated that this is indeed the case. We have shown by sequence comparison of the two 4CL genes with several cDNA clones complementary to 4CL mRNA from UV-irradiated and elicitor-treated parsley cells (Hahlbrock et al 1985), respectively, that both genes were expressed in either case (C. Douglas, H. Hoffmann, W. Schulz and K. Hahlbrock, manuscript in preparation).

It will now be interesting to find out whether one and the same gene recognizes many different stimuli or responds to only one or a few 'second messengers' which might be generated intracellularly as a result of a converging signal transduction mechanism. Possible experimental approaches to this question will be discussed below.

PHENYLPROPANOID AND SESQUITERPENOID DERIVATIVES IN NON-HOST AND RACE/CULTIVAR-SPECIFIC INTERACTIONS OF POTATO WITH PHYTOPHTHORA SPECIES

Rapid browning in association with the hypersensitive cell death in localized areas of leaf or tuber tissue has been observed in non-host interactions of potato with Pmg and incompatible interactions of R-gene carrying potato cultivars with the corresponding pathotypes of Pi. Although, as mentioned above, little is known about the biochemistry of the browning reaction, the involvement of phenylpropanoid compounds is very likely (Fritzemeier et al 1986, and references cited therein). This is in line with the observation in all cases studied that the hypersensitive response was associated with transient increases in the activities of PAL and other enzymes of phenylpropanoid metabolism.

The typical sesquiterpenoid phytoalexins, which have always been found in hypersensitively reacting potato tubers, were absent in similarly reacting potato leaves. In tubers, the accumulation of these compounds occurred most rapidly in the non-host response to Pmg, more slowly in the incompatible interaction with Pi, and relatively late in the compatible interaction with Pi (Rohwer et al 1986). Thus it is possible that the speed with which sesquiterpenoid phytoalexins accumulate is an indication of the efficiency of the defense response of potato tubers. Since these phytoalexins were not found in leaves, the apparent more immediate connection of the induction of phenylpropanoid pathways with the hypersensitive resistance response was chosen as a particularly useful basis for further studies.

The browning of individual cells and the positive staining of accumulated phenolic material with Toluidine Blue could be seen as early as three hours post-inoculation, independent of whether virulent or avirulent races of Pi were used for infection of potato leaves. Subsequent differences in symptom development could be visualized by a combination of light microscopy and immunohistochemistry, using antibodies recognizing components of the fungal hyphae in situ. All mesophyll cells in close contact with hyphae attempting to penetrate the leaf tissue appeared brownish-green and necrotic in incompatible interactions, but unaffected in compatible interactions. Approximately 24 h post-inoculation, the fungus started to sporulate on the opposite side of the leaf in the compatible interaction, whereas the incompatible interaction had confined all fungal hyphae within the leaf tissue and completely prevented sporulation (B. Cuypers and K. Hahlbrock, manuscript in preparation). Thus, despite the great similarity of the very early effects in compatible and incompatible interactions, the subsequent cytological appearance (Fig. 3) and the final result (Fig. 4) differed greatly. It will be interesting to see whether this difference is reflected at the level of gene expression rates in individual cell types, particularly with respect to phenylpropanoid metabolism.

In this direction, initial experiments were aimed at the determination of the induction kinetics of PAL and 4CL mRNA and enzyme activities in incompatible potato-Pi interactions, and at the isolation of cDNA and genomic DNA clones for various Pi-inducible mRNAs, including PAL and 4CL, as a prerequisite for studies of gene expression rates. We have shown that the stimulation of PAL and 4CL activities, which had frequently been observed in Pi-infected tubers, also occurred in Pi-infected potato leaves, and that this effect could be closely mimicked by the infiltration of a concentrated Pi culture filtrate solution (CF elicitor) into the leaf veins. The large similarity of the effects of Pi infection (incompatible interaction) and CF elicitor treatment was demonstrated by comparison of two-dimensional gels showing similar changes in a large number of mRNA translational activities. Among these changes were very rapid, coordinated and transient increases

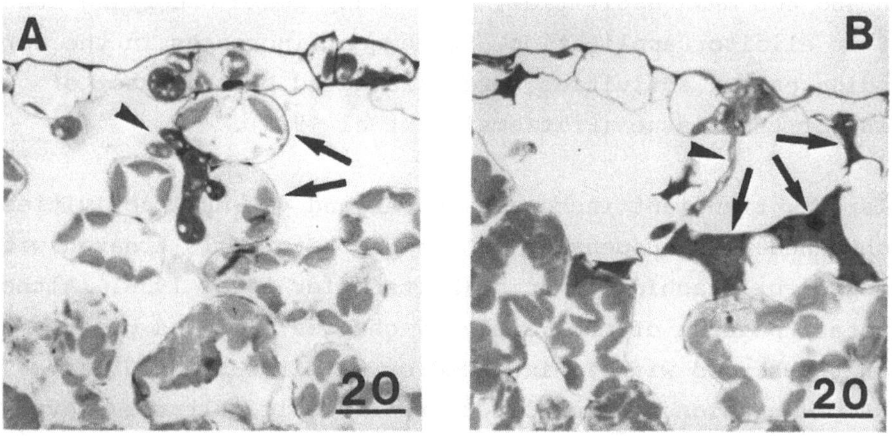

Fig. 3. Cross sections (0.5 µm) of resin-embedded potato leaves 14 h post-inoculation with Pi, stained with 0.05% Toluidine Blue O in sodium phosphate buffer, pH 6.8. Depicted are typical examples of a compatible (A) and an incompatible interaction (B) of cv. Datura (resistance gene R1) with pathotypes Pi1 and Pi4, respectively. Fungal hyphae (arrow heads) either appear intact and form haustoria on visibly undamaged mesophyll cells (A, arrows) or appear degenerated and are surrounded by heavyly stained, necrotic cells (B, arrows). Mag. bars in µm.

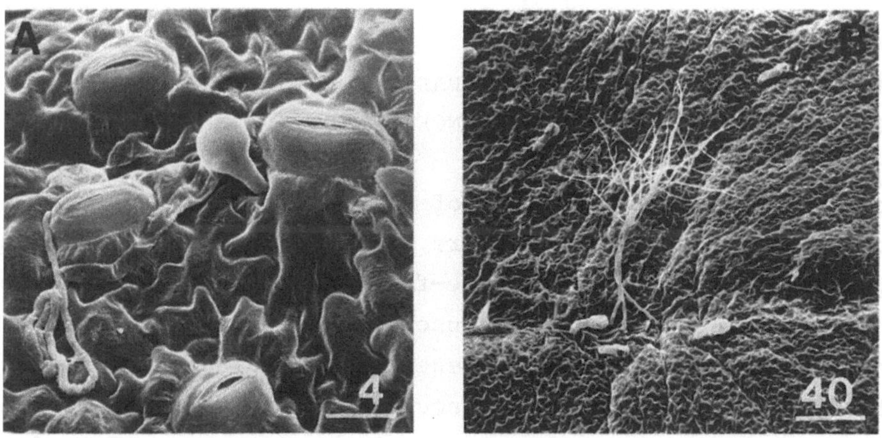

Fig. 4. Scanning electron micrographs of fungal hyphae growing on the lower epidermis of a potato leaf (A) and sporulating in compatible interactions approximately 4 days post-inoculation on the upper epidermis (B). The initial appearance of hyphal growth and tissue penetration was not significantly different between compatible and incompatible interactions, whereas sporulation was not observed in non-host or host incompatible interactions under the experimental conditions used (B. Cuypers and K. Hahlbrock, manuscript in preparation). Mag. bars in µm.

in PAL and 4CL mRNA activities, with sharp peaks occurring about 2 h after elicitor application. Subsequent increases in the corresponding enzyme activities were paralleled by the onset of browning of the tissue (Fritzemeier et al 1986).

Even larger, transient increases in PAL and 4CL mRNA activities were obtained with suspension-cultured potato cells treated with CF elicitor or arachidonic acid (Fritzemeier et al 1986). Although the overall pattern of mRNA activity changes differed greatly from that observed with Pi-infected or elicitor-treated leaves, an mRNA preparation from elicitor-treated cell cultures proved to be the best available source of PAL and 4CL mRNAs. Poly$(A)^+$ mRNA preparations from this source as well as from Pi-infected leaves (both incompatible and compatible interactions) have recently been used to generate cDNA libraries for PAL and 4CL mRNAs and several other, as yet biochemically unidentified mRNAs, whose induction during infection was taken as an indication of their relation to pathogen defense (C. Cretin, H. Fritzemeier, F. Rohwer, J. Taylor and K. Hahlbrock, unpublished results).

HYPOTHETICAL CONSIDERATIONS AS A BASIS FOR THE FUNCTIONAL IDENTIFICATION OF DISEASE RESISTANCE GENES

To date, the molecular functions of all resistance genes that have been operationally defined are unknown. However, there are a large number of cases where gene-for-gene relationships between resistance genes of potentially susceptible plants and corresponding virulence or avirulence genes of the pathogens have been established. The simplest way of accounting for this phenomenon in molecular terms is to assume that the resistance gene product and the virulence or avirulence gene product interact directly with each other (Ellingboe 1982). Another prerequisite for the following considerations is the well-founded exclusion of genes encoding enzymes of phytoalexin biosynthesis or other putative defense-related pathways as possible candidates for resistance genes (Ellingboe 1982).

It should be noted that the proposed hypothesis is meant to serve as a basis for the design of experiments towards the functional identification and characterization of disease resistance genes, and therefore is a working hypothesis with little ambition to make correct predictions in every detail. For instance, it is clear from much circumstantial evidence that a genetically identifiable gene-for-gene relationship represents but one of an unknown number of molecular interactions between plants and pathogens (partly covered by the poorly defined terms 'basic compatibility' and 'basic incompatibility'). It is further clear that great variations in the basic mechanism of interaction of the gene products (if one exists) might occur in different plant species or even in different cultivars of one species.

Finally, the present lack of evidence for the existence of race/cultivar-specific fungal elicitors - with one notable exception (De Wit et al 1985) - indicates the possibility that, at least in some cases, rather unspecific elicitors are complemented by other signal molecules mediating the observed specificity.

Several related versions of the proposed working hypothesis are depicted in Fig. 5. In principle, the resistance gene is postulated to encode an extracellularly, or less likely intracellularly, located effector-binding protein (e. g.: Re, receptor; e, endogenous signal molecule; rp, regulatory protein acting in the nucleus; etc) whose counterpart is either a virulence gene-encoded suppressor (S) or an avirulence gene-encoded elicitor (E_3). Versions II and V are the preferred and simplest explanations of the two formally distinct cases of gene-for-gene relationships of a given resistance gene-carrying plant cultivar with the corresponding pathotype carrying either a virulence or an avirulence gene. Two of various possible alternative mechanisms involving intracellular interactions are illustrated as versions III and IV. They are shown as modifications of version II, but essentially analogous alternatives might be considered for version V.

Details of the proposed intracellular signal transduction mecha-

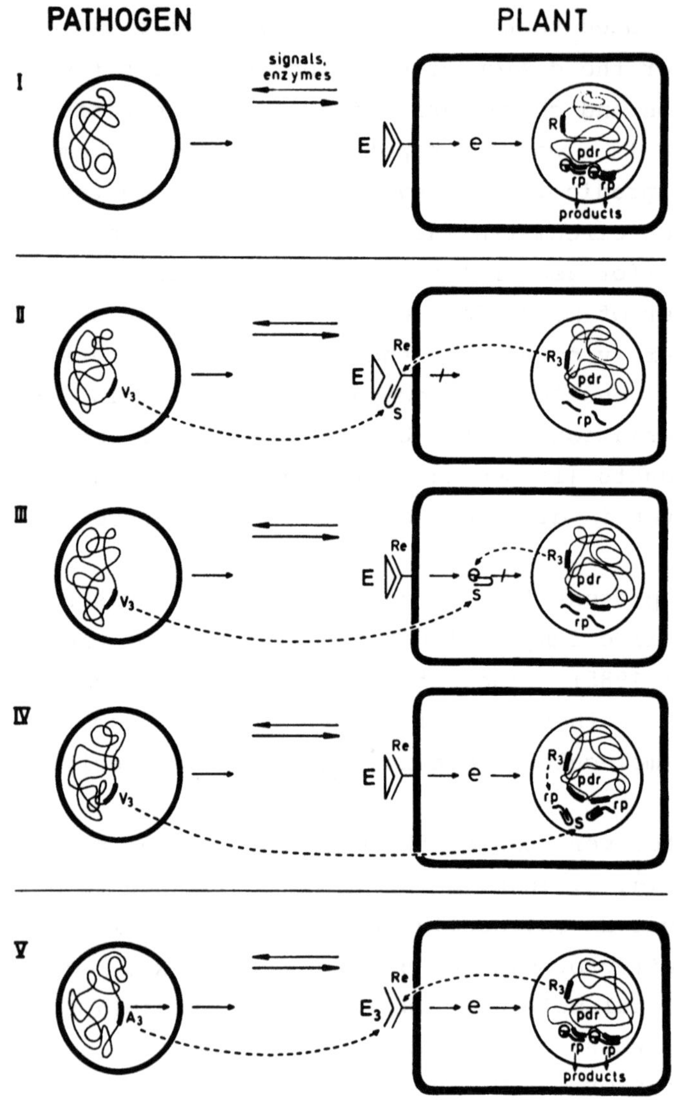

··· etc. (VI - n) and combinations

Fig. 5. Schematic representation of a working hypothesis on the mechanisms of gene activation in plant-pathogen interactions, with special emphasis on genetically defined gene-for-gene relationships. E = Elicitor (triangle: no race/cultivar specificity, open triangle: race/cultivar specificity), e = endogenous signal molecule (e.g., 'second messenger' or modified regulatory protein), rp = regulatory protein, pdr = pathogen-defense-related gene, R = resistance gene, Re = effector-binding protein ('receptor'), S = suppressor, V = virulence gene, A = avirulence gene. Note the putative involvement of additional signal molecules and enzymes in the overall complexity of biochemical interactions.

nism are not given, due to the lack of any experimental basis. However, analogous to other eukaryotic systems (Sibley and Lefkowitz 1985), it is possible that the putative membrane-associated, effector-responsive system is composed of two or more entities, as indicated in Fig. 6.

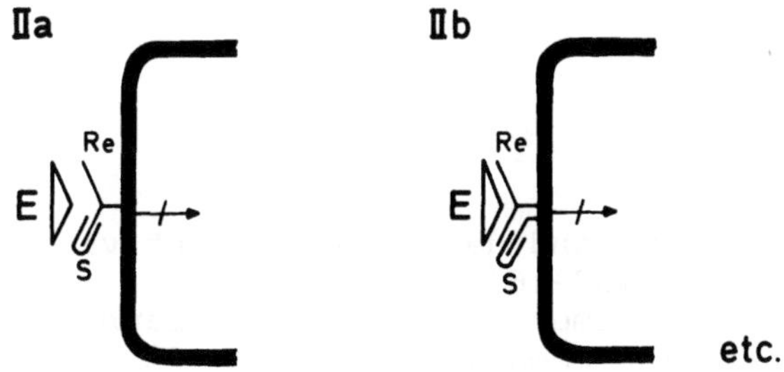

Fig. 6. Possible alternatives of version II (Fig. 5) for the proposed signal transduction mechanism at the plant cell membrane. Note the staggered pattern of lines and arrows. Analogous modifications might be considered for version V.

With regard to future experimental approaches, the recently established coordination in the activation of pathogen defense-related (pdr) genes (Chappell and Hahlbrock 1984, Somssich et al 1986) suggests the action of common regulatory proteins (rp) mediating this effect (see, for instance, Bienz 1985). One way of testing this hypothesis will be to use the regulatory sequences of the available pdr genes, e. g. PAL and 4CL, for affinity studies to determine whether such regulatory proteins can be identified. Fig. 5 would then propose functional connections of these proteins with the R gene product, either through a second messenger (endogenous signal molecule) or directly (e. g., if 'e' is the activated form of 'rp'). Though not explicitly included in the design of Fig. 5, this latter possibility should seriously be considered as long as experimental evidence, for or against, is lacking.

An alternative or additional approach would be to search directly for the putative interacting proteins that specify the complementing genotypes of the two organisms. Monoclonal antibodies might be valuable tools for such experiments.

Whatever the most successful way to the molecular identification of a disease resistance gene, a complete analysis of its mechanism of action will certainly have to include the whole spectrum of events from signal perception to the regulation of gene expression and the accumulation of defense-related products. The preceding sections of this chapter have outlined a few very early steps towards this goal.

REFERENCES

Bienz M (1985) Transient and developmental activation of heat-shock genes. TIBS 10:157-161

Bollwell GP, Bell JN, Cramer CL, Schuch W., Lamb CJ, Dixon RA (1985) L-Phenylalanine ammonia-lyase from *Phaseolus vulgaris*. Eur J Biochem 149:411-419

Chappell J, Hahlbrock K (1984) Transcription of plant defence genes in response to UV light or fungal elicitor. Nature 311:76-78

De Wit PJGM, Hofman AE, Velthuis GCM, Kuć JA (1985) Isolation and characterization of an elicitor of necrosis isolated from intercellular fluids of compatible interactions of *Cladosporium fulvum* (Syn. *Fulvia fulva*) and tomato. Plant Physiol 77:642-647

Dickman MB, Patil SS (1986) Cutinase deficient mutants of *Colletotrichum gloeosporioides* are nonpathogenic to papaya fruit. Physiol Mol Plant Physiol 28:235-242

Ellingboe AH (1982) Genetical aspects of active defence. In: Active Defense Mechanisms in Plants, Wood RKS, ed, Plenum Press, New York, pp 179-192

Fritzemeier KH, Rohwer F, Cretin C, Scheel D, Hahlbrock K (1986) Biochemical reactions of different tissues of potato (*Solanum tuberosum*) to zoospores or elicitors from *Phytophthora infestans*. Transient induction of phenylalanine ammonia-lyase and 4-coumarate:CoA ligase mRNAs. Submitted for publication

Hahlbrock K, Chappell J, Jahnen W, Walter M (1985) Early defense reactions of plants to pathogens. In: Molecular Form and Function of the Plant Genome, Vloten-Doting L, Groot GSP, Hall TC, eds, Plenum Publishing Corporation, pp 129-140

Hahlbrock K, Scheel D (1986) Biochemical responses of plants to pathogens. In: Non-Conventional Approaches to Plant Disease Control, Chet I, ed, Wiley and Sons, New York, in press

Kistler HC, VanEtten HD (1984) Regulation of pisatin demethylation in *Nectria haematococca* and its influence on pisatin tolerance and virulence. J Gen Microbiol 130:2605-2613

Knobloch , Hahlbrock K (1985) Isoenzymes of p-coumarate: CoA ligase from cell suspension cultures of *Glycine max*. Eur J Biochem 52:311-320

Kolattukudy PE (1985) Enymatic penetration of the plant cuticle by fungal pathogens. Ann Rev Phytopathol 23:223-250

Kombrink E, Scheel D, Schmelzer E, Somssich I, Hahlbrock K (1986)

Biochemical responses of non-host plant cells to fungi and fungal elicitors. NATO Advanced Research Workshop "Biology and Molecular Biology of Plant-Pathogen Interactions", Bailey JA, ed, in press

Rohwer F, Fritzemeier KH, Scheel D, Hahlbrock K (1986) Biochemical reactions of different tissues of potato (*Solanum tuberosum*) to zoospores or elicitors from *Phytophthora infestans*. Submitted for publication

Sibley DR, Lefkowitz RJ (1985) Molecular mechanisms of receptor desensitization using the ß-adrenergic receptor-coupled adenylate cyclase system as a model. Nature 317:124-129

Somssich I, Schmelzer E, Bollmann J, Hahlbrock K (1986) Rapid activation by fungal elicitor of genes encoding "pathogenesis-related" proteins in cultured parsley cells. Proc Natl Acad Sci USA 83:2427-2430.

STIMULATION OF PHYTOALEXIN FORMATION IN FUNGUS-INFECTED PLANTS AND ELICITOR-TREATED CELL CULTURES OF PARSLEY

Dierk Scheel, Karl Dietrich Hauffe, Willi Jahnen and Klaus Hahlbrock
Max-Planck-Institut für Züchtungsforschung
Abteilung Biochemie
D-5000 Köln 30
Federal Republic of Germany

Resistance of plants to pathogens usually involves a great variety of mechanisms preventing penetration, growth, and multiplication of the invading organism. Pathogen-induced reactions are an important part of the plant's defense mechanisms. A prerequisite for induction of these responses is the perception of appropriate signals by the plant cell. The molecular mechanisms of recognition and induction processes are difficult to study in intact plant/pathogen systems, because only a limited number of plant cells are challenged by and respond to the pathogen, and because multiple infection events usually do not occur synchronously.

In our studies we therefore use plant cell cultures in combination with fungal elicitors, as a system of reduced complexity, in addition to the intact non-host/pathogen system of parsley (Petroselinum crispum) and the soybean pathogen Phytophthora megasperma f.sp. glycinea (Pmg). Cultured parsley cells respond to treatment with an elicitor prepared from hyphal cell walls of this fungus with rapid but transient activation of genes coding for phenylalanine ammonia-lyase (PAL) and 4-coumarate:CoA ligase (4CL), two enzymes of general phenylpropanoid metabolism (Chappell and Hahlbrock 1984), and for so-called "pathogenesis-related" proteins of unknown function (Somssich et al 1986), with increases in activity of hydrolytic enzymes such as 1,3-ß-glucanase and chitinase (Kombrink and Hahlbrock 1986), and with the formation of coumarins, the putative phytoalexins of this species (Tietjen et al 1983; Hauffe et al 1986). From the complex mixture of coumarin derivatives, which is excreted into the culture fluid, the six major components have been identified as umbelliferone, marmesin, psoralen, xanthotoxin, bergapten, and isopimpinellin by mass and proton magnetic resonance spectroscopy in comparison with authentic compounds. An almost identical pattern of coumarin derivatives accumulated in infection

NATO ASI Series, Vol. H 4
Recognition in Microbe-Plant Symbiotic and Pathogenic Interactions. Edited by B. Lugtenberg

droplets of Pmg-infected parsley cotyledons. Maximum amounts of 160 µmol/l (equivalent to 1 µmol per g cell fresh weight) were extracted from the culture fluid of elicitor-treated cells and of 10 µmol/l (equivalent to 0.03 µmol per g cell fresh weight) from infection droplets (Fig. 1). Untreated cell cultures and healthy plants released only trace amounts of the compounds. Cultured cells first excreted psoralen, followed by umbelliferone, marmesin, and xanthotoxin, before finally bergapten and isopimpinellin accumulated in the culture fluid. Maximum concentrations of the single compounds were observed between 15 and 70 h after treatment. Psoralen, umbelliferone, and xanthotoxin disappeared from the growth medium at later stages. This phenomenon may be due to turnover rather than re-uptake by the cells, because the amounts of cell-associated coumarins remained low during the entire period of experimentation, while the excreted levels increased drastically. Synchronous accumulation of coumarins was observed in infection droplets with maximum levels around 30 h after inoculation. Concentrations of all compounds were drastically reduced after 50 h. In this case the compounds may be degraded and/or accumulate in fungal spores, which at this stage show the characteristic fluorescence of coumarins after UV irradiation under the fluorescence microscope.

The autofluorescence of these compounds also enabled us to localize their site of accumulation in Pmg-infected cotyledons by fluorescence microscopy. Necrotic lesions around the penetration site of the fungus lighted up after excitation between 395 and 425 nm (Fig. 2A), whereas the blue fluorescence of coumarins appeared after excitation at 365 nm (Fig. 2B). This latter fluorescence was associated with cells surrounding the lesions and with the intercellular spaces, from where it was eluted by simply dipping the upper parts of intact seedlings into water.

Parsley plants contain furanocoumarins as common constituents (Zaynoun et al 1985; Knogge et al 1986) of their vascular bundles (Jahnen 1986). No additional increases in coumarin concentrations were detectable in methanolic extracts of Pmg-infected cotyledons compared to healthy tissue (Knogge et al 1986). Despite the high levels of furanocoumarins present in healthy plants, the rapid local accumulation of these compounds in infection droplets of Pmg-infected cotyledons clearly defines the coumarins indentified here as phytoalexins of this species. Furanocoumarins at concentrations of 10 to 100 µmol/l have been shown to inhibit spore germination and growth of a variety of plant pathogenic fungi including Pmg (Johnson et al 1973; Juntila 1976; Tietjen et al 1983; Knogge et al 1986). In our case, the

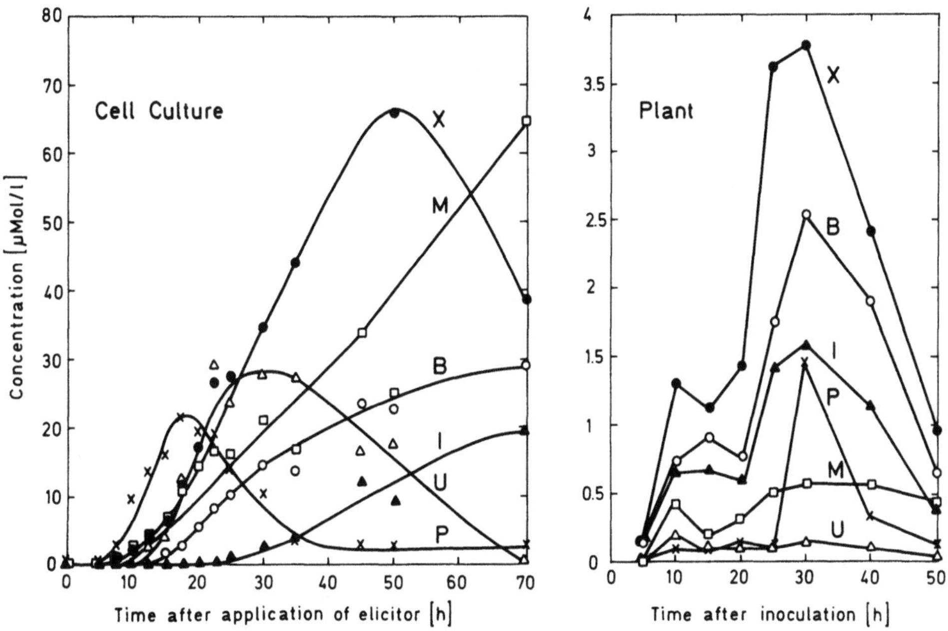

Fig. 1. Time courses of accumulation of coumarins in culture fluid of elicitor-treated cells and in infection droplets of Pmg-infected cotyledons of parsley. Coumarin derivatives were quantified on HPLC by comparison with known amounts of authentic compounds (X=xanthotoxin, M=marmesin, B=bergapten, I=Isopimpinellin, U=umbelliferone, P= psoralen).

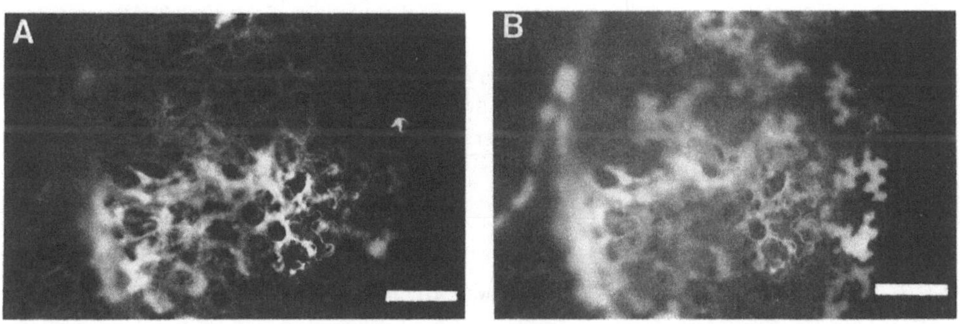

Fig. 2. Fluorescence microscopy of Pmg-infected cotyledons of parsley. Necrotic lesions of cotyledons 30 h after inoculation with Pmg zoospores after excitation with light of 395 to 425 nm (A) and of 365 nm (B) (Bars 100 µm).

level of coumarin derivatives in infection droplets was calculated on the basis of the amount of spore suspension applied to the cotyledons. Therefore the content of 10 µmol/l is rather an underestimation, because evapora-

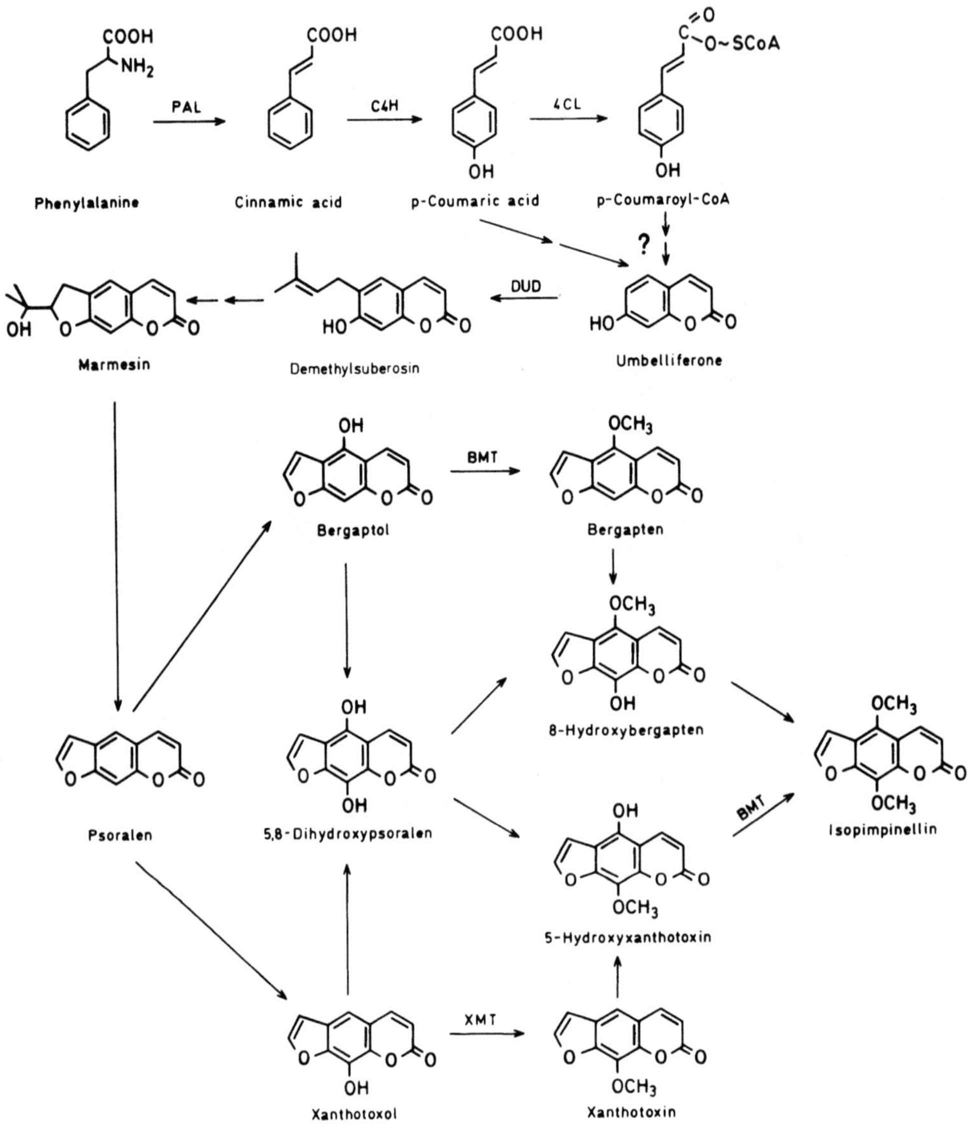

Fig. 3. Hypothetical biosynthetic pathway of furanocoumarins.

tion of water during the 50-h period of experimentation was not taken into account. All six coumarins identified here are synthesized via the phenylpropanoid and furanocoumarin pathways (Murray et al 1982) (Fig. 3). The activities of the enzymes of general phenylpropanoid metabolism, PAL, cinnamate 4-hydroxylase (C4H), and 4CL, and of three enzymes of the furanocoumarin pathway, dimethylallyl diphosphate:umbelliferone dimethylallyltransferase (DUD), SAM:bergaptol O-methyltransferase (BMT), and SAM:xanthotoxol

O-methyltransferase (XMT), have been shown to be stimulated by elicitor application to parsley cells (Hahlbrock et al 1981; Hagmann et al 1983; Tietjen and Matern 1983; Hauffe et al 1986).

The activity increases of PAL and 4CL were found to be due to transient activation of the respective genes (Chappell and Hahlbrock 1984; Schmelzer et al 1985). The previously observed sequential activity changes in elicitor-treated parsley cells of PAL, C4H, and 4CL on the one hand, and DUD on the other hand, have now also been demonstrated for XMT and BMT relative to PAL and 4CL (Fig. 4). XMT and BMT, specifically catalyzing final methyl-

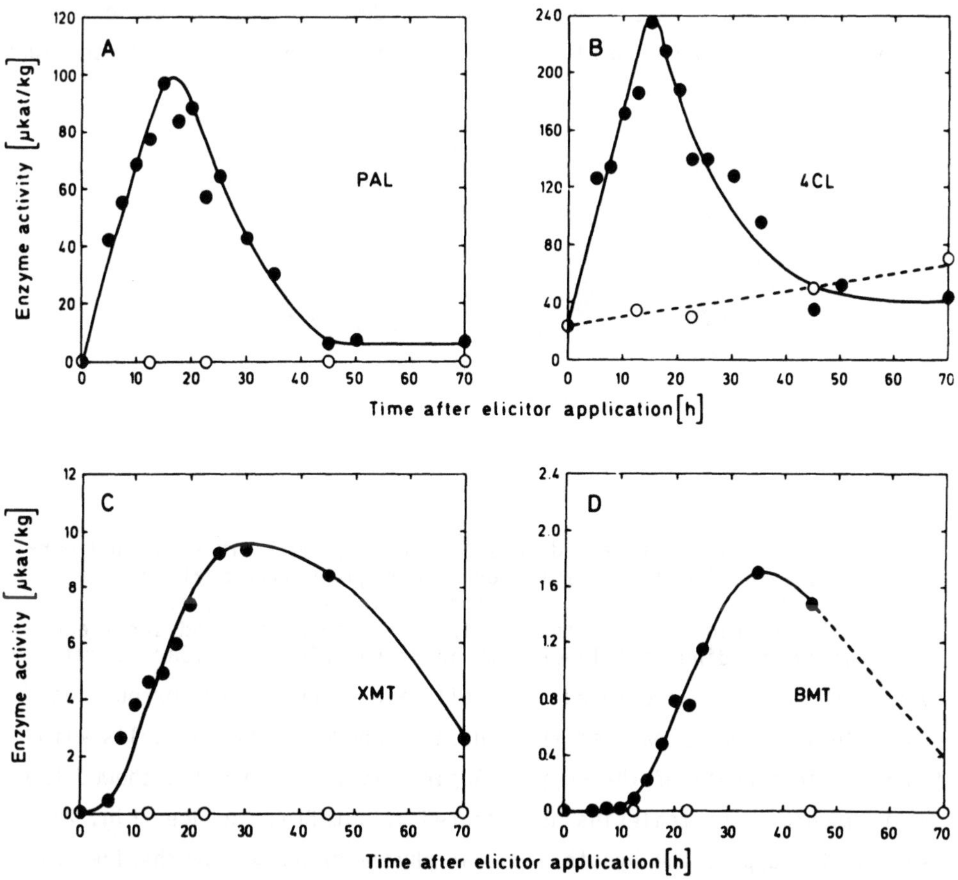

Fig. 4. Time courses of elicitor-stimulated changes in activity of enzymes related to furanocoumarin biosynthesis. Specific enzyme activities of PAL (A), 4CL (B), XMT (C), and BMT (D) were monitored at the indicated times after elicitor treatment (closed circles) in comparison with untreated cultures (open circles).

ation steps in the syntheses of xanthotoxin, bergapten, and isopimpinellin (Hauffe et al 1986), were purified 1,200- and 650-fold, respectively. Polyclonal antisera raised against both proteins in rabbits showed no crossreaction between the two methyltransferases and the antisera. These antisera are now being used to study the stimulation of both enzymes at the level of mRNA activity.

Similar studies with parsley cotyledons showed that PAL, 4CL, XMT, and BMT were present at high levels in healthy tissue (Knogge et al 1986). No additional activity increases were detected after Pmg infection using this averaging method. Polyclonal antisera against PAL from cultured parsley cells (Schröder et al 1979) were employed to detect PAL in infected seedlings by indirect immunoperoxidase staining (Sternberg 1979) on cryosections of infection sites.

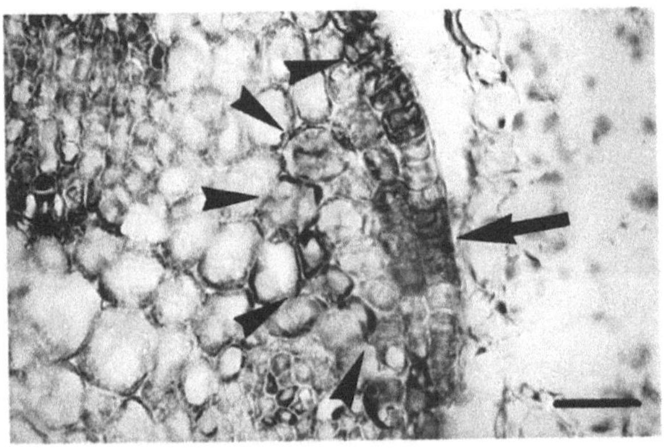

Fig. 5. Section of a Pmg-infected parsley seedling. PAL antiserum and peroxidase-labeled second antibody were sequentially applied to a section of an infection site of a primary leaf petiole. Arrows indicate dead cells in the center of the infection and arrowheads the surrounding cell layer with increased PAL level (Bar 25 µm).

Elevated PAL contents were found in a cell layer spherically surrounding necrotic lesions, but separated from these by another layer of cells without detectable amounts of the enzyme. Apparently the same cells showed the highest amounts of phytoalexins as detected by autofluorescence (Fig. 2). These results support the hypothesis that phytoalexins are synthesized by active live cells around necrotic lesions rather than being released from cells killed by the pathogen. The antisera against XMT and BMT will now be used to further investigate this question.

Chappell J, Hahlbrock K (1984) Transcription of plant defence genes in response to UV light or fungal elicitor. Nature 311:76-78

Hagmann M-L, Heller W, Grisebach H (1983) Induction and characterization of a microsomal flavonoid 3'-hydroxylase from parsley cell cultures. Eur J Biochem 134:547-554

Hahlbrock K, Lamb CJ, Purvin C, Ebel J, Fautz E, Schäfer E (1981) Rapid response of suspension-cultured parsley cells to the elicitor from Phytophthora megasperma var. sojae. Induction of the enzymes of general phenylpropanoid metabolism. Plant Physiol 67:768-773

Hauffe KD, Hahlbrock K, Scheel D (1986) Elicitor-stimulated furanocoumarin biosynthesis in cultured parsley cells: S-adenosyl-L-methionine:bergaptol and S-adenosyl-L-methionine:xanthotoxol O-methyltransferases. Z Naturforsch 41c:228-239

Jahnen W (1986) Untersuchungen zur Histochemie und Biochemie der Nichtwirt-Pathogenbeziehung von Petersilie (Petroselinum crispum, Hoffm.) und Phytophthora megasperma f.sp. glycinea. Dissertation, Universität zu Köln.

Johnson C, Brannon DR, Kuc J (1973) Xanthotoxin: a phytoalexin of Pastinaca sativa root. Phytochemistry 12:2961-2962

Juntila O (1976) Allelopathic inhibitors in seeds of Heracleum laciniatum. Physiol Plant 36:374-378

Knogge W, Kombrink E, Schmelzer E, Hahlbrock K (1986) in preparation

Kombrink E, Hahlbrock K (1986) Responses of cultured parsley cells to elicitors from phytopathogenic fungi. Timing and dose dependency of elicitor-induced reactions. Plant Physiol 81:in press

Murray RDH, Mendez J, Brown SA (1982) The Natural Coumarins. Occurrence, Chemistry and Biochemistry. John Wiley & Sons Ltd., Chichester New York Brisbane Toronto Singapore

Schmelzer E, Somssich I, Hahlbrock K (1985) Coordinated changes in transcription and translation rates of phenylalanine ammonia-lyase and 4-coumarate:CoA ligase mRNAs in elicitor-treated Petroselinum crispum cells. Plant Cell Rep 4:293-296

Schröder J, Kreuzaler F, Schäfer E, Hahlbrock K (1979) Concomitant induction of phenylalanine ammonia-lyase and flavanone synthase mRNAs in irradiated plant cells. J Biol Chem 254:57-65

Somssich I, Schmelzer E, Bollmann J, Hahlbrock K (1986) Rapid activation by fungal elicitor of genes encoding 'pathogenesis-related' proteins in cultured parsley cells. Proc Natl Acad Sci USA 83:2427-2430

Sternberger LA (1979) Immunocytochemistry, 2nd edn. John Wiley & Sons Ltd., New York Chichester Brisbane Toronto Singapore

Tietjen KG, Matern U (1983) Differential response of cultured parsley cells to elicitors from two non-pathogenic strains of fungi. 2. Effects on enzyme activities. Eur J Biochem 131:409-413

Tietjen KG, Hunkler D, Matern U (1983) Differential response of cultured parsley cells to elicitors from two non-pathogenic strains of fungi. 1. Identification of induced products as coumarin derivatives. Eur J Biochem 131:401-407

Zaynoun S, Ali LA, Tenekjian K, Kurban A (1985) The bergapten content of garden parsley and its significance in causing cutaneous photosensitization. Clin Exp Derm 10:328-331

RECOGNITION AND RESPONSE IN PLANT:PATHOGEN INTERACTIONS

Christopher J. Lamb, David R. Corbin, Michael A. Lawton,
Norbert Sauer and Vincent P.M. Wingate

Salk Institute for Biological Studies
Plant Biology Laboratory
P.O. Box 85800
San Diego
California 92138, USA

The perception of, and response to, microbes is of crucial importance in relation to the damage and reduction in crop yield attributable to plant pathogens and the potential beneficial effects arising from symbiotic plant:microbe interactions. Our research pertaining to plant:microbe recognition is based on two premises: (a) that recognitional events result in specific responses, the study of which will lead to elucidation of signal perception and transduction mechanisms and (b) that the surfaces of microbial and plant cells are crucially involved in recognitional events. In this chapter, recent studies on the molecular mechanisms governing expression of specific plant defense responses and the molecular architecture of the surface of the phytopathogenic bacterium *Pseudomonas syringae* pv *glycinae* are discussed.

As a model for analysis of the molecular mechanisms underlying activation of plant defense responses we have examined the interaction between (*Phaseolus vulgaris* L.) and the fungus *Colletotrichum lindemuthianum*, causal agent of anthracnose. This study has involved analysis both of the response of suspension-cultured cells to various fungal elicitor preparations and also the response of intact bean hypocotyl tissue to infection by *C. lindemuthianum*. The specific attributes of these complementary experimental systems have been

NATO ASI Series, Vol. H 4
Recognition in Microbe-Plant Symbiotic and

discussed in detail elsewhere (Lamb et al 1983,1986a)

Our initial focus was on the accumulation of phaseollin, kievitone and structurally-related phytoalexins since they appear to have a key role in expression of disease resistance in bean : C. lindemuthianum interactions and moreover, many steps in the biosynthesis of these phytoalexins have been well characterized at the enzyme level (Dixon et al 1983). More recently we have extended our analysis to other defense responses including induction of the hydrolytic enzyme chitinase, synthesis of the cell wall polymer lignin and accumulation of cell wall hydroxyproline-rich glycoproteins (HRGP). Our strategy has been to analyze at the molecular level the causally related sequence of events underlying expression of each defense response following elicitor treatment of cell cultures and infection of intact hypocotyl tissue, working back from the overall defense response toward signal perception and early transduction events. In the course of these studies a number of defense gene sequences have been cloned including those encoding proteins involved in the synthesis of lignin precursors and phytoalexins as well as the hydrolytic enzyme chitinase and HRGP apoproteins. These cloned sequences have been used to analyze the organization and structure of plant defense genes in relation to their activation by biological stress signals. Recent findings have been reviewed (Lamb et al 1986 a,b) and are briefly outlined below to provide the basis for discussion of signal perception and transduction mechanisms.

Elicitor treatment of cell cultures and infection of intact hypocotyl tissue causes massive changes in the patterns of RNA synthesis and protein synthesis correlated with the activation of specific defense responses (Cramer et al 1985a; unpublished). With one exception, namely the deposition of the β-3 linked wall glucan callose, activation of specific plant defense responses characteristically involves transcriptional activation of the corresponding defense genes. Transcriptional activation of the genes encoding the phenylpropanoid biosynthetic enzymes phenylalanine ammonia-lyase (PAL: lignin, phytoalexins) and chalcone synthase (CHS: phytoalexins) can first be observed within 5 minutes of elicitor treatment (Fig. 1), whereas transcriptional activation of certain other defense genes, notably those encoding HRGP apoproteins does not occur until about 1 hr after elicitor treatment (Edwards et al 1985; Lawton and Lamb, 1986; Ryder et al 1984; Showalter et al 1985).

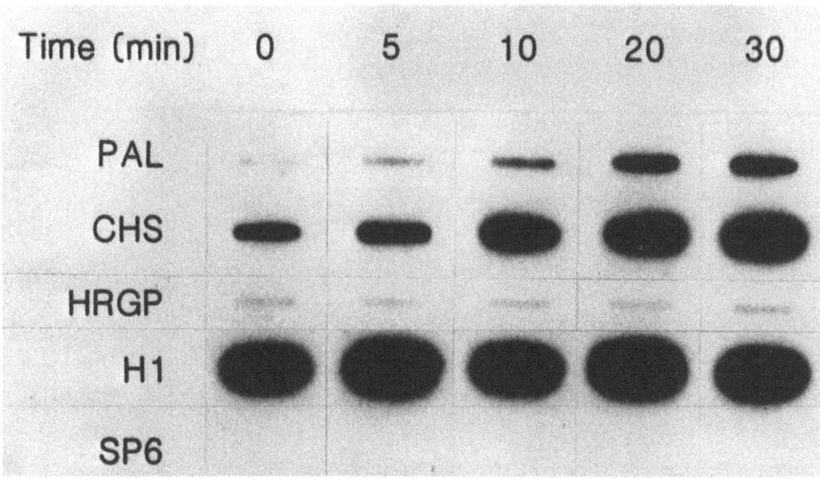

Figure 1. Rapid transcriptional activation of PAL and CHS genes by fungal elicitor. [^{32}P]-labeled in vitro run-off transcripts from nuclei isolated from bean cells at various times after elicitor treatment were hybridized to immobilized PAL, CHS, HRGP and H1 sequences. H1 gene transcription is unaffected by elicitor treatment and serves as an internal control. SP6 denotes hybdridization to vector sequences.

In infected hypocotyl tissue there are clear temporal and spatial differences between compatible and incompatible interactions in the pattern of transcriptional activation of defense genes and accumulation of the corresponding transcripts (Bell et al 1984,1986; Cramer et al 1985b). In an incompatible interaction following establishment of the fungal infection peg there is an early activation of defense genes (Fig. 2). Induction of defense genes occurs not only in the immediate neighborhood of the infected cell but also in uninfected cells in tissue at a distance from the site of infec-

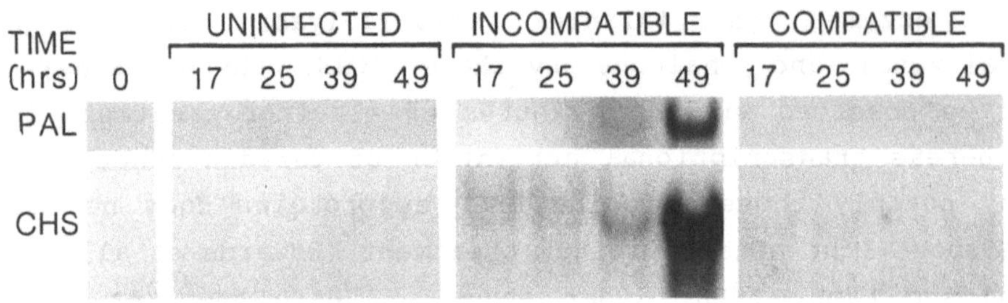

Figure 2. Accumulation of PAL and CHS mRNAs in the early stages of infection of bean hypocotyls with compatible and incompatible races of C. lindemuthianum.

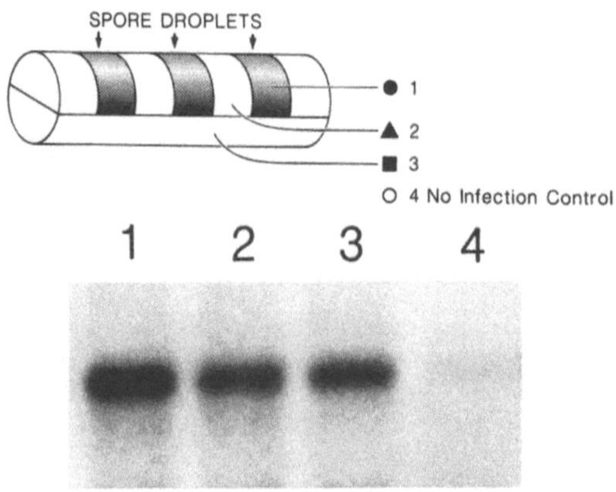

Figure 3. Induction of an HRGP transcript at a distance in the early stages of an incompatible interaction. RNA was prepared from directly infected tissue (Lane 1) and adjacent hitherto uninfected tissue (Lanes 2 and 3) 90 hr after inoculation with spores of an incompatible race of C. lindemuthianum as depicted, as well as from equivalent uninnoculated hypocotyls (Lane 4).

tion (Fig. 3). In contrast in a compatible interaction there is no early response (Fig. 2), but rather a delayed widespread response at the onset of lesion formation (Bell et al 1984,1986; Cramer et al 1985b).

A number of defense gene systems, characteristically those thought to be key regulatory elements in the expression of specific defense responses, appear to be highly polymorphic at the gene, RNA and polypeptide levels. These polymorphic forms exhibit complex, differential patterns of regulation in different biological stress situations (unpublished).

These observations have a number of implications for the molecular mechanisms underlying perception of pathogens and signalling of biological stress. Thus elicitor induction of PAL and CHS gene transcription in bean cells represents the most rapid response hitherto observed that can be causally related to expression of specific defense responses, with kinetics comparable to the most rapid gene activation systems in animal cells. The rapidity of PAL and CHS induction by elicitor implies that there are few intervening steps between elicitor binding to the putative plant cell receptor and specific transcriptional activation of these genes. Hence PAL and CHS gene activation might involve direct interaction of an elicitor-receptor complex with cis-acting regulatory DNA sequences as postulated for rapid transcriptional activation by steroid hormones in animal cells (Hollenberg et al 1985). Alternatively, elicitor binding to a receptor putatively located on the cell surface might activate a short signal transduction cascade leading to modulation of trans-acting nuclear regulatory proteins that interact with the target cis-acting regulatory DNA sequences.

The rapid transcriptional activation of PAL and CHS genes implies that the signal perception and transduction systems are in place prior to the onset of biological stress, although of course these systems may be further induced as part of the subsequent program of responses. Following isolation and structural characterization of PAL and CHS genes, it should now be possible by a combination of comparative nucleotide sequence analysis, analysis of elicitor-induced changes in chromatin structure and gene transfer techniques to identify these cis-acting regulatory DNA sequences and hence develop an assay to isolate and characterize the corresponding trans-acting regulatory proteins.

The magnitude and rapidity of elicitor stimulation of defense gene activation in suspension cultured cells strongly argues that the resultant defense responses play a primary role in disease resistance. Likewise induction of PAL and CHS mRNA accumulation within 39 hr after innoculation of bean hypocotyls with the spores of an incompatible race of C. lindemuthianum (Fig. 2) represents an early event in relation to the ingress of the pathogen, since following spore innoculation there is a period of 30-40 hr during which the spores germinate and produce an infection peg before fungal hyphae come in contact with the first host epidermal cell (Bailey 1982). In relation to the timing of this initial cell:cell contact at which the putative recognition event occurs, induction of defense mRNAs in an incompatible interaction is extremely rapid and occurs well before the first signs of hypersensitive flecking and expression of hypersensitive resistance. Moreover, there is no induction of defense gene transcripts at the comparable stage in the compatible interaction, even though the fungus has invaded plant cells at that stage (Fig. 2). This implies that activation of defense genes in directly infected tissue in the early stages of the incompatible interaction is a direct consequence of molecular recognition specified by the plant disease resistance genes and pathogen avirulence genes which underly race:cultivar specificity, rather than a response to non-specific reactions arising from fungal ingress.

These observations do not appear to be consistent with the hypothesis that the induction of defense mRNAs is causally preceded by hypersensitive cell death and initial expression of hypersensitive resistance. Thus our working hypothesis is that disease resistance genes encode proteins involved in molecular recognition of pathogens leading to rapid activation of defense genes and hence expression of localized hypersensitive resistance. From the recent kinetic analysis of gene activation, we propose that these trans-acting proteins are either the products of specific disease resistance genes as defined by genetic analysis of host:pathogen interactions (Ellingboe 1981) or are closely associated at the biochemical level with the products of such genes. Hence we conclude that disease resistance genes are expressed in the target tissue prior to elicitation or infection. Furthermore, we propose that induction of defense mRNAs in the later stages of a compatible interaction in hitherto uninfected cells ahead of the invading pathogen provides a plausible mechanism for the process of attempted lesion limitation, which under appropriate physiological conditions can restrict the size of the lesion and prevent complete rotting of the organ and hence plant death, even though the interaction is genetically compatible (Bailey 1982). Likewise the preactivation of defense genes in hitherto uninfected cells distant from the initial site of infection by an incompatible pathogen may represent a molecular mechanism by which induced systemic resistance (Dean and Kuc 1985) is established.

The temporal and spatial patterns of defense gene activation in elicited cells and infected hypocotyls implies the existence of a complex set of biological stress signalling systems. Thus the different kinetics for elicitor stimulation of HRGP gene transcription and mRNA induction relative to PAL and CHS gene activation (Edwards et al 1985; Lawton and Lamb 1986; Ryder et al 1984; Showalter et al 1985) presumably reflect two distinct stimuli or a single stimulus leading to either sequential effects or divergent pathways (Fig. 1). Operation of similarly complex signal systems in infected tissue can likewise be inferred from analysis of the kinetics of accumulation of defense gene transcripts (Bell et al 1986; Showalter et al 1985). Moreover, within the syndrome of biological stress, different regulatory circuits appear to be involved in defense gene activation in different circumstances. Perhaps the clearest example to date is a preferential activation of two HRGP genes in, respectively, the early stages of an incompatible interaction and the later stages of a compatible interaction, implying the operation of distinct signal transduction systems in the two situations. Likewise, the observation of defense gene activation in tissue distant from the site of infection (Fig. 3) implies intercellular transmission of endogenous elicitation signals in addition to localized transduction of exogenous signals in directly infected cells (Bell et al 1986; Showalter et al 1985).

Clearly, identification of the molecular mechanisms underlying these signal transduction systems represents a major challenge for future research. In principle, by further analysis of the molecular mechanisms underlying defense gene activation in different circumstances it should be possible to piece together the mechanisms underpinning signal perception and transduction of exogenous and endogenous signals of biological stress.

In parallel we have initiated an alternative approach to elucidate the molecular mechanisms underlying plant:pathogen recognition in which the surface architecture of plant pathogens is analyzed using monoclonal antibody (MCA) approaches. Our initial study has been concerned with Pseudomonas syringae pv glycinae (Psg), a bacterial pathogen of soybean (Glycine max) Murine MCAs have been generated using various physiological races as immunogens. The nature of the immunogen and the method and schedule of immunization were crucial in determining both the frequency of generation of antibody secreting cell lines and the pattern of immune response to different classes of surface epitope. Both qualitative and quantitative racial differences in surface architecture have been discerned, although from this analysis no clear correlation with the pattern of race:cultivar specificity has emerged.

To dissect the molecular architecture of Psg cell surfaces in relation to race:cultivar specificity required a less variable background. Therefore, we have generated MCA libraries against surface antigens of race 4 transconjugated with plasmid pLAFR1 carrying a 27.2 kb insert containing cosmid clone pPg6L3 of race 6 DNA which confers the race 6 phenotype on transconjugated race 4 (Staskawicz et al 1984). The properties of the epitopes reactive with these MCAs have subsequently been investigated by comparative radioimmunoassay of binding to the surfaces of wild type and transconjugated race 4 and by the pattern of immunoreactivity in protein blots following fractionation of bacterial lysates by SDS-polyacrylamide gel electrophoresis. While the majority of MCAs bound to roughly equal extents to wild type and transconjugated race 4, a small number of MCAs gave clear differential binding exhibiting very little affinity for the surface of wild type race 4 bacteria, but strong binding to the surface of transconjugated race 4 bacteria which show the race 6 phenotype, as well as to the surface of race 6 bacteria (Fig. 4).

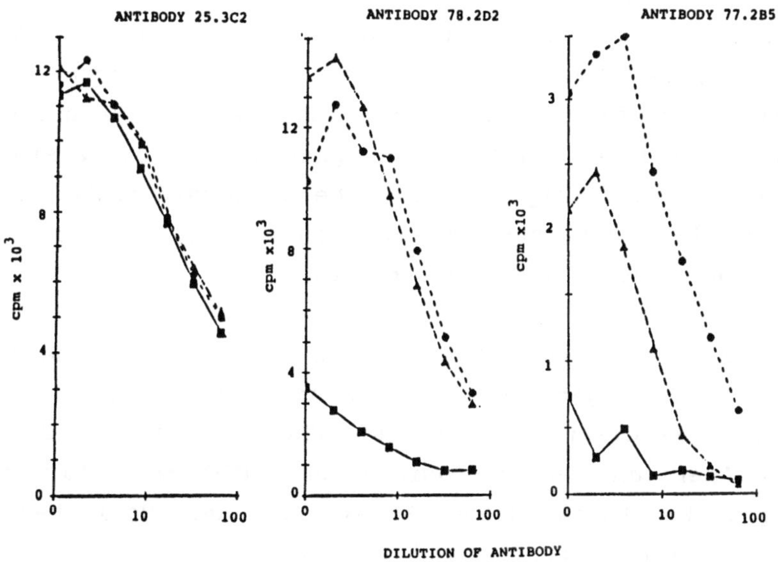

Figure 4. Titration of MCA reactivity with surface epitopes of Psg: race 4 (—■—); race 6 (—●—); race 4 transconjugated with pPg6L3 (—▲—).

To date we have not been able to identify the surface feature in which the epitopes reactive with these specific MCAs reside. However, by protein blot analysis, we have identified a number of other MCAs that react with epitopes present on a species Mr 35 kDa present in races 4 and 5 and a polypeptide Mr 36 kDa in race 6. Interestingly, race 4 transconjugated to show the race 6 phenotype shows reactivity with both species. Furthermore, while immunization with race 6 and/or transconjugated race 4 led to about 30% of the hybridomas giving MCAs reactive with the epitope(s) distributed in this pattern, indicating that the epitope(s) in this configuration or molecular background was immunodominant, immunization with wild type race 4 failed to produce MCAs with this pattern of reactivity. So although this epitope(s) is present in all races the molecular configuration and antigenicity is highly correlated with the race 6 phenotype in an otherwise constant genetic background.

Although we have not yet established a functional relationship between the features of the Psg surface reactive with these MCAs and the avirulence phenotype of race 6, the present analysis demonstrates that specific alterations in surface architecture are very closely associated with a specific bacterial avirulence gene and phenotypic expression of race:cultivar specificity. It will be of considerable interest to identify plant cell surface components which recognize these particular features of the Psg surface.

Research was supported by grants to CJL from the Samuel Roberts Noble, McKnight and Galster Foundations, Tan Sri Tan and USDA Competitive Grants Program (#84-CRCR-1-1464). DRC and NS thank the American Cancer Society and EMBO respectively for Fellowships. We thank Brian Staskawicz and Noel Keen for providing bacterial strains and for many helpful discussions. We thank Evelyn Wilson for preparation of the typescript.

REFERENCES

Bailey JA (1982) Physiological and biochemical events associated with the expression of resistance to disease. In: Wood RKS (ed) Active defence mechanisms in plants. Plenum, New York, p 39

Bell JN, Ryder TB, Wingate VPM, Bailey JA, Lamb CJ (1986) Differential accumulation of plant defense gene transcripts in a compatible and an imcompatible plant: pathogen interaction. Mol Cell Biol 6: 1615

Cramer CL, Ryder TB, Bell JN, Lamb CJ (1985a) Rapid switching of plant gene expression by fungal elicitor. Science 227: 1240

Cramer CL, Bell JN, Ryder TB, Bailey JA, Schuch W, Bolwell GP, Robbins MP, Dixon RA, Lamb CJ (1985b) Coordinated synthesis of phytoalexin biosynthetic enzymes in biologically-stressed cells of bean (Phaseolus vulgaris L.). EMBO J 4: 285

Dean RA, Kuc J (1985) Induced systemic resistance protection in plants. Trends in Biotechnol 3: 125

Dixon RA, Dey PM, Lamb CJ (1983) Phytoalexins: Enzymology and molecular biology. Adv Enzymol 55: 1

Edwards K, Cramer CL, Bolwell GP, Dixon RA, Schuch W, Lamb CJ (1985) Rapid transient induction of phenylalanine ammonia--lyase mRNA in elicitor treated bean cells. Proc Natl Acad Sci USA 82: 6731

Ellingboe AH (1981) Changing concepts in host-pathogen genetics. Ann Rev Phytopathol 19: 125

Hollenberg SM, Weinberger C, Ong ES, Cerelli G, Oro A, Lebo R, Thompson EB, Rosenfeld MG, Evans RM (1985) Primary structure and expression of a functional human glucocorticoid receptor cDNA. Nature 318: 635

Lamb CJ, Bell JN, Norman PM, Lawton MA, Dixon RA, Rowell PM, Bailey JA (1983) Early molecular events in the phytoalexin defense response. In: Ciferri O, Dure L III (eds) Structure and function of plant genomes. Plenum, New York, p 313

Lamb CJ, Bell JN, Cramer CL, Dildine SL, Grand C, Hedrick SA, Ryder TB, Showalter AM (1986a) Molecular response of plants to infection. In: St John JB (ed) Biotechnology for solving agricultural problems. Beltsville Agricultural Research Center Symposia, Vol X Roman and Allenheld, Totowa, N.J., in press

Lamb CJ, Bell JN, Corbin DR, Lawton MA, Mehdy MC, Ryder TB, Sauer N, Walter MH (1986b) Activation of defense genes in response to elicitor and infection. In: Arntzen CJ, Ryan CA Jr (eds) Molecular mechanisms of crop protection. Allan R Liss, New York, in press

Lawton MA, Lamb CJ (1986) Transcriptional activation of defense genes by fungal elicitor, wounding and infection. Submitted

Ryder TB, Cramer CL, Bell JN, Robbins MP, Dixon RA, Lamb CJ (1984) Elicitor rapidly induces chalcone synthase mRNA in Phaseolus vulgaris cells at the onset of the phytoalexin defense response. Proc Natl Acad Sci USA 81: 5724

Showalter AM, Bell JN, Cramer CL, Bailey JA, Varner JE, Lamb CJ (1985) Accumulation of hydroxyproline-rich glycoprotein mRNAs in response to fungal elicitor and infection. Proc Natl Acad Sci USA 82: 6551

Staskawicz BJ, Dahlbeck D, Keen NT (1984) Cloned avirulence gene of Pseudomonas syringae pv. glycinae determines race-specific incompatibility of Glycine max (L.) Merr. Proc Natl Acad Sci USA 81: 6024

PHYTOALEXIN SYNTHESIS IN SOYBEAN FOLLOWING INFECTION OF ROOTS WITH *PHYTOPHTHORA MEGASPERMA* OR TREATMENT OF CELL CULTURES WITH FUNGAL ELICITOR

Jürgen Ebel, Hans Grisebach, Anne Bonhoff, Dorit Grab, Christina Hoffmann, Georg Kochs, Hannelore Mieth, Walter Schmidt and Margit Stäb

Biologisches Institut II
Universität Freiburg
D-7800 Freiburg
FRG

ABSTRACT

Phytoalexin production is one of a number of inducible plant defense reactions which is thought to confer disease resistance against microbial infections in plants. Soybean (*Glycine max*) tissues produce and accumulate isoflavonoid phytoalexins (glyceollins) following either inoculation with a soybean pathogen, the fungus *Phytophthora megasperma* f.sp. *glycinea*, or treatment with a ß-glucan elicitor isolated from the fungal cell walls. Studies on the regulation of phytoalexin biosynthetic enzymes suggested that the phytoalexin defense response in soybean is controlled by temporary gene activation. The activities of several of the glyceollin biosynthetic enzymes have been studied during race-cultivar specific interactions between *P. megasperma* zoospores and intact soybean primary roots, a natural site of attack by the fungus. Only in the incompatible (host-resistant) interaction, is there an early enhancement of the enzyme activities starting at 2-4 h after inoculation, which correlates with the onset of glyceollin accumulation and expression of the hypersensitive response. Elicitor treatment of soybean cell suspension cultures causes major changes in the population of total translatable mRNA, which indicates large metabolic changes associated with phytoalexin synthesis and possibly other defense responses of the challenged cells.

Key events in the elicitor-mediated phytoalexin response are the interaction of the elicitor with the presumed primary target site and the signal transduction mechanism in the defense gene activation. Using a (^{3}H)ß-glucan elicitor fraction as ligand, saturable, high-affinity, and specific elicitor binding has been demonstrated in plasma-membrane enriched fractions from soybean roots. Substances which might be involved in signal transmission to the nucleus following elicitor binding to the suspected plasma membrane-located binding

NATO ASI Series, Vol. H 4
Recognition in Microbe-Plant Symbiotic and Pathogenic Interactions. Edited by B. Lugtenberg

site are unknown at present. Analysis of the presence and the metabolism of polyphosphoinositides showed that the elicitor did not affect the turnover of individual phosphoinositides in soybean cell cultures. It is concluded that polyphosphoinositides are not involved in signal transduction. Since Ca^{2+} ions stimulate the elicitor-mediated phytoalexin response in soybean cell cultures, changes in cellular Ca^{2+} pools could possibly be involved in signal transmission.

INTRODUCTION

Defense mechanisms conferring disease resistance in plants are effective at different levels in host-parasite interactions and include preformed defense barriers as well as defenses triggered by the invader (Horsfall and Cowling 1980). One inducible defense response is the production of phytoalexins at the site of attempted infection. There is a close correlation between phytoalexin accumulation and hypersensitive (necrogenic) resistance to fungal and bacterial infections (Bailey and Mansfield 1982). It appears that the speed, magnitude, and site of phytoalexin accumulation following penetration by the microorganism determine disease resistance in a number of plant-microbe interactions.

Recent progress in the analysis of the phytoalexin response in soybean (Glycine max) following the inoculation with a soybean fungal pathogen causing root and stem rot, the oomycete Phytophthora megasperma f.sp. glycinea, will be summarized. The relationship between changes in the levels of phytoalexin biosynthetic enzyme activity and phytoalexin production at early stages of race:cultivar-specific interactions between P. megasperma and soybean will be discussed. Fungal phytoalexin elicitor has been used for the investigation of the molecular mechanism underlying phytoalexin production in different soybean tissues, including cell suspension cultures.

BIOSYNTHESIS OF SOYBEAN PHYTOALEXINS

Following either infection or elicitor treatment, soybean tissues including cotyledons, hypocotyls, roots, and cell suspension cultures accumulate a number of structurally related pterocarpan phytoalexins (glyceollins and glycinol), which belong to the group of isoflavonoids shown in Fig. 1. Glyceollin biosynthetic pathway consists of a series of reactions which are catalyzed by enzymes of general phenylpropanoid metabolism and of flavonoid/isoflavonoid biosynthesis as well as of enzymes specifically involved in later steps of pterocarpan phytoalexin biosynthesis (Ebel 1986) (Fig. 1). The enzymes of general phenylpropanoid metabolism (phenylalanine ammonia-lyase, cinnamate 4-hydroxylase, and 4-coumarate:CoA ligase), acetyl-CoA carboxylase, and the enzymes of the common terpenoid pathway are involved in provi-

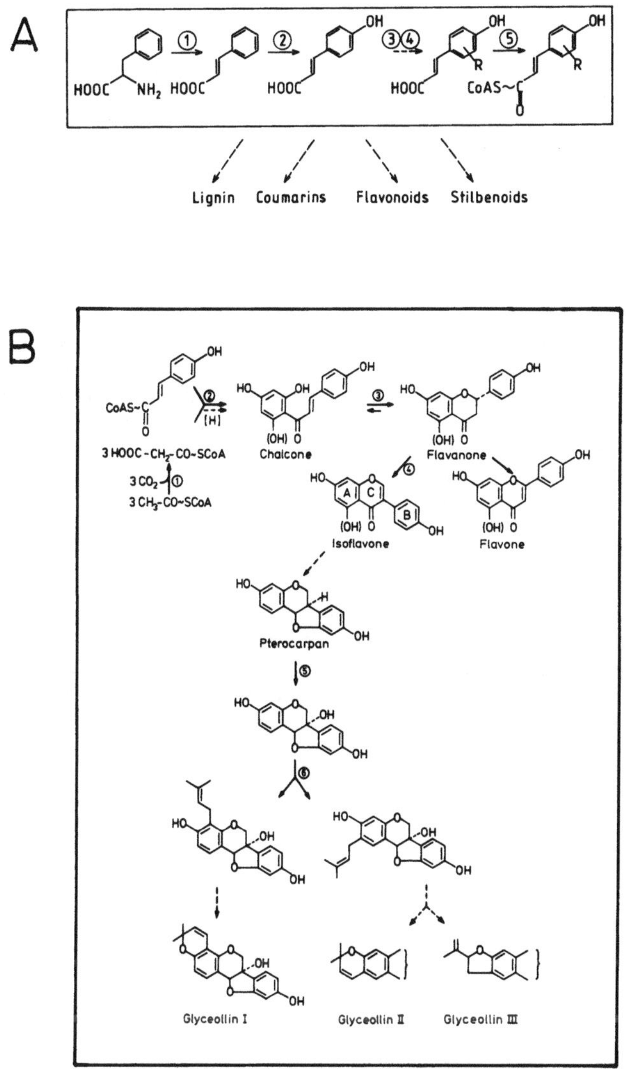

Fig. 1. Scheme illustrating the reactions of general phenylpropanoid metabolism (A) and of the proposed biosynthetic pathway of the soybean phytoalexin, glyceollin (B). The enzymes marked by numbers in A are: (1) phenylalanine ammonialyase (PAL); (2) cinnamate 4-hydroxylase; (3) 3- and 5-hydroxylase(s); (4) O-methyltransferase; (5) 4-coumarate:CoA ligase; R = H, OH, or OCH_3. The enzymes marked by numbers in B are: (1) acetyl-CoA carboxylase; (2) chalcone synthase (CHS); (3) chalcone isomerase; (4) "isoflavone synthase" (IFS); (5) 3,9-dihydroxypterocarpan 6a-hydroxylase (D6aH); (6) dimethylallylpyrophasphate:3,6a, 9-trihydroxypterocarpan dimethylallyltransferase (TDT) (modified after Ebel 1986)

ding the substrates, 4-coumaroyl-CoA, malonyl-CoA, and dimethylallylpyrophosphate for pterocarpan phytoalexin biosynthesis. The reaction linking general phenylpropanoid metabolism to flavonoid/isoflavonoid biosynthesis is catalyzed by chalcone synthase, which uses 4-coumaroyl-CoA and malonyl-CoA as substrates. The central reaction of isoflavonoid biosynthesis involves a 2,3-aryl migration with a flavanone as substrate. This reaction has been resolved into two steps. In a first step soybean microsomal preparations catalyze the NADPH- and dioxygen dependent conversion of a flavanone to give an intermediate, which in a second reaction, formally a dehydration, yields an isoflavone (Kochs and Grisebach 1986). Two enzymic reactions that utilize pterocarpanoid substrates have been identified in soybean cell-free extracts. A microsomal pterocarpan 6a-hydroxylase converts 3,9-dihydroxypterocarpan stereospecifically to 3,6a(S),9-trihydroxypterocarpan (glycinol) in the presence of NADPH and dioxygen (Hagmann et al 1984). A dimethylallylpyrophosphate:3,6a,9-trihydroxypterocarpan dimethylallyltransferase in a particulate fraction catalyzes the formation of the 4- as well as 2-dimethylallyltrihydroxypterocarpan, the suspected intermediates in the biosynthesis of the various glyceollin isomers (Zähringer et al 1981) (Fig. 1). Identification of enzymes related to glyceollin biosynthesis was largely facilitated by the fact that their respective activities are greatly enhanced upon challenge of soybean tissues by either elicitor or infection. The results of various experiments indicate that phytoalexin production in soybean depends on a general enhancement of the activity levels of all of the phytoalexin biosynthetic enzymes.

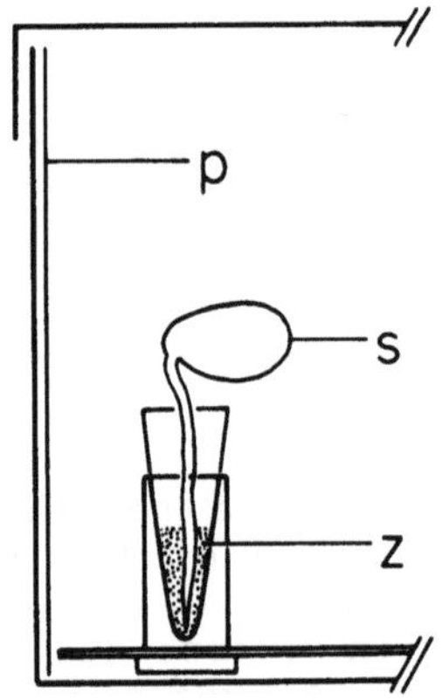

Fig. 2. Dip-inoculation of soybean seedling roots with zoospores of P. megasperma f. sp. glycinea. Two-d-old seedlings (s) were placed in test tubes containing 100 µl zoospore suspension (z) (about 10^4 zoospores) and incubated in covered glass dishes lined with wet filter paper (p) (from Hahn et al 1985)

PHYTOALEXIN PRODUCTION IN SOYBEAN ROOTS FOLLOWING INFECTION WITH P. MEGASPERMA

The natural infective propagules of P. megasperma are zoospores, which frequently infect the soybean root (Fig. 2). Using a radioimmunoassay for the major root phytoalexin, glyceollin I, and an immunofluorescent stain for the fungal hyphae, the quantitative and temporal pattern of glyceollin in relation to P. megasperma hyphae was examined in detail by Hahn et al (1985). As shown in Fig. 3, the timing of diffe-

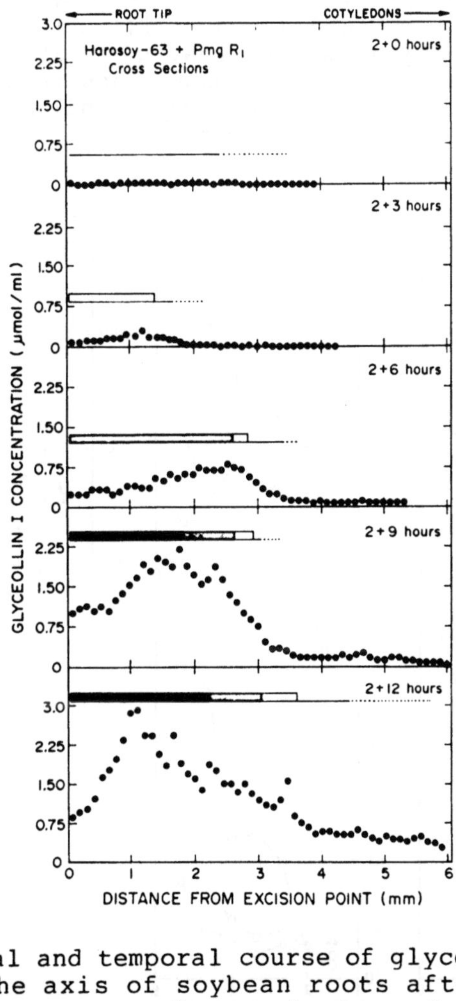

Fig. 3. Spatial and temporal course of glyceollin I accumulation along the axis of soybean roots after infection with race 1 of P. megasperma f. sp. glycinea. Seedlings were dip-inoculated as described in Fig. 2. Glyceollin I was quantitated in extracts of cryotome cross-sections by radioimmunoassay. The extent of fungal colonization, determined immunohistochemically, is indicated by the bars: (■) > 90% of cross-section colonized; (▦) 50 to 90% colonized; (□) 25 to 50% colonized (from Hahn et al 1985)

rential accumulation and the spatial distribution of glyceollin I correlated well with the restricted (incompatible interaction; race I) growth of the hyphae in the root tissue. Glyceollin I was first quantifiable in cross-sections 5 h after infection with race 1. At this time it was present at concentrations approaching the in vitro EC_{50} of this phytoalexin (0.17 µmol/ml) (Lazarovits and Ward 1982). At later times, the phytoalexin concentration in resistant roots reached or exceeded the in vitro EC_{90} (0.6 µmol/ml), dropping sharply at the leading edge of the infection. In cross-sections from roots inoculated with the compatible race 3 only low concectrations ($<$ 0.08 µmol/ml) of glyceollin I were detected in some sections of roots 11 and 14 h following inoculation. The analysis showed that glyceollin I accumulation is an early response of soybean roots to infection with an incompatible race of P. megasperma, which is correlated with the inhibition of fungal growth.

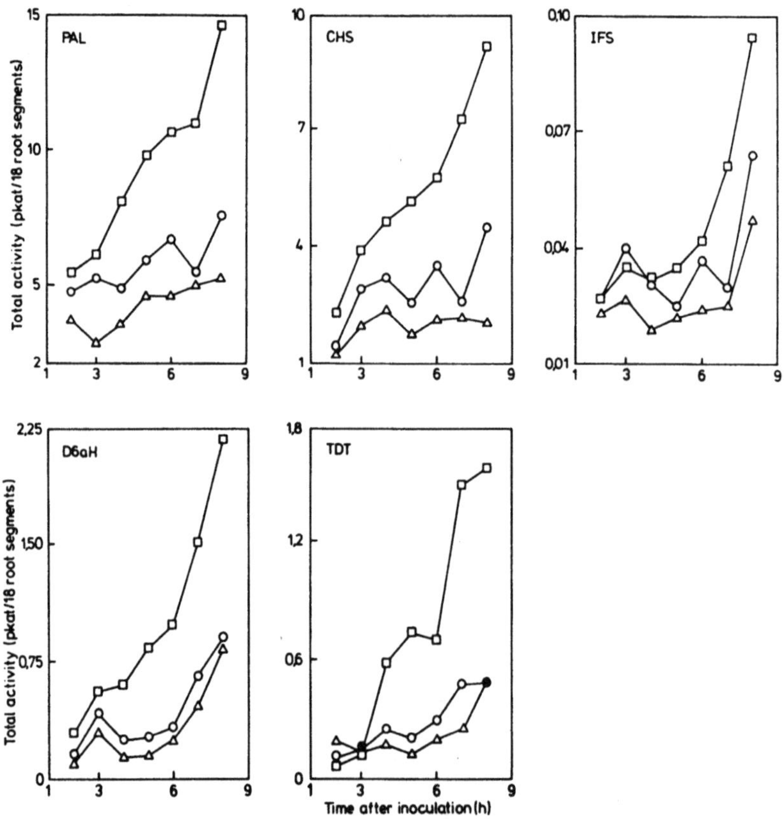

Fig. 4. Changes in total phytoalexin biosynthetic enzyme activities in soybean roots after dip-inoculation with zoospores of P. megasperma race 1 (□) or race 3 (○) and in water controls (△). For abbreviations see Fig. 1

Using the same inoculation procedure, the activities of glyceollin biosynthetic enzymes were studied at various times after infection of soybean roots with different P. megasperma races (Bonhoff et al 1986a,b). Marked differences in the enzyme activity patterns were found between the incompatible and compatible interactions (Fig. 4). At early stages (2-8 h) of infection with race I large activity increases were observed for phenylalanine ammonia-lyase, chalcone synthase, 6a-hydroxylase, and prenyltransferase, concomitant with the onset of glyceollin accumulation (Fig. 3). Following infection with race 3, the activity levels of these enzymes remained low throughout the period investigated. In contrast to the other enzymes, the isoflavone synthase activity patterns in the differently infected and non-infected roots were quite similar.

Interestingly, the activities of NADPH-cytochrome reductase and hydroxymethylglutaryl-CoA reductase, enzymes indirectly involved in glyceollin biosynthesis, and the activities of isocitrate dehydrogenase, glucose 6-phosphate dehydrogenase, and glutamate dehydrogenase were not significantly different in roots from infected and non-infected soybean seedlings (Bonhoff et al 1986a,b). In contrast, in experiments with wounded soybean hypocotyls infected with mycelium of P. megasperma, marked increases were found in the activities of glucose 6-phosphate dehydrogenase and glutamate dehydrogenase and in the activities of phytoalexin biosynthetic enzymes irrespective of the P. megasperma race used for inoculation (Börner and Grisebach 1982).

Thus, in soybean roots the biochemical responses as measured by phytoalexin accumulation and induction of phytoalexin biosynthetic enzymes largely reflect the known differences in the physiology of the race:cultivar-specific interaction of P. megasperma and soybean. The results with this experimental system further support the hypothesis that phytoalexin production is one early defense response of soybean against P. megasperma (Hahn et al 1985; Yoshikawa et al 1978).

MODE OF ACTION OF A P. MEGASPERMA GLUCAN ELICITOR

The initial event in the elicitor-mediated induction of phytoalexin production is its interaction with the presumed primary target site(s). Various sites, including the cell wall, the plasmalemma, and some intracellular components have been suggested as putative targets for the primary elicitor action. The hypothesis was tested whether soybean membrane fractions contain binding sites for the P. megasperma glucan elicitor (Schmidt and Ebel 1986). To help in the identification of binding sites, a glucan elicitor fraction with an average degree of polymerization of 20 to 25 was prepared by partial acid hydrolysis of P. megasperma cell walls and low- and high-resolution gel-filtration on Bio-Gel P-4 of the acid solubilized material. The elicitor fraction was composed entirely of glucose and possessed a glucosyl-linkage composition typical of the P. megasperma cell-wall branched ß-glucan consisting of terminal, 3-, 6-, and 3,6-linked glucosyl residues (Ayers et al 1976b, c; Sharp et al 1984). The elicitor fraction showed high phytoalexin elicitor activity when tested in a cotyledon bioassay (Ayers et al 1976a; Hahn et al 1981) (Table 1). The radioactive elicitor ligand utilized in the binding studies was obtained by reduction of the oligoglucan fraction with sodium boro(^{3}H)hydride to yield (^{3}H)oligoglucosyl-glucitols with a specific radioactivity of about 74 GBq/mmol.

Membrane binding of the glucan elicitor was demonstrated by incubating soybean root microsomal fractions with 80 nM of (^{3}H)glucan elicitor and separating free from bound elicitor by centrifugation of the reaction mixture through a sucrose layer. Elicitor binding fulfilled the criterion for specific (saturable) binding in that competition for the binding site(s) could be demonstrated in the presence of increasing concentrations of unlabelled glucan elicitor (Fig. 5). Scatchard-analysis of the data (Fig. 5) showed that elicitor binding was of high affinity with an apparent K_D value of 0.2 µM and that root microsomal fractions contained a maximal concentration of binding site(s) of 0.5 pmol/mg protein.

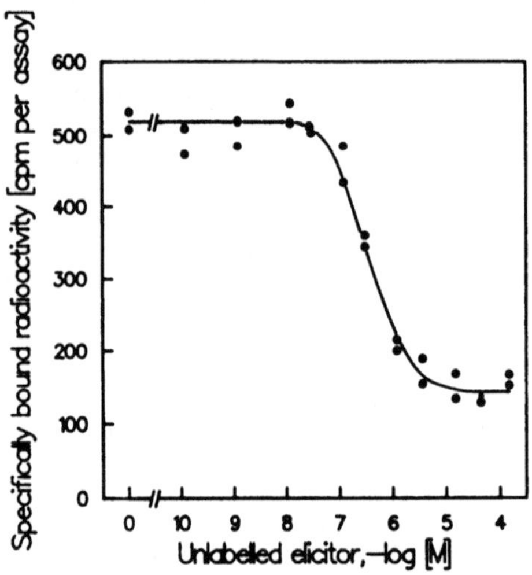

Fig. 5. Displacement of P. megasperma (^{3}H)glucan elicitor by unlabelled glucan elicitor. Soybean root microsomes were incubated at 4°C with 80 nM labelled ligand, in the absence (O) or presence of various concentrations of unlabelled ligand. After 120 min, the samples were centrifuged and the membrane pellets analyzed for radioactivity

Highest binding activity was associated with a plasma membrane-enriched fraction, as we were able to demonstrate after separating the membranes of cell-free root extracts by centrifugation on a linear sucrose density gradient and comparing the distribution in the gradient fractions of elicitor binding and of marker enzyme activities. The ligand specificity of the glucan elicitor binding was analyzed by the displacing ability of a number of polysaccharides (Table 1).

Table 1. Comparison of various polysaccharides for their displacing ability of (^{3}H)glucan elicitor binding and their phytoalexin elictor activity

Substance*	Chemical characteristics (Preparation)	Replacement of ß-glucan IC_{50} †	Phytoalexin synthesis EC_{50} †
		µg/ml (µM)	µg/ml (µM)
Phytophthora megasperma Glucan elicitor	ß-1,3(55%)-,1,6(41%)-glucan (partially acid hydrolyzed)	1 (0.3)	0.3(0.1)
Glucan elicitor	30% carbohydrate (water extracted)	3	2
Phytophthora spp. Mycolaminaran	ß-1,3(94%)-,1,6(6%)-glucan	200 (40)	200 (40)
Alternaria carthami Elicitor	50% carbohydrate (water extracted)	300	180
Laminaria digitata Laminarin	ß-1,3(appr. 90%)-,1,6-glucan	50	30
Laminarin	(reduced)	50	80

* No replacement and no phytoalexin synthesis found with: Mannan, glucan (S. cerevisiae), dextran, chitosan, xylan, arabinogalactan, pullulan, pustulan, lichenan, nigeran, and oligogalacturonide.

† IC_{50}: 50% inhibitory concentration as calculated from the dose response curves of competitors on specific (^{3}H)glucan binding; EC_{50}: 50% stimulatory concentration as calculated from the dose response curves of substances for phytoalexin production in the cotyledon bioassay.

IC_{50} (50% inhibitory concentration) values for a number of polysaccharides were calculated from the dose response curves of their displacement of (^{3}H)glucan binding. The displacing ability was compared with EC_{50} (50% stimulatory concentration) values for phytoalexin elicitor activity in a cotyledon bioassay. Interestingly, the differential effects of a number of ß-glucans such as mycolaminaran from Phytophthora spp, a fraction released from the cell wall of Alternaria carthami, and laminarin from Laminaria digitata on radiolabelled ligand binding (IC_{50} values) corresponded closely with their relative efficiency in enhancing phytoalexin production in the cotyledon bioassay (EC_{50} value; Table 1). A larger number of polysaccharides that were unable to compete for binding also failed to elicit phytoalexin production in the concentration range tested (Table 1).

SIGNAL TRANSDUCTION IN ELICITOR ACTION

Assuming that the triggering of phytoalexin production by P. megasperma or by the P. megasperma glucan elicitor is a membrane-mediated process, one has to raise important questions concerning signal generation and transduction. One such question is related to the possible involvement of metabolic control substances. Hahn and Grisebach (1983) demonstrated the presence of adenosine-3′,5′-cyclic monophosphate (cAMP) in soybean hypocotyls by a radioimmunoassay.

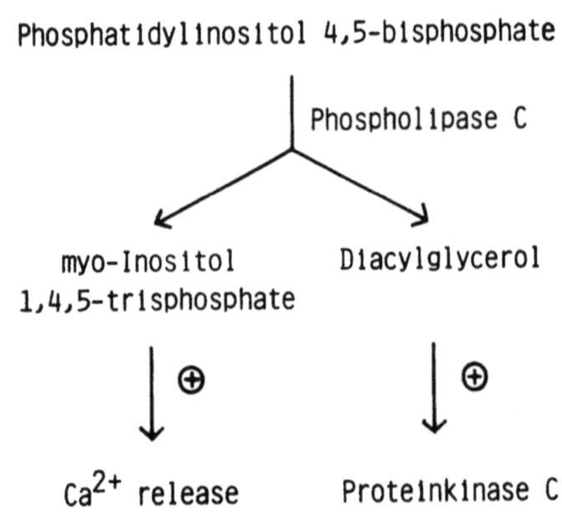

Fig. 6. Proposed second messenger role of inositol trisphosphate and diacycglycerol in vertebrate cells. Following agonist stimulation, phosphatidylinositol 4,5-bisphosphate is hydrolyzed to give myo-inositol 1,4,5-trisphosphate and diacylglycerol. Inositol trisphosphate acts to mobilize intracellular calcium, whereas diacylglycerol stimulates proteinkinase C

The cAMP level was, however, not affected by infection of soybean hypocotyls with P. megasperma, nor was cAMP found in soybean cell cultures. Hahn and Grisebach (1983) concluded that it is unlikely that cAMP, which serves as a second messenger in animal systems, plays an analogous role in the response of soybean to infection with P. megasperma or in the response of cell cultures to the glucan elicitor. In vertebrate cells, inositol trisphosphate and diacylglycerol have been shown to function as messengers in cellular signal transduction (Berridge 1984) (Fig. 6). Phosphatidylinositol monophosphate and phosphatidylinositol bisphosphate have recently been identified as lipids from soybean cell-suspension cultures by chromatographic comparison with reference substances (Strasser et al 1986). The polyphosphoinositide turnover in response to the glucan elicitor was studied. Soybean cell-suspension cultures were labelled with either myo-(2-^{3}H)inositol or (^{32}P)orthophosphate and subsequently challenged with elicitor for various lengths of time. It was found that the elicitor did not significantly affect the incorporation of the precursors into polyphosphoinositides of soybean cells between 3 and 20 min after elicitor addition (Fig. 7). Similar results were obtained in studies on phospholipid turnover with parsley cell-suspension cultures challenged with an elicitor from Alternaria carthami for 0.5 to 20 min (Strasser et al 1986). The fastest elicitor-mediated

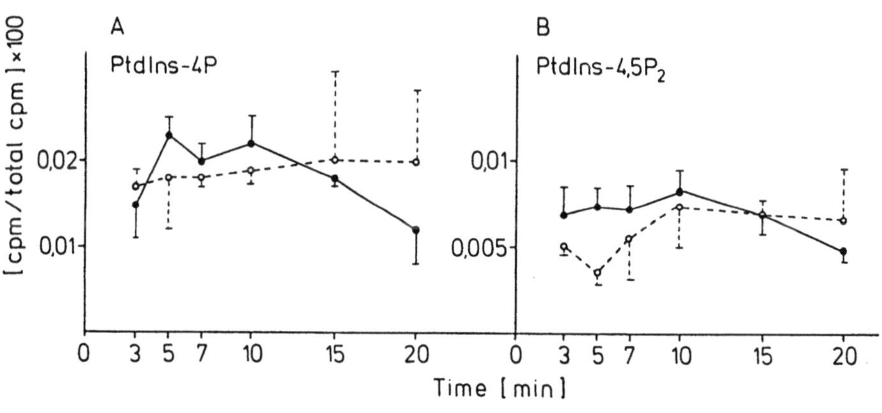

Fig. 7. Radioactivity incorporated from (^{32}P)orthophosphate into phosphatidylinositol 4-phosphate (A) and phosphatidylinositol 4,5-bisphosphate (B) in soybean cell cultures at various times after elicitor addition. The phospholipids were separated by thin-layer chromatography. The amounts of radioactivity in the phosphoinositides are indicated as % of radioactivity in the total phospholipids. (●) elicitor-treated; (○) untreated control samples; results are averages from three determinations ± SD

effects on plant cell cultures measured so far have been reported to occur within two to five min after elicitor application (Somssich et al 1986; Strasser et al 1983). Any possible modulation of the plant cell phospholipid turnover in response to elicitor would have been expected to result during a similar period following onset of challenge. It was therefore concluded that polyphosphoinositides are not involved in signal transduction in elicitor-challenged soybean cell cultures (Strasser et al 1986).

Another hypothesis is that the biological action of elicitors is mediated by Ca^{2+} ions. Ebel et al (1985) reported recently that Ca^{2+} ions stimulate the *P. megasperma* glucan elicitor activity, as measured by the enhancement of phenylalanine ammonia-lyase and chalcone synthase activities and glyceollin accumulation in soybean cell cultures. As shown in Fig. 8, verapamil, a Ca^{2+} channel blocker in animal cells, and trifluoperazine, an inhibitor of calmodulin-dependent reactions, at concentrations below 0.1 mM had little effect on the glucan elicitor activity as measured by the increases in phytoalexin biosynthetic enzyme activities. However, La^{3+} ions markedly affected elicitor induction of enzyme activities and glyceollin production at concentrations ≤ 0.1 mM (Stäb and Ebel 1985). Effective La^{3+} concentrations for 50% inhibition were 40 µM for phenylalanine ammonia-lyase activity, 0.1 mM for chalcone synthase activity, and 30 µM for glyceollin production. These results suggest that an as yet unknown cellular Ca^{2+} pool and changes in Ca^{2+} pool sizes are important in the action of the glucan elicitor.

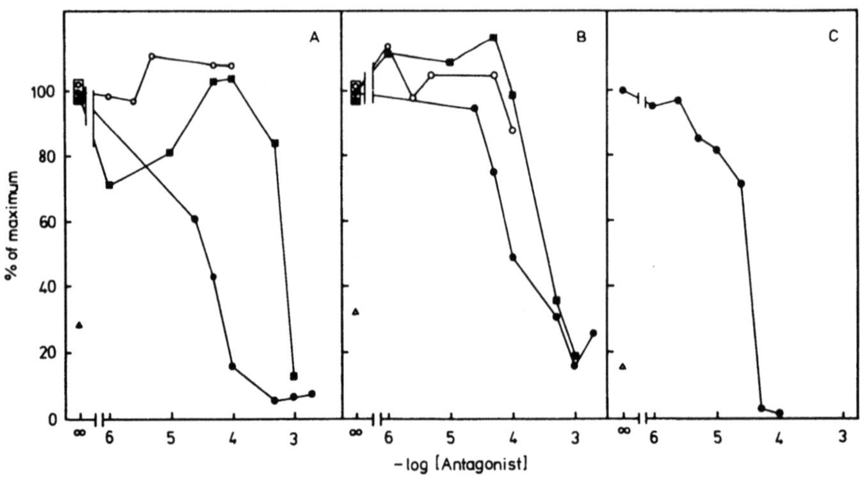

Fig. 8. Effects of verapamil (■), trifluorperazine (○), and La^{3+} (●) on the elicitor enhancement of phenylalanine ammonia-lyase (A) and chalcone synthase (B) activities and of glyceollin accumulation (C) in soybean cell cultures. (Δ) values of untreated control cells

The cell wall of soybean cell cultures could be a source of both Ca^{2+} ions and other endogenous substances (Darvill and Albersheim 1984) which might participate in the induction of phytoalexin production by the microbial elicitor. In an attempt to study these possible functions, Mieth et al (1986) tried to exploit protoplasts as an experimental system. The results showed that under the methods employed enzymatically isolated soybean protoplasts displayed a stress response which resulted in phytoalexin production and were not capable of responding to the _P. megasperma_ glucan elicitor. At least two components of the protoplast isolation medium, the osmoticum and the fungal endopolygalacturonase, had the potential to initiate the observed phytoalexin synthesis during protoplast isolation. Osmotic stress-mediated phytoalexin synthesis again points to the possible involvement of changes in ion gradients in the mechanism of elicitor action. Further investigations are required to demonstrate whether any of such changes is causally related to elicitor induction of phytoalexin production.

REGULATION OF PHYTOALEXIN BIOSYNTHETIC ENZYMES

Control of synthesis is probably one important regulatory mechanism of phytoalexin production in response to both infection and different types of elicitor (Moesta and Grisebach 1981a,b). As discussed earlier in this contribution, infection with _P. megasperma_ or treatment with the glucan elicitor derived from the fungus induced rapid and large transient increases in the activities of all enzymes as yet known to be involved in the phytoalexin biosynthetic pathway in different soybean tissues, including cotyledons, hypocotyls, roots, and cell cultures. These increases occurred concomitantly with the onset of accumulation of glyceollin and structurally related isoflavonoid compounds (Fig. 1). _In vivo_ labelling studies and _in vitro_ translation of mRNA have demonstrated that changes in the activities of three of the biosynthetic enzymes, phenylalanine ammonia-lyase, 4-coumarate:CoA ligase, and chalcone synthase, were preceded by transient increases in their rates of synthesis (Börner and Grisebach 1982; Ebel et al 1984; Grab et al 1985; Hille et al 1982; Schmelzer et al 1984). In elicitor-challenged soybean cell cultures, chalcone synthase mRNA activity increased as early as 40 to 60 min after the onset of induction and reached a maximum at about 5 h (Ebel et al 1984; Grab et al 1985) and was correlated with an increased amount of mRNA (Schmelzer et al 1984). Similar results were obtained for soybean seedlings following infection with the incompatible race 1 of _P. megasperma_ (Schmelzer et al 1984).

In vivo labelling studies and _in vitro_ translation of mRNA showed also that elicitor treatment of soybean cell cultures caused major changes in the rates of synthesis of many unidentified proteins as well as of phytoalexin biosynthetic enzymes (Grab et al 1985) (Fig. 9). The induced meta-

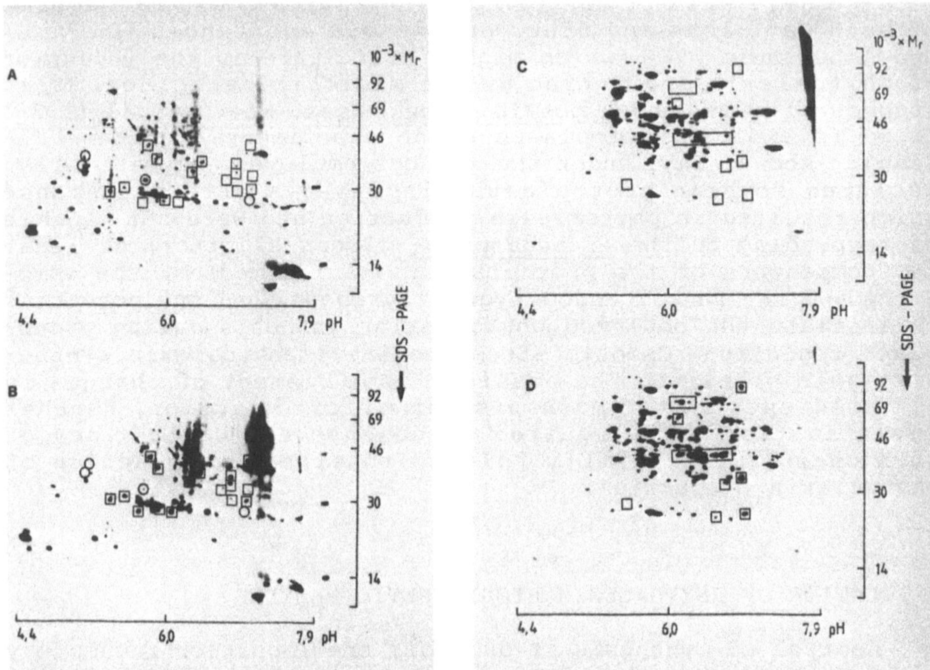

Fig. 9. Fluorographs of proteins synthesized in the presence of (^{35}S)methionine in vitro (A, B) from total cellular RNA or in vivo (C, D) and resolved by two-dimensional electrophoresis. Soybean cells were exposed for 4 (B) or 6.5 h (D) to the elicitor or incubated in the absence of elicitor (A, C). Reference notations: (□) proteins for which rates of synthesis were increased by the elicitor; (○) proteins for which rates of synthesis were decreased by the elicitor; arrows denote proteins for which rates of synthesis remained unchanged during elicitor treatment.

bolic changes involved decreases in a small number and increases in a large number of mRNA activities. The induced changes may involve mRNAs encoding hydroxyproline-rich glycoprotein, as observed in stressed bean tissues (Showalter et al 1985), and a number of proteins classified as pathogenesis-related proteins, as demonstrated for challenged parsley cell cultures (Somssich et al 1986). Studies of plant cell responses to elicitors may thus help to identify other plant defense reactions besides the accumulation of phytoalexins (Ebel 1986).

Acknowledgements. Research in our laboratories was supported by the Deutsche Forschungsgemeinschaft (SFB 206), Fonds der Chemischen Industrie, and BASF (Ludwigshafen).

REFERENCES

Ayers AR, Ebel J, Finelli F, Berger N, Albersheim P (1976a) Host-pathogen interactions IX. Quantitative assays of elicitor activity and characterization of the elicitor present in the extracellular medium of cultures of Phytophthora megasperma var sojae. Plant Physiol 57:751-759

Ayers AR, Ebel J, Valent B, Albersheim P (1976b) Host-pathogen interactionx X. Fractionation and biological activity of an elicitor isolated from the mycelial walls of Phytophthora megasperma var sojae. Plant Phyisol 57:760-765

Ayers AR, Valent B, Ebel J, Albersheim P (1976c) Host-pathogen interactions XI. Composition and structure of wall-released elicitor fractions. Plant Physiol 57:766-774

Bailey JA, Mansfield JW (eds) (1982) Phytoalexins. Blackie, Glasgow

Berridge MJ (1984) Inositol trisphosphate and diacylglycerol as second messengers. Biochem J 220:345-360

Börner H, Grisebach H (1982) Enzyme induction in soybean infected by Phytophthora megasperma f sp glycinea. Arch Biochem Biophys 217:65-71

Bonhoff A, Loyal R, Ebel J, Grisebach H (1986a) Race:cultivar-specific induction of enzymes related to phytoalexin biosynthesis in soybean roots following infection with Phytophthora megasperma f sp glycinea. Arch Biochem Biophys 246:149-154

Bonhoff A, Loyal R, Feller K, Ebel J, Grisebach H (1986b) Further investigations of race:cultivar-specific induction of enzymes related to phytoalexin biosynthesis in soybean roots following infection with Phytophthora megasperma f sp glycinea. Biol Chem Hoppe-Seyler, in press

Darvill AG, Albersheim P (1984) Phytoalexins and their elicitors - A defense against microbial infection in plants. Annu Rev Plant Physiol 35:243-275

Ebel J (1986) Phytoalexin synthesis: the biochemical analysis of the induction process. Annu Rev Phytopathol 24:235-264

Ebel J, Schmidt WE, Loyal R (1984) Phytoalexin synthesis in soybean cells: Elicitor induction of phenylalanine ammonia-lyase and chalcone synthase mRNAs and correlation with phytoalexin accumulation. Arch Biochem Biophys 232:240-248

Ebel J, Stäb MR, Schmidt WE (1985) Induction of enzymes of phytoalexin synthesis in soybean cells by fungal elicitor. In: Neumann K-H, Barz W, Reinhard E (eds) Primary and secondary metabolism of plant cell cultures. Springer, Berlin Heidelberg New York Tokyo, p 247-254

Grab D, Loyal R, Ebel J (1985) Elicitor-induced phytoalexin synthesis in soybean cells: Changes in the activity of chalcone synthase mRNA and the total population of translatable mRNA. Arch Biochem Biophys 243:523-529

Hagmann ML, Heller, W, Grisebach H (1984) Induction of phytoalexin synthesis in soybean. Stereospecific 3,9-dihydroxypterocarpan 6a-hydroxylase from elicitor-induced soybean cell cultures. Eur J Biochem 142:127-131

Hahn MG, Bonhoff A, Grisebach H (1985) Quantitative localization of the phytoalexin glyceollin I in relation to fungal hyphae in soybean roots infected with Phytophthora megasperma f sp glycinea. Plant Physiol 77:591-601
Hahn MG, Darvill AG, Albersheim P (1981) Host-pathogen interactions XIX. The endogenous elicitor, a fragment of a plant cell wall polysaccharide that elicits phytoalexin accumulation in soybeans. Plant Physiol 68:1161-1169
Hahn MG, Grisebach H (1983) Cyclic AMP is not involved as a second messenger in the response of soybean to infection by Phytophthora megasperma f sp glycinea. Z Naturforsch 38c:578-582
Hille A, Purwin C, Ebel J (1982) Induction of enzymes of phytoalexin synthesis in cultured soybean cells by an elicitor form Phytophthora megasperma f sp glycinea. Plant Cell Rep 1:123-127
Horsfall JC, Cowling EB (eds) (1980) Plant disease, vol 5. Academic Press, London New York
Kochs G, Grisebch H (1986) Enzymatic synthesis of isoflavones. Eur J Biochem 155:311-318
Lazarovits G, Ward EWB (1982) Relationship between localized glyceollin accumulation and metalaxyl treatment in the control of Phtophthora rot in soybean hypocotyls. Phytopathology 72:1217-1221
Mieth H, Speth V, Ebel J (1986) Phytoalexin production by isolated soybean protoplasts. Z Naturforsch 41c:193-201
Moesta P, Grisebach H (1981a) Investigation of the mechanism of phytoalexin accumulation in soybean induced by glucan or mercuric chloride. Arch Biochem Biophys 211:39-43
Moesta P, Grisebach H (1981b) Investigation of the mechanism of glyceollin accumulation in soybean infected by Phytophthora megasperma f sp glycinea. Arch Biochem Biophys 212:462-467
Schmelzer E, Börner H, Grisebach H, Ebel J, Hahlbrock K (1984) Phytoalexin synthesis in soybean (Glycine max). Similar time courses of mRNA induction in hypocotyls infected with a fungal pathogen and in cell cultures treated with fungal elicitor. FEBS Lett 172:59-63
Schmidt WE, Ebel J (1986) Specific binding of a fungal glucan phytoalexin elicitor to membrane fractions from soybean (Glycine max). Proc Natl Acad Sci USA, submitted for publication
Sharp JK, McNeil M, Albersheim P (1984) Purification and partial characterization of a ß-glucan fragment that elicits phytoalexin accumulation in soybean. J Biol Chem 259:11312-11320
Showalter AM, Bell JN, Cramer CL, Bailey JA, Varner JE, Lamb CJ (1985) Accumulation of hydroxyproline-rich glycoprotein mRNAs in response to fungal elicitor and infection. Proc Natl Acad Sci USA 82:6551-6555
Somssich IE, Schmelzer E, Bollmann H, Hahlbrock K (1986) Rapid activation by fungal elicitor of genes encoding "pathogenesis-related" proteins in cultured parsley cells. Proc Natl Acad Sci USA, 83:2427-2430

Stäb, MR, Ebel J (1985) Role of Ca^{2+} in the induction of the phytoalexin defense response in soybean cells. In: Molecular and cellular aspects of Ca^{2+} in plant development. NATO workshop. 15-20 July. Edinburgh, abstract A9

Strasser H, Matern U, Hoffmann C, Grisebach H (1986) Are polyphosphoinositides involved in signal transduction of elicitor-induced phytoalexin synthesis in cultured plant cells? Z Naturforsch, in press

Strasser H, Tietjen KG, Himmelspach K, Matern U (1983) Rapid effect of an elicitor on uptake and intracellular distribution of phosphate in cultured parsley cells. Plant Cell Rep 2:140-143

Yoshikawa M, Yamauchi K, Masogo H (1978) Glyceollin: its role in restricting fungal growth in resistant soybean hypocotyls infected with Phtophthora megasperma var sojae. Physiol Plant Pathol 12:73-82

Zähringer U, Schaller E, Grisebach H (1981) Induction of phytoalexin synthesis in soybean. Structure and reactions of naturally occurring and enzymatically prepared prenylated pterocarpans from elictor-treated cotyledons and cell cultures of soybean. Z Naturforsch 36c:234-241

ELICITATION OF INDOLE ALKALOID BIOSYNTHESIS IN PERIWINKLE

U. Eilert[2], V. DeLuca[1], F. Constabel[1] and W.G.W. Kurz[1]
[1]Plant Biotechnology Institute
National Research Council of Canada
Saskatoon, Saskatchewan, S7N 0W9, Canada, NRCC No. 25554
[2]Inst. Pharmazeut. Biologie, Braunschweig, FRG

INTRODUCTION

Until very recently (Eilert et al 1986), transfer of cells to alkaloid production medium was the only method available to induce indole alkaloid accumulation and enzymes involved in their biosynthesis. However, the slow rates of enzyme induction observed presented difficulties in performing detailed biochemical investigations. Eilert et al (1986) reported that elicitor treatment of Catharanthus cell suspension cultures stimulated production of indole alkaloids within 12-24 h after treatment and the alkaloid profile obtained was qualitatively and quantitatively identical to that obtained by cultivation of cells in alkaloid production medium.

In following up these results we now demonstrate that elicitor-stimulated indole alkaloid accumulation is linked to the induction of biosynthetic pathway enzymes. The induction of tryptophan decarboxylase (TDC) activity which produces tryptamine, an indole alkaloid precursor, and of strictosidine synthase (SS) activity which produces strictosidine, the first monoterpene indole alkaloid of this pathway, will be described.

RESULTS AND DISCUSSION

Cells grown in 1B5 medium were treated with Pythium aphanidermatum elicitor at 2,4,6 and 8 days after transfer to fresh 1B5 medium. They responded with rapid increase in TDC activity (Fig. 1) which varied between 3.5 to 12-fold over control cells

NATO ASI Series, Vol. H 4
Recognition in Microbe-Plant Symbiotic and Pathogenic Interactions. Edited by B. Lugtenberg

and did not correlate with the age of the culture (Fig. 1). SS activities were also stimulated by elicitor treatment (Fig. 1B). For this enzyme, culture age was an important factor of cell response. Two to 6 day-old cells were found to be highly responsive, showing 5- to 10-fold increased SS activities compared to controls within 12-24 h of Pythium treatment, whereas older cells did not respond to such treatment.

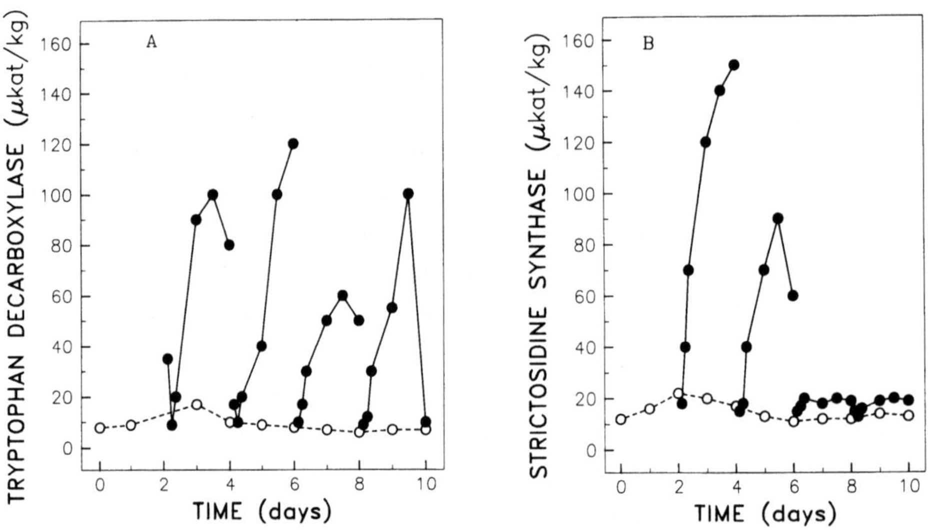

Fig. 1. Effect of fungal elicitor on the TDC (A) and SS (B) activities from cells cultured on 1B5 medium. At different times (2,4,6 or 8 days) after transfer, cells were treated with fungal elicitor and approximately 1 gram fresh weight of cells was harvested for enzyme assay at the given times after elicitation (—●—). Controls (—○—). Cells were prepared and enzymes were assayed as described previously (Sasse et al 1983; Mizukami et al 1979).

The previous experiment was repeated with cells transferred to alkaloid production medium. Both TDC (Fig. 2A) and SS (Fig. 2B) specific activities increased with time after transfer of cells to alkaloid production medium. Strictosidine synthase specific activity began to increase approximately 60 h before that of TDC and had already peaked approximately 50 h prior to the TDC peak. Treatment of cells with Pythium homogenate at

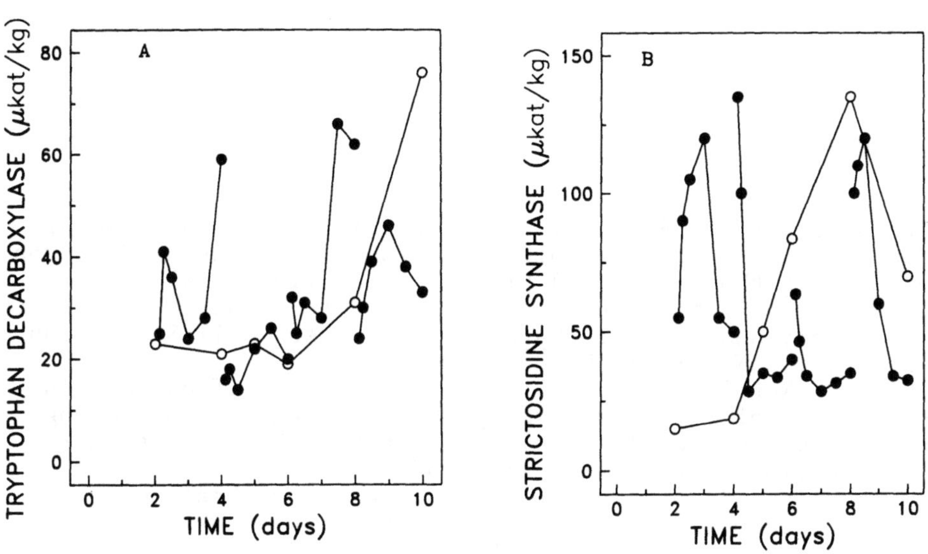

Fig. 2. Effect of fungal elicitor on TDC (A) and (SS) activities from cells cultured on APM medium. Experiment was conducted as in Figure 1. Symbols: (—o—), control cells; (—●—), elicited cells.

2,4,6 and 8 days after transfer to alkaloid production medium resulted in rapid changes in the levels of both enzymes (Fig. 2B), which varied with the age of the culture.

The induction of TDC activity has been proposed to be a key step in the activation of the pathway for the biosynthesis of indole alkaloids in Catharanthus roseus cell suspension cultures (Noe and Berlin 1986). The early induction of SS activity compared to that of TDC observed in this report allowed for the depletion of tryptamine pools by channeling of this amine into indole alkaloids (not shown). When the tryptamine level decreased by two-thirds (from 61 to 19 mg/g fresh weight between day 4 to 6), TDC activity began to rise. Therefore, decarboxylation of tryptophan may indeed be a "bottleneck" as suggested previously (Noe and Berlin 1986). Tryptamine, however, seems to regulate TDC activity and thus may prevent excessive channeling of shikimic acid pathway intermediates rather than to activate indole alkaloid biosynthesis by tryptamine formation. We suggest, therefore, that the channeling of tryptamine to indole alkaloids must be controlled at some other point in this biosynthetic pathway.

REFERENCES

Eilert U, Constabel F, Kurz, W (1986) Elicitor-stimulation of Monoterpene Indole Alkaloid Formation in Suspension Cultures of Catharanthus roseus. J Plant Physiol (in press)

Sasse F, Buchholz M and Berlin J (1983) Site of Action of Growth Inhibitory Analogues in Catharanthus roseus Cell Suspension Cultures. Z Naturforsch 38c: 910-915

Mizukami H, Nordlov H, Lee S, Scott A (1979) Purification and Properties of Strictosidine Synthase (an enzyme condensing tryptamine and secologanin) from Catharanthus roseus cultured cells. Biochemistry 18: 3760-3763

Noe W, Berlin J (1986) Induction of de-novo synthesis of tryptophan decarboxylase in cell suspension cultures of Catharanthus roseus. Planta 166: 500-504

EXPRESSION OF PLANT GENES IN THE HYPERSENSITIVE REACTION OF FRENCH BEAN (PHASEOLUS VULGARIS) TO THE PLANT PATHOGENIC BACTERIUM PSEUDOMONAS SYRINGAE PV. PHASEOLICOLA

A.J. Slusarenko, A. Longland and J. Friend
Department of Plant Biology and Genetics
University of Hull, Hull HU6 7RX, UK

INTRODUCTION

The hypersensitive response (HR) is typified by a rapid, localized necrosis of host cells, beyond which the pathogen does not spread. Necrosis is only a visible clue that indicates the hypersensitive response has occurred. HR encompasses i) recognition of the pathogen, ii) induction of the response, iii) events leading to cell collapse, iv) necrosis, v) formation of antimicrobial substances and vi) limitation of the pathogen (Klement 1982). Some fungal cell wall constituents and some abiotic substances (eg $HgCl_2$) will cause, or elicit, the accumulation of antimicrobial substances called phytoalexins. These agents are called elicitors (Keen et al 1972). Physical wounding or partial freezing will also elicit phytoalexin accumulation. Bailey (1982) has pointed out that a common denominator of biotic and abiotic elicitors of phytoalexin accumulation in plants is that they cause cell injury or death. However, in the HR there are indications that it is how the cells die which is significant. Several authors have shown, by using inhibitors of eukaryotic translation, that a period of host protein synthesis is a pre-requisite for plant tissues to be able to respond hypersensitively to pathogens (Doke and Tomiyama 1975; Keen et al 1981; Lyon and Wood 1977; O'Brien and Wood 1972). Host cells which die during a HR may do so as a result of a controlled sequence in the disorganization of the cellular machinery, a kind of "organized disorganization" as opposed to a simple necrosis due to some toxic principle. This phenomenon, termed apoptosis, is well documented as a mechanism of cell turnover in animal tissues (Wyllie 1981). Apoptosis re-

NATO ASI Series, Vol. H 4
Recognition in Microbe-Plant Symbiotic and Pathogenic Interactions. Edited by B. Lugtenberg

quires translation and probably de novo transcription, although an absolute requirement for the latter has not been demonstrated (Idem). According to Bailey (1982), the plant cells which die during HR release constitutive elicitors which stimulate the surrounding healthy cells to produce phytoalexins which then accumulate in the dead cells which act as a sink. Elicitors mimic this process but usually in a non-specific manner; steps i), ii) and iii) (see above) are missing and the process begins at stage iv). The organized disorganization of host cells in the HR might regulate the changes in the host's metabolism in the adjacent healthy cells which leads to the plethora of resistance responses often found in plants e.g. lignification, accumulation of hydroxyproline-rich glycoproteins, induction of chitinase and phytoalexin biosynthesis. Indeed, in French bean hypocotyls challenged by an avirulent race of Colletotrichum lindemuthianum the, as yet unidentified, elicitation signal was shown to be transmitted intercellularly prior to the appearance of local necrosis in the HR (Bell et al 1986). Thus the importance of studying the HR as a whole tissue response is clear.

Our aim was to follow the course of host gene expression in leaf tissue expressing a HR and compare this with gene expression in a susceptible reaction. It was hoped this would show more clearly whether or not an apoptotic mode of cell death might be involved in the HR, and form a basis for further research to determine the mechanisms involved in hypersensitive cell collapse at the molecular level.

EXPERIMENTAL APPROACH

Pseudomonas syringae pv. phaseolicola causes halo blight of French bean (Phaseolus vulgaris). Several races of the pathogen can be identified using differential bean cultivars. The French bean cultivar Red Mexican (RM) is resistant to race 1 isolates of the pathogen but susceptible to races 2 and 3. Cultivar Tendergreen is resistant to race 3 isolates but is susceptible to races 1 and 2. Thus the reciprocal check criterion common to gene for gene interactions (Keen & Holliday 1982) is satisfied by this host-pathogen combination. The resistance trait in both cultivars is associated with a hypersensitive reaction and is inherited as though it were conditioned by a single dominant gene (Taylor et al 1978; Taylor, personal communication). Primary leaves from 7 to 9 day old Red Mexican seedlings were infiltrated under reduced pressure with suspensions of bacteria or buffer. Leaves were not detached from the plants and the inoculation method minimises wounding which is unavoidable if inoculum is injected with a syringe or forced through stomata by an atomizer. HR cell collapse visible to the naked eye occurred from 20 to 25 hours after inoculation. Light microscopy of trypan blue--stained leaf discs 22 hours after inoculation showed a mosaic of healthy and stained cells in the incompatible combination, whereas no stained cells were visible in the compatible combination. Thus, in our system there was ample opportunity for communication between dying and healthy cells, which if Bailey's hypothesis is correct (Bailey 1982) is of fundamental importance.

Total cellular and polysomal RNA was isolated from RM leaves at various times after inoculation and was used as a template in an in vitro translation system derived from a rabbit reticulocyte lysate (Pelham and Jackson 1976). After separation by polyacrylamide gel electrophoresis (Laemmli 1970; O'Farrell 1975) the polypeptides were visualized by fluorography (Laskey and Mills 1975).

Comparison of mRNA activities from leaf tissues inoculated with avirulent race 1, virulent races 2 or 3, or buffer showed a cascade of changes in gene activity which was characteristic for tissues undergoing a HR (Slusarenko and Longland 1986; Slusarenko 1986). Coordinated increases and decreases in certain mRNA activities were observed. Some increases in mRNA activity were transient, with two pronounced bands with M_r 74 800 and 68 400 peaking at 6 hours after inoculation with avirulent race 1 cells of P.s. pv. phaseolicola. Messenger RNA activities for polypeptides with M_r 40 600 and 43 200 were detected from 12 and 21 hours after inoculation, respectively, in the HR. Some mRNA activities, notably those encoding polypeptides with M_r 50 400 and 27 800, decreased rapidly in tissues expressing a HR. Interestingly, these decreases in mRNA activity were also found in RM leaf tissues inoculated with virulent race 2 or 3 isolates of P.s. pv. phaseolicola. However, the rates of decrease of these mRNA activities were markedly greater in the HR. All the above changes in mRNA activities were reflected in both total cellular and polysomal RNA populations. Thus, they represent real changes in gene expression since the polysomes contain transcripts of genes actively being expressed at the time of sampling. The agreement between total and polysomal mRNA activities suggests that transient changes in transcription of the relevant genes might be involved (Lawton et al 1983). However, use of specific cDNA probes to quantify any given mRNA might reveal both transcriptional and translational control steps as has been reported for induction of phenylalanine ammonia lyase and chalcone synthase in bean hypocotyls in response to an avirulent race of Colletotrichum lindemuthianum (Bell et al 1986).

For ease of comparison of relative changes in mRNA activities in the several time course experiments 1-dimensional SDS polyacrylamide gels were used to resolve the products of in vitro translation. When 2-dimensional gel electrophoresis was used several more changes in mRNA activity associated with the HR were discernible (Figure 1).

Similar co-ordinated changes in mRNA activities have been observed in tissues of Brassica campestris undergoing a HR to a non-pathogen of Brassica, Xanthomonas campestris pv. vitians (Collinge and Daniels, personal communication).

This approach appears to show that specific changes in gene activity, co-ordinated temporally, are a feature of tissues undergoing a hypersensitive response. The problem of assigning functions to the products of the genes in question must also be addressed, and it would also be desirable to locate particular mRNA activities within dying or surrounding healthy cells. To this end we shall speculate on some of the possibilities.

WHAT EVENTS MIGHT LEAD TO CELL COLLAPSE IN THE HR?

a) Plasma membrane dysfunction

Loss of electrolytes from plant cells has been observed by

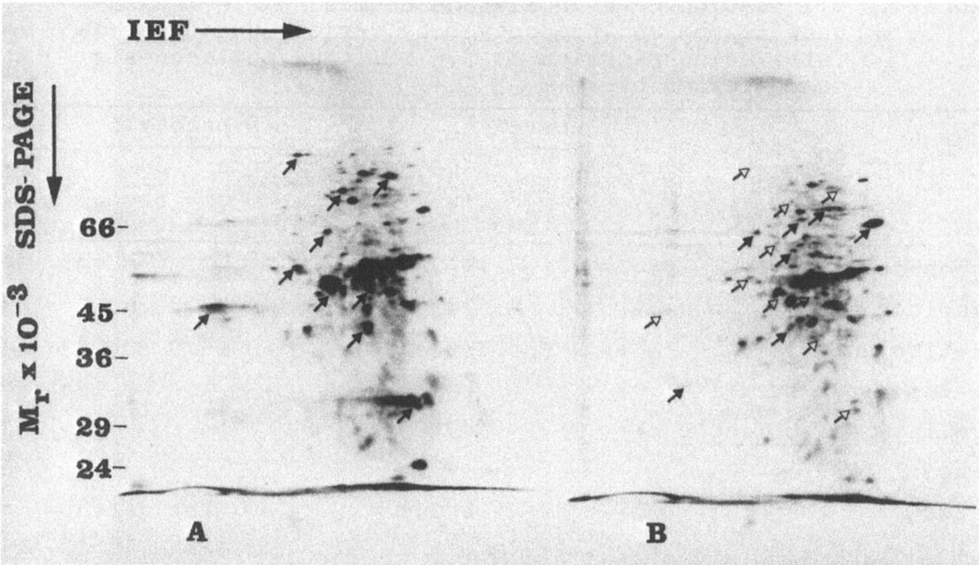

Figure 1 Two-dimensional electrophoresis of the in vitro translation products of total cellular mRNA from leaves of RM after inoculation with P.s. pv. phaseolicola race 1 (HR). A at the time of inoculation, B 6 h after inoculation. The solid arrows in A (⟶) indicate polypeptides which decrease in intensity after 6 hours and are shown in B with hollow arrowheads (⟶▷). The solid arrows in B indicate polypeptides which have increased in intensity by 6 hours after inoculation.

many authors as one of the earliest symptoms of the HR (Lyon and Wood 1975a; Atkinson et al 1985). French bean cells exposed to an avirulent race of Colletotrichum lindemuthianum were changed in the way they plasmolyse before any other microscopical effects were discernible (Bailey, personal communication). Both of these phenomena have been interpreted as indications of changes in the plasmalemma of cells undergoing a HR. Keppler and Novacky (1986) suggested that peroxidation of host membrane lipids was a key step in producing the electrolyte leakage and tissue collapse seen in the HR. In view of this we propose that some of the mRNA activities we detected increasing in the HR might give rise to polypeptides involved in an active oxygen generating system. For example lipoxygenase will oxidize unsaturated fatty acids to lipid peroxides and produce hydroxyl radicals as a by-product of the reaction (Hitchcock and Nichols 1971). We feel this may be the mechanism by which exogenously applied arachidonic and linoleic acids cause cell death and elicit phytoalexins in French Bean

TABLE 1. Elicitation of necrosis and coumestrol in French bean leaves and hypocotyls by various unsaturated fatty acids supplied at 3.3 mM. Maximum necrosis is indicated by ++++

Treatment	Leaves		Hypocotyls	
	Symptoms	Coumestrol ($\mu g\ g^{-1}$ dry weight)	Symptoms	Coumestrol ($\mu g\ g^{-1}$ dry weight)
Arachidonic acid	+++	52.8	++	0.32
Linoleic acid	++++	26.3	+	0.35
α-Linolenic acid	±	Not detectable	±	0.183
γ-Linolenic acid	+	trace	±	0.108
P.s. pv *phaseolicola* race 1 (HR)	+++	49.2		
Water	-	trace	-	Not detectable

leaves (Table 1). Soybean lipoxygenase oxidizes arachidonic and linoleic acids at a greater rate than α- or γ-linolenic acid (Holman et al 1969); if French bean lipoxygenase has similar specificity this could account for the difference in elicitor activity of the fatty acids tested. The lipid peroxy or hydroxyl radicals generated by lipoxygenase activity could damage membrane lipids by causing *in situ* peroxidation or cause cell death by damaging membrane transport proteins and disrupting the ionic homeostasis of the cells. Indeed, damage to membrane transport proteins has been suggested as a primary cause of radical-induced cell death (Wolff et al 1986). There have been many recent reports implicating various active oxygen species in causing cell damage and phytoalexin accumulation in host-pathogen interactions (Doke 1985; Doke and Chai 1985; Epperlein et al 1986).

Many membrane proteins have a rapid turnover rate and half-lives measured in hours. One hypothesis we are testing is that mRNA activities which decrease rapidly in the HR encode for membrane proteins. If there were a reduction or abolition of transcription of the genes for some of these membrane proteins and mRNA decayed at its normal rate, then protein lost from membranes by natural turnover would not be adequately replaced, leading to loss of membrane integrity and function. Such a mechanism might contribute to the membrane damage observed in the HR.

b) Pectic enzymes

In the absence of a plasmolyticum pectic enzymes (both polygalacturonase and pectate lyase) kill plant cells and this cell death is associated with increased permeability of the plasma membrane to ions (Wood 1976). It is possible that in the HR the host cells are stimulated to produce pectic enzymes whose activity might lead to cell death. The role of pectic enzymes in host-pathogen interactions is complex. Oligomers of uronic acid released from pectic polysaccharides by the action of pectic enzymes elicit resistant reactions such as phytoalexin accumulation (Nothnagel et al 1983; Davis et al 1984). However, pectic enzyme production is a trait possessed by several successful virulent plant pathogens, particularly those which cause soft rots. P.s. pv. phaseolicola does not cause a soft rot but endopolygalacturonase was isolated from infected leaves (Longland et al, unpublished). This enzyme was a potent elicitor of phytoalexin accumulation in RM leaves. Indeed, compared on the basis of specific enzyme activity, c. 30 X more of a commercial polygalacturonase was required to elicit an equivalent response in bean leaves. However, in infected material the enzyme may be of pathogen origin and we found more in the compatible than the incompatible combination. Endopolygalacturonase activity peaked about 70 h after inoculation and may have been responsible for the accumulation of phytoalexins in the susceptible reaction which accompanied cessation of bacterial multiplication (Lyon and Wood 1975b; Longland et al, unpublished). Certainly there appears to be a physical interaction between P.s. pv. phaseolicola and pectic polysaccharides in bean cell walls (Slusarenko and Wood 1983; Ebrahim-Nesbat and Slusarenko 1983) but the role of pectic enzymes in this host pathogen combination remains unclear.

RESISTANT RESPONSES IN HYPERSENSITIVELY REACTING TISSUES

It seems probable that some of the mRNA activities which increased in the HR were for enzymes on the biosynthetic pathway of phytoalexin production (Bell 1986; Lawton et al 1983; Lamb, this volume). Similarly, mRNA activity leading to the production of hydroxyproline-rich glycoproteins which accumulate in plant cell walls would be expected to increase (Bolwell et al 1985; Esquerré-Tugayé, this volume).

Antiserum containing specific antibodies to French bean chitinase (kindly given to us by Thomas Boller) precipitated ^{35}S-labelled chitinase from in vitro translations. Messenger RNA activity for chitinase was first detected at 15 hours after inoculation in the HR but later than 20 hours after inoculation in the compatible combination (Voisey and Slusarenko, unpublished). Chitinase accumulates in the vacuole (Boller and Vogeli 1984), but it may be released when cells collapse in the HR; and since it shows lysozyme activity, may have a role in limiting the pathogen (Idem).

FUTURE PROSPECTS

In situ hybridization of specific cDNA probes might help to identify which changes in mRNA activity occur in dying and healthy cells. A high priority of future research must be to establish the nature of the intercellular signal transmitted from the dying to the surrounding, healthy cells.

REFERENCES

Atkinson MM, J-S Huang, JA Knopp 1985 Hypersensitivity of suspension-cultured tobacco cells to pathogenic bacteria. Phytopathology 75: 1270-1274

Bailey JA 1982 Physiological and biochemical events associated with the expression of resistance to disease. In: RKS Wood (Ed) Active Defense Mechanisms in Plants, Plenum Press, New York pp 39-65

Bell JN, TB Ryder, VPM Wingate, J Bailey, CJ Lamb 1986 Differential accumulation of plant defense gene transcripts in a compatible and an incompatible plant-pathogen interaction. Molecular and Cellular Biology 6: 1615-1623

Boller T, U Vogeli 1984 Vacuolar localization of ethylene-induced chitinase in bean leaves. Plant Physiology 74: 442-444
Bolwell GP, MP Robbins, RA Dixon 1985 Metabolic changes in elicitor-treated bean cells. Enzymic responses associated with rapid changes in cell wall components. European Journal of Biochemistry 148: 571-578
Davis KR, GD Lyon, AG Darvill, P Albersheim 1984 Endopolygalacturonic acid lyase from Erwinia carotovora elicits phytoalexin accumulation by releasing plant cell wall fragments. Plant Physiology 74: 52-60
Doke N 1985 NADPH-dependent O_2^- generation in membrane fractions isolated from wounded potato tubers inoculated with Phytophthora infestans. Physiological Plant Pathology 27: 311-322
Doke N, HB Chai 1985 Activation of superoxide generation and enhancement of resistance against compatible races of Phytophthora infestans in potato plants treated with digitonin. Physiological Plant Pathology 27: 323-334
Doke N, K Tomiyama 1975 Effect of blasticidin S on hypersensitive death of potato leaf petiole cells caused by infection with an incompatible race of Phytophthora infestans. Physiological Plant Pathology 6: 169-175
Ebrahim-Nesbat F, AJ Slusarenko 1983 Ultrastructure of the interaction of cells of Pseudomonas phaseolicola with cell walls of a resistant and susceptible bean cultivar. Phytopathologische Zeitschrift 108: 148-159
Epperlin MM, AA Noronha-Dutra, RN Strange 1986 Involvement of the hydroxyl radical in the abiotic elicitation of phytoalexins in legumes. Physiological and Molecular Plant Pathology 28: 67-77
Hitchcock C, BW Nichols 1971 Biological degradation of plant fatty acids. In: Plant Lipid Biochemistry, Academic Press, London pp 201-235
Holman RT, PO Egwim, WW Christie 1969 Substrate specificity of soybean lipoxidase. Journal of Biological Chemistry 244: 1149-1151
Keen NT, T Ersek, M Long, B Bruegger, M Holliday 1981 Inhibition of the hypersensitive reaction of soybeans to incompatible Pseudomonas spp. by blasticidin S, streptomycin or elevated temperature. Physiological Plant Pathology 18: 325-337
Keen NT, MJ Holliday 1982 Recognition of Bacterial Pathogens by Plants. In: MS Mount and GH Lacy (Eds) Phytopathogenic Prokaryotes Vol 2 Academic Press, New York pp 179-217
Keppler LD, A Novacky 1986 Involvement of membrane lipid peroxidation in the development of a bacterially induced hypersensitive reaction. Phytopathology 76: 104-108
Klement Z 1982 Hypersensitivity. In: MS Mount and GH Lacey (Eds) Phytopathogenic Prokaryotes Vol 2 Academic Press, New York pp 149-177
Laemmli UK 1970 Cleavage of structural proteins during the assembly of the head of bacteriophage T4. Nature 227: 680-685
Laskey RA, AD Mills 1975 Quantitative film detection of ^3H and ^{14}C in polyacrylamide gels by fluorography. European Journal of Biochemistry 56: 335-341
Lawson MA, RA Dixon, K Hahlbrock, C Lamb 1983 Elicitor induction of mRNA activity. Rapid effects of elicitor on phenylalanine ammonia-lyase and chalcone synthase mRNA activities in bean cells. European Journal of Biochemistry 130: 131-139

Lyon FM, RKS Wood 1975a The hypersensitive reaction and other responses of bean leaves to bacteria. Annals of Botany 40: 479-491

Lyon FM, RKS Wood 1975b Production of phaseollin, coumestrol and related compounds in bean leaves inoculated with Pseudomonas spp. Physiological Plant Pathology 6: 117-124

Lyon FM, RKS Wood 1977 Alteration of the response of bean leaves to compatible and incompatible bacteria. Annals of Botany 41: 359-367

Nothnagel EA, M McNeil, P Albersheim, A Dell 1983 A galacturonic acid oligosaccharide from plant cell walls elicits phytoalexins. Plant Physiology 71: 916-926

O'Brien FM, RKS Wood 1972 Role of ammonia in infection of Phaseolus vulgaris by Pseudomonas spp. Physiological Plant Pathology 3: 315-325

O'Farrell PH 1975 High resolution two-dimensional electrophoresis of proteins. Journal of Biological Chemistry 250: 4007-4021

Pelham HRB, RJ Jackson 1976 An efficient mRNA dependent translation system from rabbit reticulocyte lysates. European Journal of Biochemistry 67: 247-256

Slusarenko AJ 1986 Gene expression and resistance of French bean to Pseudomonas phaseolicola. In: PR Day and GJ Jellis (Eds) Genetics and Pathogenesis, Blackwell Scientific Publications, Oxford (in press)

Slusarenko AJ, A Longland 1986 Changes in gene activity during expression of the hypersensitive response in Phaseolus vulgaris cv. Red Mexican to an avirulent race 1 isolate of Pseudomonas syringae pv. phaseolicola. Physiological and Molecular Plant Pathology (in press)

Slusarenko AJ, RKS Wood 1983 Agglutination of Pseudomonas phaseolicola by pectic polysaccharide from leaves of Phaseolus vulgaris. Physiological Plant Pathology 23: 217-227

Taylor JD, NL Innes, CL Dudley, TA Griffiths 1978 Sources and inheritance of resistance to halo-blight of Phaseolus beans. Annals of Applied Biology 90: 101-110

Wolff SP, A Garner, RT Dean 1986 Free radicals, lipids and protein degradation. Trends in Biochemical Sciences 11: 27-31

Wood RKS 1976 Killing of Protoplasts. In: J Friend and DR Threlfall (Eds) Biochemical Aspects of Plant Parasite Relationships, Academic Press, London pp 105-116

Wyllie AH 1981 Cell death: A new classification separating apoptosis from necrosis. In: ID Bowen and RA Lockshin (Eds) Cell Death in Biology and Pathology, Chapman and Hall, London pp 9-34

HYDROXYPROLINE-RICH GLYCOPROTEINS (HRGPs) IN INFECTED PLANTS : SIGNALING, ACCUMULATION AND GENE EXPRESSION

D. Rumeau, D. Mazau and M.T. Esquerré-Tugayé
Centre de Physiologie Végétale, UA 241 CNRS,
118, route de Narbonne, 31062 Toulouse,
France

INTRODUCTION

Defense responses are initiated in plants as a result of recognition of self and non-self . Research on cell surfaces in plant-microorganism interactions have shown during the past ten years that the cell wall plays a key role both as a source of signal molecules (Darvill and Albersheim 1984), and as a site of signal-induced defense responses. The accumulation of hydroxyproline-rich glycoproteins (HRGPs), and the enhanced deposition of lignin-like material account for such responses (Bell 1981 ; Esquerré-Tugayé et al 1979). This paper is concerned with hydroxyproline-rich glycoproteins, under the double point of view of :

1. summarizing the data in favor of its accumulation as a defense response, and
2. defining HRGPs at the chemical and molecular levels.

I ACCUMULATION OF HYDROXYPROLINE-RICH GLYCOPROTEINS IN PLANT CELL WALLS

Hydroxyproline (Hyp) was mentioned for the first time in plant cell walls in 1960 (Lamport and Northcote 1960).
It was then found to be a constituant of a hyp-rich glycoprotein, called extensin by Lamport (Lamport 1965), which is also abbreviated now as HRGP. The presence of this unusual amino acid in the protein moiety enables to measure any modification of the total amount of HRGP, and by means of ^{14}C-proline, of its biosynthesis. It thus appeared that the level of HRGP

1. increases during infection by plant pathogens,
2. can be elicited by cell wall fragments, and,
3. depends on developmental and environmental conditions.

NATO ASI Series, Vol. H 4
Recognition in Microbe-Plant Symbiotic and Pathogenic Interactions. Edited by B. Lugtenberg

Hydroxyproline in the cell walls of infected plants

Since the first report that hydroxyproline was increased in the cell wall of melons infected by Colletotrichum lagenarium (Esquerré-Tugayé and Mazau 1974), a similar response was found in several plants inoculated by either fungi, bacteria, viruses or nematodes (Giebel and Stobiecka 1974 ; Mazau and Esquerré-Tugayé 1986). Among the numerous systems which have been analyzed, significant increases were only found in dicotylenous plants. In race-cultivar specific systems, enrichment of the cell wall in hydroxyproline resembled a defense response in that it appeared earlier and to a higher extent in resistant cultivars than in susceptible ones (Hammerschmidt et al 1984 ; Mazau and Esquerré-Tugayé 1986). A clear correlation was found between the amount of Hyp in melon cell walls and the resistance against C. lagenarium : an increased Hyp level correlated with an induced resistance, and a lowered Hyp level correlated with an increased susceptibility.
Incubation of infected tissues in the presence of ^{14}C-proline (proline being the precursor of Hyp), indicated that enrichment of the cell wall in Hyp resulted from de novo synthesis of HRGP (Toppan et al 1982). The possibility that signal molecules resulting from the early stages of plant pathogen interaction could induce the synthesis of this glycoprotein was then investigated.

Hydroxyproline in the cell wall of elicitor-treated plants

It has been demonstrated that fungal cell wall components called elicitors, as well as plant cell wall endogenous elicitors, induce the synthesis of phytoalexins, a defense response of many plants against pathogens (Darvill and Albersheim 1984 ; Hargreaves and Bailey 1978). The effect of elicitors isolated from C. lagenarium and from Phytophthora megasperma on the synthesis of HRGP was measured in their respective hosts (melon, soybean). At µg amounts, these elicitors triggered the synthesis of the glycoprotein, an effect which could also be reproduced by melon endogenous elicitors (Roby et al 1985). At the cellular level, the elicitation process results probably from a cascade of complex events. Ethylene might play a role since elicitors, as well as infections, stimulate the production of ethylene

(Toppan et al 1982), and ethylene alone promotes the synthesis of HRGP (Esquerré-Tugayé et al 1979). As in the case of other defense mechanisms, elicitation of HRGP is time-(~ 10 hours) and organ-dependent. Other factors may also be involved in the regulation of the Hyp level.

Hydroxyproline in the cell wall as a function of plant development and environment

The level of Hyp in the cell wall of plants grown under standard conditions is usually low. It is, however, higher in roots than in stems, and in stems than in leaves in the plants analyzed so far, and apparently higher in secondary than in primary cell walls. It also increases in response to extreme environmental conditions (Esquerré-Tugayé et al 1985) such as tissue culture, light (red), hormonal (ethylene) treatments, temperature (Stermer and Hammerschmidt 1984), mechanical wounding, and infection by several pathogens.

It is interesting to note that other defense reactions are also induced by the same stress factors, which reinforces the hypothesis that the level of Hyp correlates with the defense of plants.

By several criteria, it thus appears that the accumulation of HRGP fulfills the characteristics of a defense response. Analysis of the glycoprotein at the biochemical and molecular levels provides further evidence to this role.

II BIOCHEMICAL AND MOLECULAR ANALYSIS OF HYDROXYPROLINE-RICH GLYCOPROTEINS

It is only within the past five years that HRGPs have been successfully purified from a few plant tissue cultures. This has allowed to approach :

1. the chemical properties and possible role of these glycoproteins, and,
2. the molecular events underlying their accumulation.

Chemical composition of HRGPs

Four different glycoproteins corresponding to two families of HRGP have been identified in melon callus and seedlings, after ion exchange and affinity chromatography (Mazau et al 1986). The two families ($HRGP_1$ and $HRGP_2$) can be distinguished on the basis of their chemical composition, structural markers, electrophoretic mobilities, immunochemical properties, and response to infection.

$HRGP_1$ is an arabino-galactan protein (HR-AGP) composed of 94 % sugars and 6 % protein. The sugar moiety contains galactose (66 %) and arabinose (34 %). In the protein moiety, hydroxyproline, serine, alanine and glycine are the most abundant amino acids residues (45, 12, 10 and 7 %, respectively). The glycosylation pattern of Hyp indicates a major hyp-oligosaccharide and a low proportion of hyp-arabinosides in $HRGP_1$.

The sugar to protein ratio of $HRGP_2$ is 64 : 36, with arabinose, galactose and glucose as the recovered sugar residues (respectively 77 %, 22 %, 1 % of the sugar moiety). The protein moiety is characterized by its high amounts of hydroxyproline (39 %), lysine (15 %) and tyrosine (7 %). These three amino acids can be considered as structural markers since they confer to the molecule its high glycosylation, its high positive charge and an assumed high cross linkage.

In an attempt to estimate the MW of $HRGP_1$ and $_2$, the two glycoproteins were deglycosilated (Edge et al 1981), and their protein moiety analyzed by SDS-PAGE. In these conditions, the protein portion of $HRGP_{2b}$ (the main component of $HRGP_2$) had an apparent MW comprised between 52 000 and 56 000, whereas it was higher in $HRGP_1$.

Antibodies raised either against $HRGP_{2b}$, or its deglycosylated counterpart, did not react with $HRGP_1$.$HRGP_1$, on the other hand, reacted strongly with a phenol-derived compound called "Yariv antigen" (Yariv et al 1962), which had no affinity for $HRGP_{2b}$.

$HRGP_1$ and $_{2b}$ are increased to a different extent during infection. Most of the HRGP which accumulates in the cell wall corresponds to $HRGP_{2b}$. This glycoprotein resembles the HRGP isolated from carrot root discs by Stuart and Varner, the P_2-HRGP isolated from tomato cells by Smith et al and the HRGP agglutinins isolated from potato by Leach et al and from tobacco by Mellon and Helgeson (Leach et al 1982 ; Mellon and Helgeson 1982 ; Smith et al 1984 ; Stuart and Varner 1980). $HRGP_1$ is also increased as

a result of infection but to a lower extent than $HRGP_{2b}$; changes in the level of arabinogalactan proteins were recently mentioned to occur during an incompatible interaction (Clarke 1986).
The purification of HRGPs has revealed their structural diversity. Knowledge of the different HRGP which are present in a given plant and account for an increase in Hyp is a prerequisite to studying these glycoproteins at the molecular level.

Molecular biology of HRGP

The increased synthesis of HRGPs in infected melons has been studied at the level of the mRNAs of their protein moiety. We have looked for cytosine-rich mRNA in view of the high amounts of Hyp (~ 40 %), and of the CC"X" nature of proline codons. Cytosine-rich mRNA (C-RNA) were isolated from poly(A)$^+$RNA by affinity chromatography on an oligo(dG) cellulose column. In vitro translation in the presence of ^{3}H-proline of the obtained C-RNA led to several proline-rich peptides, but only two of them (MW 54.5 and 56 kd) were specific of the C-RNA of infected plants. Other proline-rich peptides were synthesized, but they were not unique to infected plants or to C-RNA. Northern hybridization to a carrot genomic clone of HRGP (Chen and Varner 1985) indicated that C-RNA of infected plants contained at least 3 major RNA species which hybridize to this probe in the molecular size range of 1.6, 2.3 and 3.8 kb. Two additional mRNA species (6.3 and 9.0 kb) were found when the same experiment was performed on total poly(A)$^+$RNA at late stages of infection. A differential induction in the pattern of these five mRNA species was observed during the course of infection. Similar temporal differences were reported in other situations leading to an increase in HRGP (Showalter et al 1985, and Lamb this workshop). Cytosine-rich RNA encoding the proline precursor of HRGP were thus increased in melons as a result of infection.

DISCUSSION

Biochemical studies have allowed to calculate that the proportion of HRGP in the cell wall of healthy plant is usually low (around 1 %). Although the sites of accumulation are not yet determined, it is obvious that dramatic changes in cell wall properties are induced when the level of HRGP reaches 5 to 15 % of the wall upon infection. One may assume that, as a wall polymer, HRGP reinforces the strength of the cell surface, and as a polycation it modifies the net charge of the wall. As such, it can agglutinate negatively charged pathogens (Leach et al 1982 ; Mellon and Helgeson 1982). Each of these potent properties should now be re-appraised, in view of the apparent differential expression of HRGP genes, and of the presence of several, distinct, HRGPs in infected plants.

REFERENCES

Bell AA (1981) Biochemical mechanisms of disease resistance. Annu Rev Plant Physiol 32 : 21-81

Chen J, Varner JE (1985) Isolation and characterization of cDNA clones for carrot extensin and a proline-rich 33 k-Da protein. Proc Natl Acad Sci USA 82 : 4399-4403

Clarke AE (1986) Tissue-specific responses to recognition of self and non self. In : UCLA Symposia on molecular and cellular biology 15th Annual Meetings "Molecular strategies for crop protection". March 30-April 12 1986. J Cell Biochem Suppl 10 C : abstract J 1. Alan R Liss Inc New York

Darvill AG, Albersheim P (1984) Phytotoalexins and their elicitors. Annu Rev Plant Physiol 35 : 243-275

Edge ASB, Faltynck CR, Hof L, Reicherts LE, Weber P (1981) Deglycosylation of glycoproteins by trifluoromethanesulfonic acid. Anal Biochem 118 : 131-137

Esquerré-Tugayé MT, Mazau D (1974) Effect of a fungal disease on extensin, the plant cell wall glycoprotein. J Exp Bot 25 : 509-513

Esquerré-Tugayé MT, Lafitte C, Mazau D, Toppan A, Touzé A (1979) Cell surfaces in plant-microorganism interactions. II Evidence for the accumulation of hydroxyproline-rich glycoproteins in the cell wall of diseased plants as a defense mechanism. Plant Physiol 64 : 320-326

Esquerré-Tugayé MT, Mazau D, Pélissier B, Roby D, Rumeau D, Toppan A (1985) Induction by elicitors and ethylene of proteins associated to the defense of plants. In : Cellular and Molecular Biology of Plant Stress, Alan R Liss Inc, P 459

Giebel J, Stobiecka M (1974) Role of amino acids in plant tissue response to Heterodera rostochiensis. I. Protein proline and hydroxyproline content in roots of susceptible and resistant solanaceous plants. Nematologica 20 : 407

Hammerschmidt R, Lamport DT A, Muldoon EP (1984) Cell wall hydroxyproline enhancement and lignin deposition as an early event in the resistance of cucumber to Cladosporium cucumerinum. Physiol Plant Path 24 : 43-47

Hargreaves JA, Bailey JA (1978) Phytoalexin production by hypocotyls of Phaseolus vulgaris in response to constitutive metabolites released by damaged cells. Physiol Plant Pathol 13 : 89-100

Lamport DTA, Northcote DH (1960) Hydroxyproline in primary cell walls of higher plants. Nature 188 : 665-666

Lamport DTA (1965) The protein component of primary cell walls. Advan Bot Res 2 : 151-218

Leach JE, Cantrell MA, Sequeira L (1982) Hydroxyproline-rich bacterial agglutinin from potato. Plant Physiol 70 : 1353-1358

Mazau D, Rumeau D, Esquerré-Tugayé MT (1986) Biochemical study of hydroxyproline-rich glycoproteins in plant-pathogen interactions In : Bailey JA (ed) "Biology and molecular biology of plant-pathogen interactions" 1-6 September 1985. Plenum Press (in press)

Mazau D, Esquerré-Tugayé MT (1986) Hydroxyproline-rich glycoprotein accumulation in the cell walls of plants infected by various pathogens. Physiol Plant Pathol (in press)

Mellon JE, Helgeson JP (1982) Interaction of a hydroxyproline-rich glycoprotein from Tobacco callus with potential pathogens. Plant Physiol 70 : 401-405

Roby D, Toppan A, Esquerré-Tugayé MT (1985) Cell surfaces in plant-microorganims interactions. V. Elicitors of fungal and of plant origin trigger the synthesis of ethylene and of cell wall hydroxyproline-rich glycoprotein in plants. Plant Physiol 77 : 700-704

Showalter AM, Bell JN, Cramer CL, Bailey JA, Varner JE, Lamb C (1985) Accumulation of hydroxyproline-rich glycoprotein mRNAs in response to fungal elicitor and infection. Proc Natl Acad Sci USA 82 : 6551-6555

Smith JJ, Muldoon EP, Lamport DTA (1984) Isolation of extensin precursors by direct elution of intact tomato cell suspension cultures. Phytochem 23 : 1233-1238

Stermer BA, Hammerschmidt R (1984) Disease resistance induced by heat shock : association with ethylene production and cell wall HRGP accumulation. In : 13th Annual UCLA Symposia "Cellular and Molecular Biology of Plant Stress". March 31-April 29 1984. J Cell Biochem Suppl 8B : abstract p 257. Alan R Liss Inc New York

Stuart DA, Varner JE (1980) Purification and characterization of a salt-extractable hydroxyproline-rich glycoprotein from aerated carrot discs . Plant Physiol 66 : 787-792

Toppan A, Roby D, Esquerré-Tugayé MT (1982) Cell surfaces in Plant-Microorganism interactions. III. *In vivo* effect of ethylene on hydroxyproline-rich glycoprotein accumulation in the cell wall of diseased plants. Plant Physiol 70 : 82-86

Yariv J, Rapport MM, Graf L (1962) The interaction of glycosides and saccharides with antibody to the corresponding phenylazo glycosides. Biochem J 85 : 383-388

A MODEL GENETICALLY ENGINEERED PESTICIDE: CLONING AND EXPRESSION OF THE *BACILLUS THURINGIENSIS* SUBSP. *KURSTAKI* δ-ENDOTOXIN INTO *PSEUDOMONAS FLUORESCENS*

T. L. Graham*, L. S. Watrud, F. J. Perlak, M.T. Tran, P.B. Lavrick, M.A. Miller-Wideman, P. G. Marrone, R. J. Kaufman

Monsanto Agricultural Products Company
800 N. Lindbergh Blvd.
St. Louis, MO 63166
USA

*Present address: The Ohio State University, Dept. of Plant Pathology, 201 Kottman Hall, 2021 Coffey Road, Columbus, Ohio 43210 USA

INTRODUCTION

It is rare for a naturally occurring biological control agent to process all of the traits required for its practical use in crop systems. Genetic engineering is a potentially excellent tool for the enhancement of the effectiveness of selective biological control agents. Through the "component" approach which genetic engineering allows, one may be able to combine the best attributes of various naturally occurring organisms to produce a rationally designed biological control agent tailor made to a specific agronomic application. Conceptually, one could isolate the most effective and highly target selective pesticidal activities from a "donor" organism and move these active principles into a second organism which possesses more effective and environmentally satisfactory traits as a "delivery" system. If successful, the new organism could retain the desirable traits of both systems while effectively extending practical effectiveness, host range or both.

This paper describes a model system designed to demonstrate the feasibility of such an approach and describes the rationale for and results of the specific approach taken. The gene from *Bacillus thuringiensis* coding for the δ-endotoxin was chosen as the pesticidal principle due to its well demonstrated safety as a highly selective commercial agent for the control of lepidopteran insect

NATO ASI Series, Vol. H 4
Recognition in Microbe-Plant Symbiotic and Pathogenic Interactions. Edited by B. Lugtenberg

pests. The potential "delivery" systems chosen were selected from indigenous soil organisms using a series of strict criteria to optimize both their effectiveness and safety. The most effective organisms selected, both corn root colonizing *Pseudomonas fluorescens* isolates, were then transformed with the *B.t.* gene to determine the feasibility of 1) expression of the heterologous gene at effective levels, and 2) the effectiveness of those engineered organisms as highly targeted pesticides under practical field conditions.

MATERIALS & METHODS

Criteria for Selection of Pseudomonas Root Colonizers

Traits thought desirable for an effective and environmentally satisfactory biological control agent were used as guidelines to isolate the *Pseudomonas* corn root colonizers. These included a high degree of association specifically with the host root, active colonization of new root growth, resistance to commonly used agricultural chemicals for the corn crop system, resistance to dessicaation and other physical factors likely to be encountered in its storage and application, ability to be produced readily in large quantities, ease of genetic engineering with the available tools of molecular biology. Other traits considered essential included non-pathogenicity to beneficial organisms, sensitivity to clinical antibiotics and sterilants, limited ability to actively exchange inserted traits with other organisms, and limited ability to survive independently of the plant root in agronomic soils or limited ability to overwinter in the environment. In addition, the organisms should possess genetic markers which should allow it to be monitored in the environment. These markers should not include resistance markers for commonly used clinical or agricultural antibiotics.

Using these criteria as guidelines, selection assays were established and agronomic soils were sampled. Early in the investigation of rhizosphere associated soils, it was determined that *Pseudomonas* species were predominant among those which would most likely match the desired criteria. Two *Pseudomonas* fluorescens isolates, Ps 3732-3-7 and 112-12 were isolated which showed colonization rates as high as 10^7 CFU per fresh weight of corn roots. Through extensive laboratory and greenhouse tests, these isolates were also shown to meet the majority of the above criteria (Watrud et al 1985). Moreover, these isolates lacked indigenous plasmids and lacked the ability to support the conjugative plasmids RP4, N3 and pSa which belong, respectively, to the incompatibilty groups IncP, IncN and IncW (Jacob et al 1977). To date, only derivatives of RSF1010 plasmids have successfully been introduced into Ps 3732-3-7 or 112-12. Such plasmids, which are members of the IncQ group, are not self-transmissible. They require a conjugative plasmid to provide the transfer functions necessary for movement out of the P. fluorescens strains.

Cloning of the δ-Endotoxin Gene in E. coli

Plasmid DNA of B. thuringiensis subsp. kurstaki strain HD-1 was isolated, enzymatically restricted with BAM HI and inserted into the E. coli plasmid pBR328. Presumptive transformants were selected for ampicillin resistance and screened with a 21-mer ^{32}P-labelled synthetic nucleotide probe based on N-terminal amino acids 19-25 of the 134,000 molecular weight -endotoxin protein (Wong et al 1983). This protein had been sequenced after isolation from stationary-phase cultures of B. thuringiensis subsp. kurstaki strain HD-1 and purification in Renograffin gradients and polyacylamide gel electrophoresis (Sharpe et al 1975). Southern blots (Southern 1975) were used to screen DNA from presumptive transformants. Western blots (Towbin et al 1979) used to screen protein extracts from putative clones were carried out with a polyclonal antibody prepared against the purified protein. Subcloning fragments of original clones was accomplished using plasmid vector pUC8 in E. coli JM101 (Vieira and Messing 1982). Tests for insect toxicity were performed by diet incorporation against Manduca sexta (tobacco horworm) and Heliothus zea (corn earworm).

Subcloning into Pseudomonas Fluorescens

Subcloning of the δ-endotoxin gene into *P. fluorescens* was accomplished in several ways, including introduction by means of non-self-transmissible plasmids through triparental matings (Figurski and Helinski 1979) and into the bacterial chromosome by means of suicidal vectors (Selvary and Dyer 1983) and by a process of homologous recombination using the transposon Tn5. Tests for retention of the *B.t.* gene were as above.

Growth Chamber and Field Tests on Pseudomonas Isolates

Variants of *Pseudomonas fluorescens* Ps 3732-3-7 and 112-12 were selected which were doubly marked for resistance to rifampicin and nalidixic acid. These isolates were applied as inocula to corn seeds and examined for their ability to colonize corn roots in non-sterile soil in the growth chamber and in field plots at St. Charles, Missouri. All studies with engineered Ps 3732-3-7 and 112-12 were done in growth chambers.

RESULTS

DNA from clones pMAP1 and pMAP2 contained *Bam* HI inserts of 16 kilobases (kb) in opposite orientations which hybridized strongly with the 21-mer nucleotide probe against the N-terminal end of the *B.t.* structural gene. Since this size fragment is obviously much larger than the approximately 3.5kb needed to encode the toxin protein, subcloning in pUC8 in *E. coli* JM101 was undertaken. Selection of putative transformants in this system was made on the basis of loss of ability to utilize lactose.

Results of subcloning are shown in Figure 1. Utilization of restriction enzymes *Bam* HI and *Pst* I resulted in an 8.1kb fragment, while digestion with *Hpa* I and *Pst* I resulted in a 4.6kb fragment, both of which retained the intact structural gene as shown by DNA hybridization, insect toxicity and western gel analysis of pMAP3 and pMAP4 respectively. Western gels, shown in Figure 2, demonstrate the expression of the intact 134,000 molecular weight toxin.

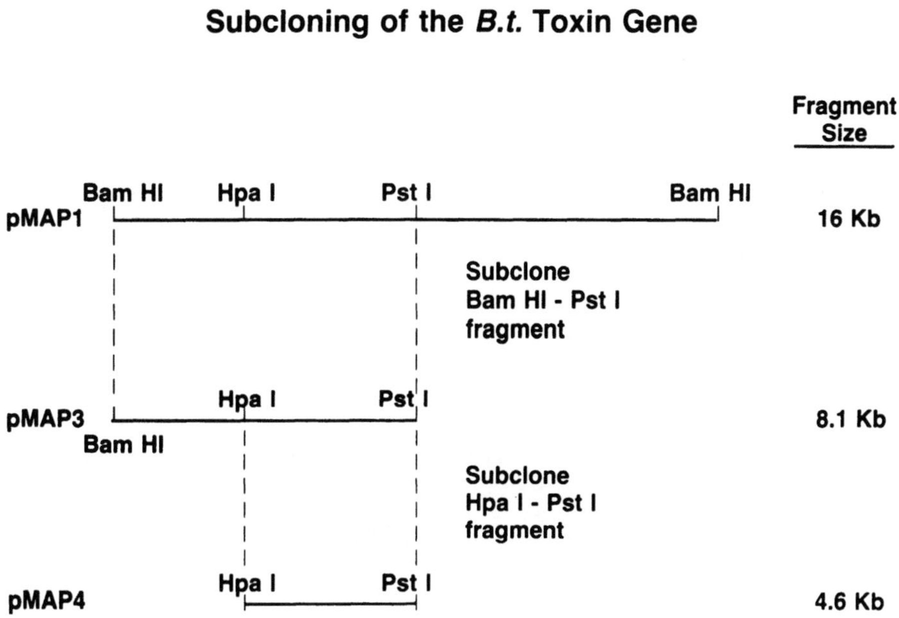

Figure 1. Subcloning of the delta-endotoxin gene from B. thuringiensis var. kurstaki into pUC8 in E. coli. Plasmid vectors pMap1, pMap3, and pMAP4 were formed which contained the delta-endotoxin gene on fragments which were, respectively, 16, 8.1 and 4.6 kb in length.

Successful subcloning and expresion in Pseudomonas fluorescens isolates Ps 3732-3-7 and 112-12 was demonstrated in a similar manner.

Survival of parental and engineered derivatives of strain Ps 3732-3-7 on the rhizoplane of corn roots grown in nonsterile agronomic soil in growth chambers is shown in Figure 3. As shown, the engineered strain showed nearly an identical survival curve as the non-engineered isolate.

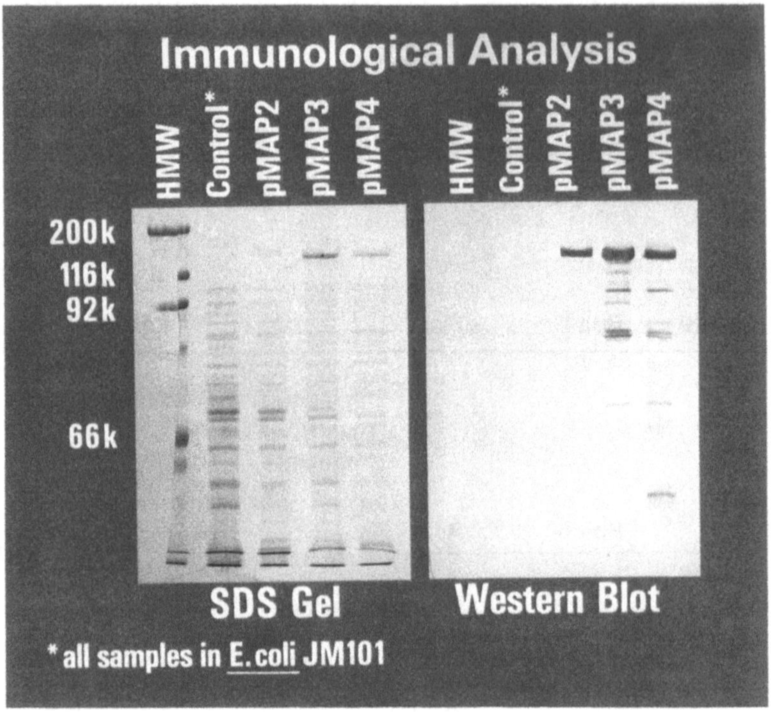

Figure 2. Use of an antibody probe to the delta-endotoxin to demonstrate production of the toxin protein by clones containing pMAP2, pMAP3, or pMAP 4. SDS, sodium dodecyl sulfate.

Recovery of the doubly marked non-engineered Ps 3732-3-7 from the rhizoplane of three cultivars of corn in field tests at St. Charles, Missouri is shown in Figure 4. These data confirm growth chamber data and demonstrate an effective translation of some colonization and survival parameters to the field environment. Importantly, the data also demonstrates the decline of the isolates late in the season.

SUMMARY AND DISCUSSION

The δ-endotoxin gene carried on a Bam HI fragment from Bacillus thuringiensis (B.t.) subsp. kurstaki strain HD-1 was cloned into E. coli using pBR328. DNA from clones pMAP1 and pMAP2

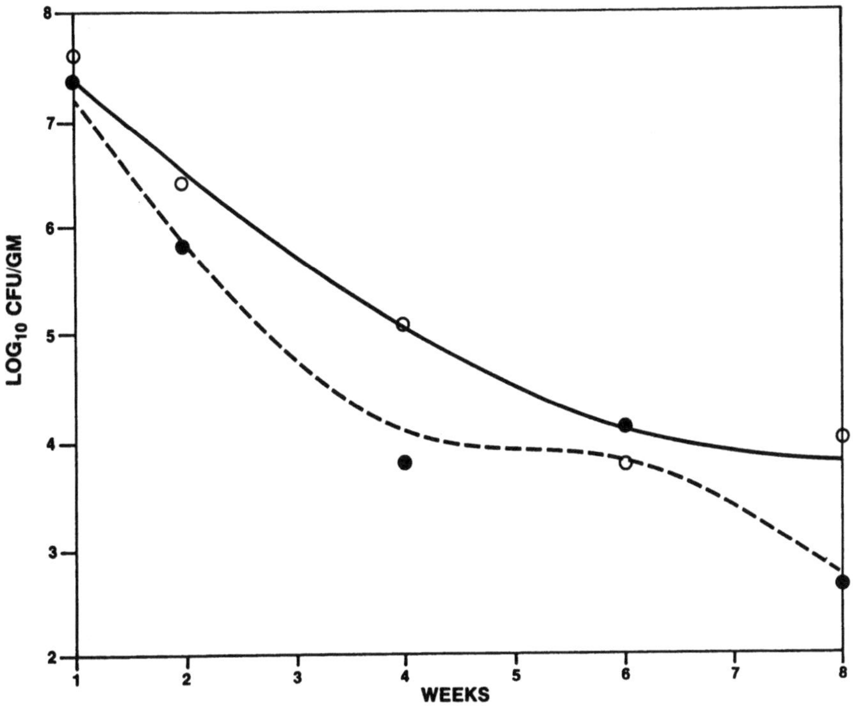

Figure 3. Determination of the survival of parental (O) and engineered (●) derivatives of P. fluorescens strain Ps3732-3-7 on the rhizoplane of corn roots grown in nonsterile agronomic soil in growth chambers.

contained Bam HI inserts of 16 kilobases (kb) in opposite orientations. Subcloning of the gene into plasmid vector pUC8 in E. coli JM101 led successfully to subclones containing DNA inserts of 8.1kb (pMAP3) and 4.6kb (pMAP4). Clones pMAP1-pMAP4 retained the pesticidal gene as determined by DNA hybridization, immunological reaction and pesticidal activity against Heliothus zea and Manduca sexta. Independently, two isolates of P. fluorescens, Ps 3732-3-7 and 112-12, were isolated from agronomic soil using strict selection criteria designed to optimize their efficacy as practical "delivery" systems for the B.t. gene within the corn rhizoshpere. Doubly marked (Rif^r Nal^r) isolates of these Pseudomonas strains were field tested in 1984 and demonstrated translation of laboratory corn root colonization characteristics on several corn

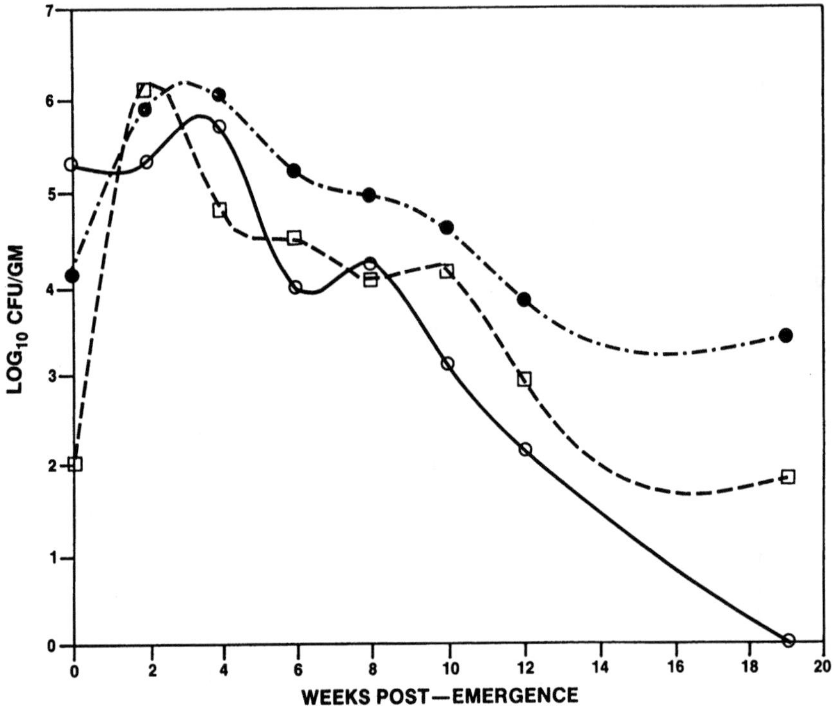

Figure 4. Recovery of a nongenetically engineered Rif^r Nal^r isolate of P. fluorescens from the rhizoplane of three cultivars of corn (PN3183, PN3732, ans PN3780A) grown in the field at St. Charles, MO., 1984.

varieties. Subcloning of the B.t. gene into isolates Ps 3732-3-7 and 112-12 was successful both through introduction on non-self-transmissable plasmids *via* triparental matings and through a process of chromosomal insertions via homologous recombination using Tn5. These engineered variants of Ps 3732-3-7 and 112-12 express pesticidal levels of the 134,000 molecular weight B.t. protein toxin. Field testing against *Agrotis ipsilon* is planned.

The above results, though preliminary, are highly encouraging. Toxic levels of expression of a heterologous gene coding for highly selective insect toxin has been achieved in an effective crop root colonizing organism.

Moreover, the delivery systems, Ps 3732-3-7 and 112-12 appear to demonstrate many of the traits of an effective field colonization microbe. A true test of the effectiveness of this model biocontrol agent will, however, require more extensive tests, particularly in the fields crop system.

REFERENCES

Figurski D, Helinski DR (1979) Replication of an origin containing derivative of plasmid RK2 dependent on a plasmid function provided in trans. Proc Natl Acad Sci USA 76:1648-1652

Jacob AE, Shapiro JA, Yamamoto L, Smith DI, Cohen SN, Berg D. (1977). Plasmid Studies in E. coli and other Enteric Bacteria p. 607-656. In: AI Bukhari, JA Shapiro, SL Adhya (eds), DNA Insertion Elements, Plasmids, and Episomes. Cold Spring Harbor Laboratories, Cold Spring Harbor, NY

Selvary G, Dyer VN (1983) Suicide plasmid vehicles for insertion mutagenesis in Rhizobium meliloti and related bacteria. J Bacteriol 156:1292-1300

Sharpe ES, Nickerson KW, Bulla LA, Aronson JN, (1975) Separation of spores and parasporal crystals and Bacillus thuringiensis in gradients of certain X-ray contrasting agents. Appl Microbiol 98:503-517

Southern EM (1975) Detection of specific sequences among DNA fragments separated by gel electrophoresis. J Mol Biol 98:503-517

Towbin HT, Stabelin T, Gordon J (1979) Electrophoretic transfer of proteins from poly-acrylamide gels to nitrocellulose sheets. Proc Natl Acad Sci USA 76:4354

Vieira J, Messing J (1982) The pUC plasmids, an m13 mp7-derived system for insertion mutagenesis and sequencing with sythetic universal primers. Gene 19:259-268

Watrud LS, Perlak, FJ, Tran MT, Kusano K, Mayer EJ, Miller-Wideman MA, Obukowicz MG, Nelson DR, Krietinger JP, Kaufman RJ (1985) Cloning of the Bacillus thuringiensis subsp kurstaki -Endotoxin Gene into Pseudomonas fluorescens: Molecular Biology and Ecology of an Engineered Microbial Pesticide. In: "Engineered Organisms in the Environment: Scientific Issues", HO Halvorson, D Pramer, M Rogul, (eds.) Publ. American Society for Microbiology, Washington, DC

Wong HC, Schnept HE, Whiteley HE (1983) Transcriptional and translational start sites for the Bacillus thuringiensis crystal protein gene. J Biol Chem 258:1960-1967

MOLECULAR ASPECTS OF PLANT GROWTH AFFECTING PSEUDOMONAS SPECIES

Bob Schippers[1], Peter A.H.M. Bakker[1], Albert W. Bakker[1], Gerard A.J.M. van der Hofstad[2], Joey D. Marugg[2], Letty A. de Weger[3], Jan G. Lamers[4], Wiel P.M. Hoekstra[2], Ben J. Lugtenberg[3], Peter J. Weisbeek[2]

[1]State University of Utrecht, Willie Commelin Scholten Phytopathology Laboratory, Javalaan 10, 3742 CP Baarn, The Netherlands

[2]State University of Utrecht, Department of Molecular Cellbiology, Padualaan 8, 3584 CH Utrecht, The Netherlands

[3]State University of Leiden, Department of Molecular Botany, Nonnensteeg 3, 2311 VJ Leiden, The Netherlands

[4]Research Station for Arable Farming and Field Productions of Vegetables (PAGV), Edelhertweg 1, 8200 AK Lelystad, The Netherlands

INTRODUCTION

Organic compounds released by roots stimulate the activity of a variety of micro-organisms in the root environment, the rhizosphere (Rovira and Davey 1974). Of these micro-organisms, soil-borne plant pathogens and nitrogen fixing micro-organisms have obtained most attention. Relatively little is known about the majority of the rhizosphere micro-organisms, the saprophytes and their influence on root functioning, plant development and crop yield.

Fluorescent Pseudomonas spp. are among the first saprophytes that establish on the root tip growing through the soil and that flourish on the adjacent young root parts. Analysis of the cell envelope proteins of many Pseudomonas fluorescens and P. putida isolates isolated from these young root parts and analysis of their interactions with other rhizosphere micro-organisms on agar media, revealed that a great diversity of metabolically different isolates of Pseudomonas spp. inhabit the rhizosphere (De Weger et al. 1986; Geels and Schippers 1983a; Geels et al. 1985). The fluorescent pseudomonads are well known for the production of a great variety of biologically active secondary metabolites, including plant growth substances, antibiotics, cyanide and iron-chelating compounds (siderophores) with a strong affinity to ferric iron (Leisinger and Margraff 1979). They are

NATO ASI Series, Vol. H 4
Recognition in Microbe-Plant Symbiotic and Pathogenic Interactions. Edited by B. Lugtenberg

therefore suspected to interfere with the physiology of young root parts and with interactions between young root parts and symbionts, parasites and non-parasitic harmful rhizosphere micro-organisms. This suspicion is supported by recent reports on positive or negative plant growth responses following root treatments with selected fluorescent pseudomonads (Geels and Schippers 1983b; Kloepper et al 1980; Schippers et al 1986c; Suslow 1982).
Little is known, however, about the underlying molecular interactions with roots and with other rhizosphere micro-organisms, and on the population dynamics and activity of the different isolates of Pseudomonas spp. in various cropping systems. This is mainly because techniques have not been available to easily distinguish isolates of Pseudomonas spp. in vitro and in situ and to identify relationships between the production of particular metabolites and responses of other micro-organisms in the rhizosphere and plant growth responses. Molecular biological techniques, however, now open unprecedented possibilities in this respect as is illustrated in this paper.

Experimental results on molecular interactions of isolates of fluorescent Pseudomonas spp. in the rhizosphere are presented and consequent hypotheses on their inpact on plant development and crop yields are discussed. They are the result of a joint research enterprise of agronomists, plant pathologists, molecular biologists, soil microbiologists and plant physiologists.

PSEUDOMONAS SIDEROPHORES AND PLANT GROWTH STIMULATION

Detailed field experiments on the effects of cropping frequencies on potato yields form the basis of this research project. These experiments manifest a considerable decrease in potato tuber yield with increasing potato cropping frequency. Physical and chemical soil factors and regular soil-borne pathogens do not fully account for the observed reductions in yield (Schippers et al 1985; Scholte et al 1985).
Yield reductions of 15-30%, obtained in short rotations of potato in the field, could be reproduced in pot experiments in soil taken from the relevant cropping frequency field plots. If potato seed tubers were dipped in cell suspensions of the selected fluorescent Pseudomonas spp. isolates P. fluorescens WCS 374 or P. putida WCS 358 before sowing, these tuber yield reductions could be prevented in the pot experiments (Geels and Schippers

1983b). Rooted potato stem cuttings lack the tuber as a source of nutrients. They totally rely on nutrient uptake from soil for their growth. The root development of such rooted stem cuttings was markedly retarded after 10 days in short potato rotation soils. If the roots were dipped in a suspension of Pseudomonas fluorescence WCS 374 or P. putida WCS 358 before planting, root growth inhibition was minimized (P.A.H.M. Bakker et al, to be published). These treatments with bacteria had no effect on root development and yields in soil obtained from field plots with no or little history of potato cropping. Because potato roots in short potato rotation soil show no distinct disease symptoms, non-parasitic harmful rhizosphere micro-organisms are suspected to be the main causal agents of the retarded growth and yields (Geels and Schippers 1983b; Schippers et al 1985).

The plant growth stimulating isolates of Pseudomonas spp. have been isolated from potato rhizosphere and selected on Fe^{3+} deficient King's Medium B for their succesful competition for Fe^{3+} ions with a variety of parasitic and saprophytic potato rhizosphere micro-organisms (Geels and Schippers, 1983a). Iron ions are indispensable for the energy metabolism of micro-organisms. The availability of iron in most soils, especially in alkaline soils such as those used in our experiments (pH 7.3-7.6), is very low because most iron is present as highly insoluble $Fe(OH)_3$ polymers.

The selected isolates of Pseudomonas spp. used in our field and pot experiments are supposed to deprive the harmful micro-organisms from iron by their production of strong Fe^{3+}-chelating siderophores resulting in improved plant growth (Geels and Schippers 1983a; Kloepper et al 1982; Schippers et al 1985; Schippers et al 1986c). Siderophore-negative (S^-) mutants of the isolate WCS 358, obtained by transposon Tn5-induced mutagenesis (Marugg et al 1985), did not improve root development of potato sprouts in short potato rotation soil in contrast to the siderophore-producing (S^+) transposon Tn5 mutants. Both S^- and S^+ mutants colonized the roots equally well (P.A. H.M. Bakker et al, to be published). Seed tuber treatment with isolate WCS 358 wild type in field plots with high potato cropping frequency (potatoes were cropped every third year) prevented a 10% reduction in seed potato tuber yield, that occurred in the untreated control plots with the same frequency of potato cropping. Treatments with the WCS 358 S^- Tn5 mutants had no such effect although this mutant colonized the roots as well as the WCS 358 S^+ Tn5 mutant, throughout the growing season (P.A.H.M. Bakker et al, to be published). This is the first time that the role of a siderophore in plant growth stimulation and yield increase has been demonstrated in the field.

PSEUDOMONAS SIDEROPHORE PRODUCTION IN THE RHIZOSPHERE

Evidence for the actual production of siderophores in the rhizosphere by plant growth-promoting pseudomonads and its consequences for other rhizosphere pseudomonads, was obtained by studying the interactions between transposon Tn5 S^- mutants of the isolates WCS 358 and WCS 374 and their wild types and with their purified siderophores. These interactions were studied on King's Medium B agar (KB) and in the potato rhizosphere. WCS 358 S^- mutants were shown to be able to use the siderophore excreted by the wild type and that of isolate WCS 374 on KB. WCS 374 S^- mutants were able to use the WCS 374 siderophore but not the siderophore of WCS 358. When roots of rooted potato sprouts were dipped in cell suspensions of the WCS 358 wild type or the WCS 374 wild type isolate and planted in soil mixed with the WCS 358 S^- mutant, ten days later this S^- mutant had colonized the developing roots significantly better than roots which had not been treated with one of the wild type isolates. When the WCS 374 S^- mutant was mixed through the soil, root colonization by this mutant was significantly stimulated if the roots had been treated with its wild type isolate before planting. Root colonization of WCS 374 S^- was comparable to that of non-treated roots, if this mutant was confronted with roots treated with the WCS 358 wild type isolate. These interactions in the rhizosphere are consistent with those on KB. This makes it very likely that the relevant siderophores are actually produced in the rhizosphere and can significantly affect the development of other isolates of Pseudomonas spp., either positively or negatively (P.A.H.M. Bakker et al, to be published). The population developments and interactions of the WCS 374 Tn5 and WCS 358 Tn5 mutants could be studied in the rhizosphere, thanks to the transposon Tn5 introduced resistance to kanamycin and streptomycin of these mutants.

P. putida isolate WCS 358 is rather unique, not only because its siderophore can only be used by very few of many isolates of fluorescent Pseudomonas spp. isolated from the potato rhizosphere, but also because it can use the Fe^{3+} siderophore complexes produced by most of these rhizosphere isolates (P.A.H.M. Bakker unpublished). The latter explains why the WCS 358 S^-

mutant colonizes the potato roots as well as its wild type and S^+ Tn5 mutant.

GENETIC ORGANIZATION OF SIDEROPHORE BIOSYNTHESIS

P. putida WCS 358 was mutagenized with transposon Tn5 using a broad-host-range RP4-dependent mobilization system (Simon et al 1983). Six classes of mutants were isolated which were completely or partially disturbed in fluorescence (Flu^- and $Flu^\pm$), and/or Fe (III)-binding capacities (Sid^- and $Sid^\pm$) (Marugg et al 1985).

Using a gene bank in cosmid pLAFRI, thirteen mutants from a total of twenty-eight were complemented to the wild type phenotype (Marugg et al 1985). The complementation results and subsequent restriction analysis revealed that the complementing cosmids could be divided in five distinct clusters. The position of the Tn5 mutational inserts was determined by hybridization of genomic DNA from the mutants with the corresponding radiolabelled cosmids. One cluster, which complemented one Flu^- Sid^+ and eight Flu^- Sid^- mutants, was studied in more detail. The mutations are located on the chromosome in a region of 30 kb. At least four transcription units are encoded by this region, the numbers of genes varying between four and fifteen. This region is being subcloned in the broad-host-range promoter-probe vector pKT240 in order to study the expression of the genes (transcription as well as gene products). In vivo protein synthesis in E. coli minicells has identified several proteins encoded by the different subclone inserts. Hybridization of total RNA isolated from wild type cells grown under iron-limiting and iron-rich conditions with nick-translated subclones or subclone-derived RNA-probes indicate that expression of at least some of the siderophore biosynthetic genes is regulated at the transcriptional level. Our aim is to isolate the iron-regulated promoters to study their structure, and use them as a tool to find regulation-defective mutants (van der Hofstad et al, this volume).

STRUCTURE OF PSEUDOBACTIN 358

The siderophore of WCS 358 was isolated as the iron-(III) complex from culture medium of stationary cultures grown in minimal succinate medium. A modification of the procedures described by Meyer & Abdallah (1978) was followed (van der Hofstad et al 1986). The yield was about 200 mg per liter of culture medium. Purity was checked by HPLC gel filtration analysis (range 1 to 300 kD) showing one major peak comprising more than 95% of A_{280}-absorbing material. The iron-(III) binding constant as determined by competition with EDTA was 2.1 x 10^{25} at pH 7.0, which is more than tenfold higher than that described for pyoverdine Pf (Meyer and Abdallah 1978). Amino acid analysis was performed on 6 N HCl and 57% HI hydrolysates of both the iron- (III)- and the free siderophore giving the following amino acid composition (in molar): Lys (2) HO-Orn (1) Ala (1) Ser (1) Thr (2) Asp (1) and HO-Asp (1). The molecular weight was determined by spectrophotometric titration of the free siderophore with $FeCl_3$. A molecular weight of 1333 D was obtained when we assume a ligand to iron-(III) molar ratio of 1 to 1. The molecular weight as calculated from the amino acid composition for the peptide part, assuming that the fluorescing group is identical to that reported for pyoverdin and pseudobactin, was found to be 1265 D, which is in agreement with the value obtained by titration.

The amino acid sequence was determined both by means of manual and automatic Edman degradation. From these results combined with our optical and fluorimetric data a tentative structure of the WCS 358 siderophore, which we designate as pseudobactin 358, is proposed. ^{1}H- and ^{13}C-NMR studies are performed to complete the structure. At this moment we are purifying and analyzing mutant-siderophores obtained from mutants defective in fluorescence and/or iron-(III) binding capacities (Marugg et al 1985) to get insight in the biosynthesis of pseudobactin 358 (van der Hofstad et al, this volume).

PLANT GROWTH INHIBITING PSEUDOMONADS

Very little is known about saprophytic rhizosphere micro-organisms that inhibit root functioning and plant growth. The harmful saprophytic bacteria, which are supposed to be involved in the reductions in potato tuber yields which increase with increasing frequency of potato cropping (Schippers et al 1985), are difficult to trace. Our attention has been focussed on cyanide producing fluorescent pseudomonads because HCN is known to inhibit cytochrome oxidase respiration and thereby the energy metabolism of plant cells (Lambers 1980). Moreover, HCN production by Pseudomonas aeruginosa was demonstrated to be dependent on the availability of Fe^{3+} ions in the environment (Castric 1976).

More than 50% of the isolates of fluorescent Pseudomonas spp. randomly isolated from the potato rhizosphere could produce HCN on KB supplemented with glycine, a constituent of potato root exudates. Concentrations of 5 µM HCN and even lower, significantly reduced the cytochrome oxidase respiration of intact potato roots (A.W. Bakker and Schippers, to be published; Schippers et al 1986a).

An impaired energy metabolism of root cells probably implies a decrease in energy demanding ion uptake (e.g. phosphorus). The lower phosphate and potassium content of potato plants in high, compared to low frequency of potato cropping agrees with this hypothesis (J. Vos and K. Groenewold, personal communications).

The biomass production of the HCN-producing fluorescent Pseudomonas spp. isolate WCS 361, and also its HCN production per unit of biomass decreased with decreasing availability of Fe^{3+} ions in a batch culture medium (A.W. Bakker and B. Schippers, to be published).

These observations lead to the hypothesis that isolates of fluorescent Pseudomonas spp. improving potato growth and seed tuber yields in short rotations of potato, deprive HCN-producing Pseudomonas spp. of Fe^{3+} ions thus limiting their population development and/or HCN production. The consequent improvement of root cell energy metabolism will enhance the uptake of nutrients from soil thereby favouring crop production (A.W. Bakker and B. Schippers, to be published; Schippers et al 1986a,b). It is interesting to note that the isolates P. fluorescens WCS 374 and P. putida WCS 358 promote potato growth and increase tuber yields, but do not have the ability to produce cyanide (A.W. Bakker and B. Schippers, to be published).

Whether the hypothesis is valid needs further examination. Harmful rhizosphere micro-organisms other than HCN-producing bacteria may be involved as well.

EPILOQUE

Many questions remain to be answered. Why do harmful micro-organisms increase with increasing potato cropping frequency? Possibly, the root exudate composition of potato roots or specific molecular interactions between potato root cells and cells of particular Pseudomonas spp. isolates favour the development and/or activity of HCN-producers or disadvantage antagonizing isolates of Pseudomonas spp. with strong Fe^{3+}-chelating capacity, which do not produce HCN.

In field experiments performed over five successive years (1981-1985), tuber bacterization with Pseudomonas spp. isolates WCS 358 and WCS 374 increased yields of *seed* tubers but seldom of *ware* potatoes which are harvested one to two months later (P.A.H.M. Bakker et al , to be published; Geels et al , to be published). Probably, the numbers of introduced Pseudomonas spp. isolates drop on newly formed roots later in the season. Root colonization studies with transposon Tn5 Sid^{+} mutants in the field sustain this supposition (P.A.H.M. Bakker et al , to be published). It is also possible, that the harmful rhizosphere Pseudomonas spp. are favoured by the changes in root exudate composition of the aging potato plants. Finally, we may not exclude the possibility, that certain harmful rhizobacteria can make use of the Fe^{3+} siderophore complex of P. putida WCS 358 and although neglectable in the early season, increase at the end of the season.

Another intriguing question is whether certain isolates of fluorescent Pseudomonas spp. are better colonizers of potato roots than others. If so, what are the underlying mechanisms? Are there differences in mobility? Is there selective stimulation by particular potato root exudate components? Do specific interactions between root cell surface polymers and polymers of the Pseudomonas cell surface occur?

By using flagella-negative mutants, mobility of P. putida WCS 374 was demonstrated to be an important factor in the succesful colonization of potato roots growing in soil (de Weger et al, this volume).

Plant growth retardation and yield reductions by harmful rhizosphere micro-organisms are not restricted to potato. A variety of micro-organisms and of mechanisms may be involved, all which of we know little about. Plant growth promotion by selected fluorescent pseudomonads has been demonstrated for a variety of crops. Fluorescent pseudomonads, their siderophores and antibiotic metabolites have also been demonstrated to be involved in the natural and manipulated suppression of several regular soil-borne plant pathogens (Schippers et al 1986c; Suslow 1982). Little is known about their influence on microbe-plant symbiotic systems. Fluorescent pseudomonads seem to be a group of rhizobacteria with great potential to influence plant development and crop yield positively as well as negatively. In order to make them useful as a tool in agriculture, knowledge is also needed on the molecular mechanisms determining their root colonization and survival. This will facilitate large scale selection of good root colonizers. Detailed information about the process of iron uptake, including its genetic organization and regulation, will allow us to manipulate these bacteria to improve their beneficial and minimize their detrimental effect on crop production. It will also lead to a better understanding of the population dynamics and activity of beneficial and of harmful isolates of Pseudomonas spp. in different cropping systems.

ACKNOWLEDGEMENT

The research described was performed in part with financial support from the Netherlands Technology Foundation (STW) and by the Bio-molecular Engineering Programme of the EEC (contract nr. GBI - 4 - 108 - NL). We thank Mrs M.A. Williamson for correcting the English.

REFERENCES

Castric P A (1981) The metabolism of hydrogen cyanide by bacteria. In: Cyanide biology (Vennesland B, Conn E E, Knowles C J, Westley J, Wissing F F (eds) Academic Press, New York, p 233

Geels F P, Schippers B (1983a) Selection of antagonistic fluorescent Pseudomonas spp. and their root colonization and persistence following treatment of seed potatoes. Phytopath Zeitschr 108: 193-206

Geels F P, Schippers B (1983b) Reduction of yield depressions in high frequency potato cropping soil after seed tuber treatments with antagonistic fluorescent Pseudomonas spp. Phytopath Zeitschr 108: 207-214

Geels F P, Schippers B (1983c) Reduction of yield depressions in high frequency potato cropping by fluorescent Pseudomonas spp. Les Coll de l'INRA 18: 231-238

Geels F P, Schmidt E D L, Schippers B (1985) The use of 8-hydroxyquinoline for the isolation and prequalification of plant growth-stimulating rhizosphere pseudomonads. Biol Fert Soils 1: 167-173

Hofstad G A J M van der, Marugg J D, Verjans G M G M, Weisbeek P J (1986) Characterization and structure analysis of the siderophore produced by the PGPR Pseudomonas putida strain WCS 358. In: Swinburne T R (ed) Iron, siderophores and plant disease. Plenum Press, New York London

Lambers H (1980) The physiological significance of cyanide resistant respiration in higher plants. Plant cell envir 3: 293-302

Leisinger T, Margraff (1979) Secondary metabolites of the fluorescent pseudomonads. Microbiol Rev 43: 422-442

Marugg J D, Spanje M van, Hoekstra W P M, Schippers B, Weisbeek P J (1985) Isolation and analysis of genes involved in siderophore-biosynthesis in the plant growth-stimulating Pseudomonas putida strain WCS 358. J Bacteriol 164: 563-570

Meyer J M, Abdallah M A (1978) The fluorescent pigment of Pseudomonas fluorescens: Biosynthesis, purification and physicochemical properties. J Gen Microbiol 107: 319-328

Rovira A D, Davey C B (1974) Biology of the rhizosphere. In: Carson E W (ed) The plant root and its environment. University Press, Charlottesville, USA, p 153

Schippers B, Bakker A W, Bakker P A H M, Weisbeek P J, Lugtenberg B (1986a) Plant growth-inhibiting and stimulating rhizosphere micro-organisms. In: Jensen V, Kjøller A, Sørensen L H (eds) Microbial communities in soil. FEMS-publ, Elsevier, Amsterdam Oxford New York

Schippers B, Geels F P, Bakker P A H M, Bakker A W, Weisbeek P J, Lugtenberg B (1986b) Methods of studying plant growth-stimulating pseudomonads: problems and progress. In: Swinburne T R (ed) Iron, siderophores and plant pathogens. Plenum Press, New York London

Schippers B, Geels F P, Hoekstra O, Lamers J G, Maenhout C A A A (1985) Yield depressions in narrow rotations caused by unknown microbial factors and their suppression by selected pseudomonads. In: Parker C A, Rovira A D, Moore K J, Wong P T (eds) Ecology and management of soilborne, plant pathogens. APS, St Paul Minnesota, p 127

Schippers B, Lugtenberg B, Weisbeek P J (1986c) Plant growth control by fluorescent pseudomonads. In: Chet I (ed) Non-conventional aproaches to plant disease control. Wiley & Sons, New York

Scholte K, Veenbaas-Rijks J W, Labruyère R E (1985) Potato growing in short rotations and the effect of Streptomyces spp., Colletotrichum coccodes, Fusarium tabacum and Verticillium dahliae on plant growth and tuber yield. Potato Research 28: 331-348

Simon R, Priefer U and Pühler A (1983) A broad host range mobilization system for in vivo genetic engineering: transposon mutagenesis in gram negative bacteria. Bio/Technology 1: 784-791

Suslow T V (1982) Role of root-colonizing bacteria in plant growth. In: Mount M S, Lacey G H (eds) Academic Press, New York London Paris San Diego San Fransisco Sao Paulo Sydney Tokyo Toronto, p 187

Weger, L A de, Boxtel R van, Burg B van der, Gruters R, Geels F P, Schippers B, Lugtenberg B (1986) Outer membrane proteins of plant growth-stimulating antagonistic root colonizing Pseudomonas spp. J Bacteriol 165: 585-594

Weisbeek P J, Hofstad G A J M van der, Schippers B, Marugg J D (1986) Genetic analysis of the iron-uptake system of two plant growth promoting Pseudomonas strains. In: Swinburne T R (ed) Iron, siderophores and plant pathogens. Plenum Press, London, etc.

THE IRON-UPTAKE SYSTEM OF THE PLANT-GROWTH-STIMULATING *PSEUDOMONAS PUTIDA* WCS358: GENETIC ANALYSIS AND PROPERTIES AND STRUCTURE ANALYSIS OF ITS SIDEROPHORE

Gerard A.J.M. van der Hofstad, Joey D. Marugg, Henk B. Nielander, Ingrid van Megen, Ans M.M. van Pelt, Kees Recourt and Peter J. Weisbeek

Department of Molecular Cell Biology and
Institute of Molecular Biology
University of Utrecht
Padualaan 8
3584 CH Utrecht
The Netherlands

INTRODUCTION

Treatment of potato seed tubers with certain root-colonizing *Pseudomonas putida* and *Pseudomonas fluorescens* strains protects the plants from deleterious microorganisms which are present in the root environment during narrow rotation cropping, resulting in an increase of tuber yield in these soils (Schippers et al 1985). The plant-growth-stimulating activity is thought to be the consequence of their very efficient iron (III)-binding capacity. Under iron-limiting conditions which exist in most soils, the pseudomonads produce and excrete large quantities of yellow-green, fluorescent siderophores which chelate iron(III) with high affinity. The pseudomonads take up the iron(III)-siderophore complex specifically and with high efficiency. The deleterious microorganisms, however, are not able to do so and thus are prevented to grow efficiently due to iron(III)-depletion.

Our research is focussed on the relationship between plant-growth-stimulation and siderophore production. We here report on the genetic organization and regulation of siderophore biosynthesis, as well as properties and structure analysis of the siderophore of the plant-growth-stimulating *P. putida* strain WCS358.

NATO ASI Series, Vol. H 4
Recognition in Microbe-Plant Symbiotic and
Pathogenic Interactions. Edited by B. Lugtenberg

RESULTS AND DISCUSSION

P. putida WCS358 was mutagenized with transposon Tn5 using a broad host range RP4-dependent mobilization system (Simon et al 1983, Marugg et al 1985). Six different classes of mutants have been obtained, the two most important of which are $Flu^{-}Sid^{-}$ (no fluorescence and no iron (III)-binding) and $Flu^{-}Sid^{+}$ (no fluorescence, but normal iron(III)-binding). These mutants are being used in studies on growth-stimulation, survival in soil and colonization of the root, and for molecular genetical and biochemical analysis of the siderophore biosynthesis (see also Schippers et al, this issue).

As a measure for growth-stimulation both root-development of potato stem cuttings and potato tuber formation (in the field) were used and a close correlation was found between the increase of growth in narrow rotation soil and the ability to produce an effective siderophore (P. Bakker et al, to be published).

Using a genomic library in cosmid pLAFR1 (Friedman et al 1982), five distinct gene clusters involved in the siderophore biosynthesis of WCS358 were isolated by complementation of the mutants (Marugg et al 1985).

One major cluster, which complemented one $Flu^{-}Sid^{+}$ mutant and eight $Flu^{-}Sid^{-}$ mutants, was studied in more detail. The mutations are located on the chromosome in a region of 30 kb. At least four transcription units are encoded by this region, and so far seven genes have been mapped. The size of the proteins belonging to these genes and the direction of transcription were determined by *in vivo* protein synthesis in *E. coli* minicells and by northern hybridization, respectively. These last experiments also indicate that expression of at least some of the genes is regulated at the transcriptional level. This information was used to identify the iron-dependent promoters and is being used for the isolation of the major regulatory gene that controls all the genes for siderophore biosynthesis.

Also the gene encoding the outer membrane receptor for the iron (III)-siderophore complex has been isolated, and appears to be located adjacent to the biosynthetic genes of the major cluster.

When grown in iron-limiting culture medium WCS358 produces and excretes the yellow-green, fluorescent siderophore pseudobactin 358. The siderophore was purified as the iron(III) complex, yielding about 200 mg per liter (van der Hofstad 1986). The molecular weight of pseudobactin 358 as determined by spectrofotometric titration with $FeCl_3$ was found to be 1330D, when we suppose a ligand to iron(III) molar ratio of 1 to 1. The iron(III) binding constant was determined by competition with EDTA and amounted 2.1×10^{25} at pH 7.0, which is more than tenfold higher than that described for pyoverdin_{Pf} (Meyer and Abdallah 1978).

The structure of pseudobactin 358 has been analyzed by means of acid degradation, manual and automatic Edman sequence analysis and 'H-NMR. It consists of a nonapeptide containing the following amino acids: two residues of lysine and one each of N^{δ}-hydroxyornithine, alanine, serine, threonine, allo-threonine, aspartic acid and *threo*-β-hydroxy-aspartic acid. The sequence of the first seven amino acids was:

α-NH_2-Lys . *threo*-β-OH-Asp . Ser . Thr . Ala . allo-Thr . Lys.

The yellow-green, fluorescent quinoline derivative was linked to the first lysine, presumably via the Σ-amino group. The three bidentate iron (III)-chelating groups of pseudobactin 358 consist of a hydroxamate group derived from N^{δ}-hydroxyornithine, an α-hydroxy acid group derived from β-hydroxyaspartic acid and an o-dihydroxy aromatic group derived from the yellow-green, fluorescent chromophore.

At this moment we are purifying and analyzing mutant-siderophores obtained from the different classes of mutants defective in fluorescence and/or iron(III) binding to get insight in the biosynthesis of pseudo-bactin 358.

ACKNOWLEDGEMENT

This investigation was supported by the Netherlands Technology Foundation (STW).

REFERENCES

Friedman AM, Long SR, Brown SE, Buikema WJ and Ausubel FM (1982) Construction of a broad host range cosmid cloning vector and its use in the genetic analysis of *Rhizobium* mutants. Gene 18: 289-296

Marugg JD, van Spanje M, Hoekstra WPM, Schippers B and Weisbeek PJ (1985) Isolation and analysis of genes involved in siderophore biosynthesis in plant-growth-stimulating *Pseudomonas putida* WCS358 J. Bacteriol. 164: 563-570

Meyer JM and Abdallah MA (1978) The fluorescent pigment of *Pseudomonas fluorescens*: biosynthesis, purification and physicochemical properties. J. Gen. Microbiol. 107: 319-328

Schippers B, Geels FP, Hoekstra O, Lamers JG, Maenhout CAAA and Scholte K (1985) Yield depressions in narrow rotations caused by unknown microbial factors and their suppression by selected pseudomonads. In: Parker CA, More KJ, Wong PTW, Rovira AD and Kollmorgen JF (eds) Ecology and management of soil-borne plant pathogens. APS, St. Paul Minnesota, p 127

Simon R, Priefer U and Pühler A (1983) A broad host range mobilization system for *in vivo* genetic engineering: transposon mutagenesis in gram-negative bacteria. Bio/Technology 1: 784-791

van der Hofstad GAJM, Marugg JD, Verjans GMGM and Weisbeek PJ (1986) Characterization and structure analysis of the siderophore produced by the PGPR *Pseudomonas putida* strain WCS358. In: Swinburne TR (ed) Iron, siderophores and plant disease. Plenum Press, in press

ROLE OF FLAGELLA OF THE PLANT GROWTH STIMULATING PSEUDOMONAS FLUORESCENS ISOLATE WCS374 IN THE COLONIZATION OF POTATO ROOTS

Letty A. de Weger[a], Lia van der Vlugt[a], Peter A.H.M. Bakker[b], Bob Schippers[b] and Ben Lugtenberg[a]

[a]Department of Plant Molecular Biology, University of Leiden, Nonnensteeg 3, 2311 VJ Leiden, The Netherlands
[b]Department of Plant Pathology, University of Utrecht and University of Amsterdam, 'Willie Commelin Scholten' Phytopathological Laboratory, Javalaan 20, 3742 CP Baarn, The Netherlands

INTRODUCTION

Although potato is an economically profitable crop, its cultivation in the Netherlands is limited to once every three years on the same soil, since diseases (e.g., caused by nematodes or pathogenic microorganisms) may occur in soil more frequently cropped to potato. Even in this rotation schedule, crops are reduced when compared to potato crops from soils cropped only once every six years. Presently, this crop reduction is often due to an increase in the population of harmful microorganisms (Schippers et al 1985, 1986). This crop reduction can be abolished by applying antagonistic Pseudomonas spp. to the seed-potatoes (Geels and Schippers 1983; Bakker et al, submitted for publication).

The plant growth promoting Pseudomonas spp. comprise a diverse family as was shown by analysis of their membrane proteins (de Weger et al 1986a) and lipopolysaccharides (de Weger et al, in preparation) by SDS polyacrylamide gel electrophoresis. Membrane protein and lipopolysaccharide patterns obtained by the former technique, were similar or identical for seven strains. These strains most probably were descendants of one ancestor as they were isolated from a potato plant, grown in a pot (de Weger et al 1986). The remaining twenty-three Pseudomonas strains

NATO ASI Series, Vol. H 4
Recognition in Microbe-Plant Symbiotic and Pathogenic Interactions. Edited by B. Lugtenberg

could all be distinguished from each other on the basis of differences in these membrane proteins and lipopolysaccharides patterns. From these strains the plant root-colonizing strain WCS374 was chosen for further study as it efficiently increases potato crop yield in narrow rotation soil (Geels and Schippers 1983).

At least two properties of these Pseudomonads, namely the abilities to colonize potato roots and to inhibit the growth of other microorganisms, are thought to be essential for this effect (de Weger et al 1986b). The antagonistic effect is explained by the production of fluorescent siderophores which complex the low concentration of Fe^{3+}-ions present in most soils. This complexed Fe^{3+} is supposed not to be available to most other microorganisms, but the producing Pseudomonas strain itself induces special outer membrane receptor proteins which function in the specific uptake of the homologous Fe^{3+}-siderophore complex. Due to the resulting lack of the essential element iron, the harmful microorganisms are no longer able to grow, and a more favourable environment for development of the potato plant is created (Bakker, submitted for publication; see also Schippers et al, this volume).

Efficient root-colonization is the second factor supposed to be very important for plant growth stimulation (de Weger et al 1986b). As it is likely that the cell surface is involved in root colonization we are planning to study the effect of several mutations affecting the bacterial cell surface on colonization. In this paper the importance of motility for efficient root colonization is described.

RESULTS

Five hundred Tn5 mutants of strain WCS374 were screened for motility on 0,3% agar plates. Four non-motile mutants were isolated: strains LWM74-4,LWM74-29, LWM74-30 and LWM74-36. Neither of the mutant cells showed any motility when judged by phase contrast microscopy. Further characterization of the mutants using anti-flagella serum in immunoblots and with electron microscopy learned that the three mutant strains LWM74-4,LWM74-29 and LWM74-30 are probably impaired in the synthesis of wild type flagellin, the subunit of the flagellum. Mutant strain LWM74-36 is able to synthesize flagellin but is unable to polymerize these subunits into a functional flagellum. Instead, the subunits are excreted into the medium.

The ability of the four mutant strains to colonize potato roots was compared with that of the wild type strain by applying the bacteria to stem cuttings with roots approximately 1 cm in length. After 12 days of growth in soil which was cropped with potato once every six years, the number of bacteria on various parts of the roots was determined (Table 1). The plants inoculated with the non-motile mutants contained 500 to 5000-fold less bacteria on the deeper parts (8 cm) of the roots than the plants inoculated with the wild type strain (Table 1).

Table 1. The logarithm of the number of bacteria determined on different parts of the roots of plants inoculated by the wild type or by a non-motile mutant. Root sections were taken from various depths as shown in the figure at the left side of the table.

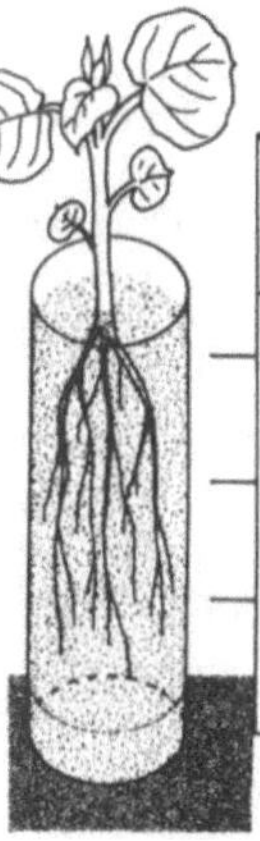

cm \ strains	wildtype	Tn5-mutant strains			
	WCS374	LWM74-4	LWM74-29	LWM74-30	LWM74-36
0	7.4	7.4	7.2	7.2	6.7
4	5.0	4.7	4.3	2.9	1.6
8	4.5	1.7	1.2	1.6	0.8

Values given are the logarithms of the mean values of 10 replicates.

CONCLUSION

Flagella of Pseudomonas fluorescens WCS374 are essential for efficient colonization of potato roots. The importance of flagella for plant growth stimulation will be tested in a greenhouse bioassay and in field experiments.

ACKNOWLEDGEMENT

This work was supported by the Netherlands Technology Foundation (STW).

REFERENCES

Geels FP and Schippers B (1983) Reduction of yield depressions in high frequency potato cropping soil after seed tuber treatments with antagonistic fluorescent Pseudomonas spp. Phytopathol. Z. 108: 207-214

Schippers B, Geels FP, Hoekstra O, Lamers JG, Maenhout CAAA and Scholte K (1985) Yield depressions in narrow rotations caused by unknown microbial factors and their suppression by selected Pseudomonads. In: C.A. Parker, K.J. Moore, P.T.W. Wong, A.D. Rovira and J.F. Kolmorgen (eds) Ecology and management of soil-borne plant pathogens. The Amer. Phyt. Soc., St Paul, Minnesota, p. 127-130

Schippers B, Lugtenberg B and Weisbeek PJ (1986) Plant growth control by fluorescent pseudomonads. In: I. Chet (ed) Non--conventional approaches to plant disease control. Wiley and Sons, New York (in press)

Weger LA de, van Boxtel R, van der Burg B, Gruters R, Geels FP, Schippers B and Lugtenberg B (1986a) Siderophores and outer membrane proteins of antagonistic, plant-growth stimulating, root-colonizing Pseudomonas spp. J. Bacteriol. 165: 585-594

Weger LA de, Schippers B and Lugtenberg B (1986b) Plant growth stimulation by biological interference in iron metabolism in the rhizosphere. In: D. van der Helm, J. Neilands, G. Winkelman (eds) Iron transport in microbes plants and animals. VCH verlagsgesellschaft (in press)

CONDITIONS IN THE RHIZOSPHERE IN RELATION TO MICROBIAL DEVELOPMENT

P.B. Tinker
Natural Environment Research Council
Polaris House, North Star Avenue, Swindon, Wilts, UK

INTRODUCTION

Interactions between microbes and higher plants take place either on the roots or shoots. The interactions in the soil-root zone are more difficult to monitor and measure, and are certainly more difficult to control, because of the difficulty of access and the extreme complexity of the environment in which the interaction occurs. It is the purpose of this paper to examine some of the specialised physical and chemical conditions immediately around plant roots, so that the potential effect of these on the microbe-plant interaction can be better understood. The normal soil situation around a root is vastly different from that observed in water culture or in other artificial situations, and needs to be taken into account in any attempt at a rational genetic modification of rhizosphere organisms for improved disease control, nitrogen fixation or other beneficial effects.

THE SOIL ENVIRONMENT

The soil habitat of microorganisms is extremely heterogeneous, due to the variable physical skeleton, and the processes of enrichment and depletion which constantly occur. Microorganisms therefore experience a varying series of conditions in time and space, superimposed upon the general background of the average soil properties measured by the normal analytical techniques. Such variation is much more extreme in natural soils, or in old pasture soils, than in arable, cultivated soils, but it is found to some extent even in the latter.

It is generally accepted that most of the microorganisms in the

NATO ASI Series, Vol. H 4
Recognition in Microbe-Plant Symbiotic and
Pathogenic Interactions. Edited by B. Lugtenberg

soil are normally in a state of starvation, with a very low metabolic rate (Paul, 1984; Brookes et al, 1983). It is very unexpected that despite this, they on average maintain a high adenylate energy change (Brookes et al, 1983), which suggests that a large part of this population has properties quite different from those normally studied in artificial culture. Equally, the work of Macdonald (to be published) has shown that the normal isolation techniques of bacteria from soil may detect only a small fraction of the total number of species.

However, the immediate root environment contains much more microbial substrates than the average soil, hence it may be more analogous to the situation in plate culture, and the organisms identified in this way may therefore be reasonably typical of the rhizosphere population.

RHIZOSPHERE POPULATION NUMBERS AND DISTRIBUTION

The population levels are not accurately known. It was customary at one time to determine "R/S ratios" (R = rhizosphere, S = soil) by direct plating and colony counting (see Martin, 1985). The numbers used were certainly underestimated by this, but the question is whether the relative increase at the root surface was correctly measured. The organisms which increase rapidly because of the root-derived substrate tend to be those which respond well in culture conditions, which may have caused the ratios to be over-estimated. Little weight is now placed on such quantitative results, and there appear to be no fully reliable and quantitative measurements of distributions of microorganisms in position and time. Newman & Watson (1977) used a mathematical model to predict that the rhizosphere contracted with time, as the root surface population increased and intercepted more exudates. In the limit, the increase was predicted to be solely on the root surface, or even in the cracks between cortical cells, from which it is assumed that exudate diffuses.

ION AND SALT CONCENTRATIONS

The absorbing root introduces concentration perturbations around it in the soil. The extent of these depends, for each chemical species, upon the demand by the plant, the initial concentration in the soil, the diffusion coefficient of that ion species, the root dimensions and the rate of water uptake. The theory of this is now well established (Nye & Tinker, 1977; Barber, 1984) and will not be discussed here.

The net result is that close to an actively absorbing root, the phosphate concentration will often be close to zero (below 10^{-6}M), potassium concentration will be greatly depressed, and other nutrients may vary widely depending upon the initial supply in the soil. Whenever an element is deficient in the sense that it restricts the growth of the plant, its concentration at the root surface must be close to zero. This is because the plant root uptake parameters (Vmax and Km) alter with the internal concentration of an element. In a deficient state, the roots are highly efficient, and can supply the plant so long as even very low concentrations (Ca 10^{-6}M for N, P and K) are maintained at the root surface (Wild & Breeze, 1981; Glass & Siddiqi, 1984). Equally, the concentrations within the root free space in such situations must be close to zero.

For ions with large diffusion coefficients, such as nitrate or chloride, it is unusual to find any major change in concentration at the root surface. On the other hand, there are situations in which slowly absorbed nutrients, such as calcium sulphate, can be carried in with the transpiration stream so rapidly that the concentration increases to above the initial value. In particular, this could occur in saline or gypseous soils in semi-arid climates. The salt concentrations immediately outside the root could thus possibly be increased 2 or 3 fold in extreme conditions.

ACIDITY CHANGES IN THE RHIZOSPHERE

Perhaps the most important local perturbation around roots is the change in pH. It has been shown by several means (Nye,

1986) that pH changes of 1 pH unit or so often occur. These changes were supposed to be caused by CO_2 release, exudation of organic acids or by cation-anion imbalance. The first two have now been eliminated, except for reports of high citric acid release in lupin species. The major mechanism is undoubtedly the excess of cations or anions in nutrient uptake, in that the over-riding requirement for electro-neutrality is met by excretion of protons or hydroxyl/bicarbonate ions. The most usual situation for crops is that nitrate is taken up to an extent which causes a raised pH in the rhizosphere.

The consequences of this for microbe interactions have scarcely been examined yet. The best known example is that of Gaeumannomyces graminis, which causes take-all in cereals, is acid-sensitive, and is therefore favoured by nitrate rather than by ammonium as the nitrogen source (Smiley, 1978). When plants obtain their nitrogen from an N_2-fixing symbiont, there is no net charge transfer, and the pH will normally then rise.

This flux of acidity or alkalinity generates a rapid pH change at the root surface which spreads out through the soil away from the root. The buffering by the soil and the cylindrical geometry of the system rapidly dampens the change, but if roots are very close to each other, then the whole soil pH will gradually change.

The theory of this process is quite complicated (Nye, 1986). In principle the equations are exactly analogous to those describing the flux of nutrients towards a root, but the effective diffusion coefficient for acidity is highly pH dependent. Several ionic species carry 'acidity': H^+, OH^-, HCO_3^-, Al^{+++}, and the behaviour of each individually varies with pH. In practice the effective diffusion coefficient is least at around pH 5.3, and the largest gradients will occur if the initial soil pH is near this value. Roughly speaking, the disturbance zone rarely exceeds 3 mm from the root.

These equations can in principle be used to calculate pH changes around roots, even if it is difficult in practice to measure all the physico-chemical quantities required. The most difficult part is however to determine the net flux of acidity or alkalinity out of the root. This can be done reasonably accurately as an average for the plant as a whole. However, it is now clear that this flux varies over the root system, and that different parts can at the same time excrete H^+ or OH^- fluxes (Römheld & Marschner, 1984). This is so even when root systems are growing in uniform media, but the reasons are so far unknown. Our own observations have shown intense acidification around single roots of leeks, whereas most of the rest were producing moderately alkaline conditions around them (Wong, Stribley & Tinker, unpublished). It seems possible that this differential behaviour of single main axes in monocot root systems may result from differences in vascular connections. Similar behaviour has been seen in water uptake and in sulphate uptake. It is thus impossible to predict from theory what the pH change will be around each root in soil, despite our theoretical understanding of the process.

This raises the question of the pH at the 'rhizoplane' or actual surface of the root, and that within the free space. The pH at the root surface cannot be measured, but is assessed by extrapolation of the pH gradient found in the soil. The smaller the pH micro-electrode available, the more nearly can a surface pH measurement be attained, but to do so in soil, without extreme disturbence of the system, is exceedingly difficult.

The pH of the free space is even more difficult to determine. With a steady outward flux of, for example, H^+ ions, the buffering power of the cell walls must rapidly become saturated, and the only factor limiting the drop in pH will be the rate of H^+ diffusion. If H^+ flows uniformly out of all cortical cells, it seems possible that the pH in the free space could differ by up to at least 1 pH unit from that at the surface. The free space pH could therefore reach extreme and unexpected values. This free space is occasionally inhabited by bacteria (the "endorhizosphere"), though it is not clear at what stage of development of a root they make their entrance to the intercellular spaces.

There is also another, and highly specific mechanism of rhizosphere acidification. In both phosphate deficiency (Grinsted et al, 1982) and in iron deficiency (Römheld & Marschner, 1984) the root surface pH can be depressed. This is not associated with the general cation/anion balance of the plant.

ORGANIC SUBSTRATES

There is a great deal of uncertainty at present about the form and quantity of the organic materials deposited in the rhizosphere (Rovira et al, 1979). Direct measurements on roots in solution culture show at most 1-2% of the total photosynthetically fixed carbon appearing in solution (Newman, 1985); sometimes it is as small as 0.1%. A great deal of work using ^{14}C labelling and axenic conditions has made it clear that much more than this is lost from the root in total (Whipps & Lynch, 1985).

The general consensus is that some 15-40% of the total fixed carbon is lost from the root, including respired CO_2. Helal & Sauerbeck (1984) assumed that microbial and root respiratory losses were equal, in their experiments with maize. On this basis, they concluded that some 7.5% of fixed carbon was deposited as organic compounds in the rhizosphere, of which one quarter remained after 25 days.It seems fair to assume, from solution culture work, that of this, perhaps 1% was as soluble, low-molecular compounds, and the rest polysaccharide gels or insoluble debris of various types (Rovira et al, 1979). This is the basis for the microbiological population increase in the rhizosphere. Any microorganism which is intended to multiply in the rhizosphere must be a good competitor for such materials.

SPECIFIC ROOT EXUDATES

A number of compounds released from roots are expected to have an effect on organisms in the external soil, including Rhizobia and nematodes. This may include chemotaxis or specific physiological responses. The behaviour of such small molecular weight compounds in soil requires consideration before assuming that they must display some activity. They may be sorbed on many different soil materials; thus hydroxycarboxylic acid or dicarboxylic acids are strongly sorbed on aluminium or iron oxy/hydroxy surfaces; phenyl compounds may be sorbed by expanding clays, and amino acids and proteins by both clays and humic acids. The diffusion of these compounds out of the root can in theory be dealt with by standard equations. In all cases diffusion rate depends on sorption, and if the latter is strong, then the compounds will move correspondingly slowly. In addition, bacterial breakdown must happen. Grinsted et al (1982) attempted to extract root-derived organic acids from the rhizosphere soil around rape roots, but were unable to detect any significant difference in the amounts of acids in rhizosphere and non-rhizosphere soils, except for carboxylic acids ionizing at up to pH 9. The latter groups may well be on solid debris. The implication of this is that assumptions about the effects of compounds diffusing out from the root should not be made without some evidence for the ability of the system to produce and sustain the necessary concentrations at some distance from the root.

REFERENCES

Barber SA (1984) Soil nutrient availability: a mechanistic approach. John Wiley, New York

Brookes PC, Tate KR, Jenkinson DS (1983) The adenylate energy charge of the soil microbial biomass. Soil Biol Biochem 15:9-16

Glass AD, Siddiqi MY (1984) The control of nutrient uptake rates in relation to the inorganic composition of plants. In: Tinker PB, Läuchli A (eds) Advances in plant nutrition, vol I. Praeger Publishers, New York, p 103

Grinsted MJ, Hedley MJ, White RE, Nye PH (1982) Plant-induced changes in the rhizosphere of rape (Brassica napus var Emerald) seedlings I. pH change and the increase in P concentration in the soil solution. New Phytol 91:19-29

Helal HM, Sauerbeck DR (1984) Influence of plant roots on C and P metabolism in soil. In: Tinsley J, Darbyshire JF (eds) Biological processes and soil fertility. Martinus Nijhoff/Dr W Junk, The Hague, p 175

Macdonald RM (to be published) Sampling soil microfloras, I. Dispersion of soil by ion exchange and extraction of specific microorgansims from suspension by elutriation. Soil Biol Biochem

Martin JK (1985) Biology of the rhizosphere, In: Soils, an Australian viewpoint. Division of Soils, CSIRO/Academic Press

Newman EI (1985) The rhizosphere: carbon sources and microbial populations. In: Fitter AH (ed) Ecological interactions in soil. Blackwell, Oxford (Sp Publ No 4 of British Ecological Society)

Newman EI, Watson A (1977) Microbial abundance in the rhizosphere: a computer model. Plant Soil 48:17-56

Nye PH (1986) Acid-base changes in the rhizosphere. In: Advances in Plant Nutrition, vol II. Praeger Scientific, New York, p 129

Nye PH, Tinker PB (1977) Solute movement in the soil-root system. Blackwell, Oxford

Paul EA (1984) Dynamics of organic matter in soils. In: Tinsley J, Darbyshire J (eds) Biological processes and soil fertility. Martinus Nijhoff/Dr W Junk, The Hague, p 275

Römheld V, Marschner H (1984) Plant-induced pH changes in the rhizosphere of 'Fe-efficient' and 'Fe-inefficient' soybean and corn cultivars. J Plant Nutrn 7:623-630

Rovira AD, Forster RC, Martin JK (1979) Note on terminology: origin, nature and nomenclature of the organic materials in the rhizosphere. In: Harley JL, Russell RS (eds) The soil-root interface. Academic Press, London, p 1

Smiley RW (1978) Colonization of wheat roots by Gaeumannomyces graminis inhibited by specific soils, microorganisms and ammonium nitrogen. Soil Biol Biochem 10:175-180

Whipps JM, Lynch JM (1985) Energy losses by the plant in rhizodeposition. Sum Proc Phytochem Soc Eur 26:59-71

Wild A, Breeze VG (1981) Nutrient uptake in relation to growth. In: Johnson CB (ed) Physiological processes limiting plant productivity. Butterworths, London, p 331-344

CONSENSUS AND PERSPECTIVE ON RHIZOBIUM

Sharon R. Long
Department of Biological Sciences
Stanford University
Stanford, CA 94305
U.S.A.

The discussions on Rhizobium during the workshop provided us with a consensus on several important questions and pointed to new lines of research. However, it was clear that we lack fundamental knowledge of and tools in other central areas.

First, the consensus: the work of many laboratories allows us to assign names to genes required for or involved in nodulation of plant hosts: nod A through J are defined in at least one species of Rhizobium. One caution must be added: because of the distinctive arrangement, and occasionally differing phenotypes, for the various nod genes, we require a convention for assigning equivalent names to loci in different species. Until a better way of defining genes comes along, we should use sequence homology as a way of defining allelism, and gene names should be assigned very cautiously.

Of the nod genes so far defined, nodDABC appear absolutely to be required for nodulation in all systems, plus nodH in R. meliloti and possibly others in Bradyrhizobium japonicum and other species. Other genes delay or reduce nodulation, and/or alter host range characteristics when mutated or when transferred to heterologous species. More study of these subtle phenotypes is required: kinetics of nodulation should be observed, along with microscopy of cell divisions, root hair curling and infection thread initiation and growth.

Several reports indicate that the job of locating nod genes is not over, and suggest a need for more searches. Examples include the demonstration that some chv- equivalent Rhizobium trifolii genes are required for nodulation, and that His^- B. japonicum mutants are also Nod^-. Preliminary results indicate that the slow-growing bacteria may have different nod genes working with the "ABCD" group, rather than the EF...etc. series present in fast-growing species.

NATO ASI Series, Vol. H 4
Recognition in Microbe-Plant Symbiotic and
Pathogenic Interactions. Edited by B. Lugtenberg

Besides the non-nodulating phenotype, the "empty nodule" (morphogenesis without infection) phenotype has been studied, leading to the observation that exopolysaccharides are essential for normal infection. Extracellular and lipopolysaccharides are still mysterious structures in *Rhizobium*, and while work on some systems is progressing, there was a general call for further work: first, it must be recognized that the surface and extracellular domains bear not a single polysaccharide, but a complex mixture of oligomers and polymers, and these various entities all need to be defined and analyzed. The possible changes of these molecules in response to *nod* gene regulation, or during phases of growth in culture, need to be analyzed.

Whether the *nod* and host range genes affect surface polysaccharides is an important question which will receive direct attention in the near future. This is, however, only a small part of the global question, what do the *nod* genes do? The availability of DNA sequence data has revealed some homologies to known proteins: specifically, Shearman et al. have found that *nodF* of *R. leguminosarum* displays considerable homology to *Escherichia coli* acyl carrier protein. Other sequence analyses, and in two cases protein localization, also point to a membrane localization for *nod* gene products.

Future research proposals focus partly on more definitive localization for the *nod* gene proteins. To test function, it would be tremendously helpful to have in vitro systems or at least to have access to material highly enriched for infection threads and/or newly dividing cells. Several possible methods of achieving this were proposed, including J. Dénarié's suggestion that delayed -Nod mutants which are Inf^{++} might be useful as sources of growing threads. Such cellular tests might be combined with new analyses of proteins and other components on bacterial surfaces. This might be approached by chemical analysis or by antibody or phage testing which would highlight surface alterations.

In vivo phenotypes and *in vitro* biochemical functions of *nod* genes can be studied in a new light due to the breakthrough on *nod* gene regulation. The discovery by four research groups that various flavones and flavanones induce nodulation genes greatly refines the recent finding that total plant exudates influence *nod* gene transcription and *nod*-gene mediated root growth effects ("tsr"). This pushes the field forward into specific experiments on plants and bacteria.

Several questions now need to be answered: are inducer molecules taken up into the cell? Are they metabolized or modified in the cell or at its surface? What is (are) the receptor(s)? It has been shown the nodD is required for inducer molecules to affect nod gene expression. Does nodD interact directly with inducer? And, are various Rhizobium species "fine-tuned" to interact with particular plants by preferential interactions with certain flavones? In several systems a conserved sequence has been observed upstream of nod genes. The Szeged group proposes to call this the "nod box". The function of this sequence needs to be established by deletion and alteration of sequences, in conjunction with transcript mapping and gene-fusion expression assays. Among the fundamental questions we must answer are these: why is the "nod box" so large? Where does it lie with respect to the transcript start? Does expression of nod genes require altered RNA polymerase, or positive activator proteins, or both?

Turning, as a last subject, away from nod genes and towards later stages of nodulation, we find that "recognition" - that is, specific interaction - occurs at many stages. There are more genes out there to be found for control of bacterial release into target cells, bacterial differentiation, and nitrogen fixation. Research by Brewin and colleagues suggests the exchange of molecules between bacteroid surface and peribacteroid membranes. This study points the way to further analysis of macromolecular assemblies and exchange between the symbiotic partners, which will doubtless be subject to proper recognition. Whether large molecules or small, the signals which lead to the final differentiation of plant and bacteroid remain to be defined: what triggers the expression of nitrogen fixation in the bacteroid? Does it work through inducing nifA? What bacterial signal cause the expression of unique plant genes, nodulins, early and late in nodule formation? What are the functions of these plant proteins in receiving or sending signals and in nodule construction?

Recognition is defined as a "clear perception of something as having been seen before." Although Rhizobium bacteria and their plant hosts cannot be said to have "seen each other before" their interaction, we might say that their genes have. This fanciful analogy reminds us of a helpful approach: as we look for the many levels at which these organisms interact, it will be useful to refer back to basic level, that of the gene.

PROSPECTS FOR NOVEL PLANT DISEASE CONTROL RATIONALES

N. T. Keen
Department of Plant Pathology
University of California
Riverside, CA 92521, U.S.A.

The practical control of plant diseases generally uses methods which were implemented in the last century. Plant pathologists have improved the application and deployment of these strategies, but conceptually they are little changed. Where available, the most popular and effective means of plant disease control is by genetically inherited resistance. Such disease resistance genes have been used in cultivar improvement since early in this century, but we still do not know their structures or what they do. The importance of resistance genes in practical agriculture and the fact that they are inherited as single Mendelian traits makes them lucrative objectives for the improvement of crop plants by biotechnology. To attain the objective, however, it will be necessary to clone disease resistance genes from one or more plant species. This is a technically imposing task and it would also be necessary to introduce the cloned gene into a cultivar of the same plant lacking the allele in question or into another plant species and test for expression by inoculation with an appropriate pathogen. While this exercise is of general interest to plant biologists, there are several other approaches to improving pathogen resistance. Some of these are:

(i) Introduce genes to increase the number of phytoalexins made by a plant. The generalization is emerging that successful pathogens of a certain plant are able to degrade the phytoalexins of that plant but not the different phytoalexins of another plant. Since genes encoding biosynthetic enzymes to several phytoalexins have been cloned, it should eventually be possible to introduce the requisite phytoalexin synthetic genes from one plnat into a different plant so that it can make an additional phytoalexin(s). It would then be of interest to see if improved disease resistance occurred.

(ii) Transform genes into plants that encode secreted proteins which are antagonistic to pathogens or insects. These may include the production of lysozyme or other proteins which antagonize bacteria, ribonuclease or other proteins to antagonize RNA plant viruses, chitinase or glucanase to antagonize insects and fungi and Bacillus thuringiensis proteins which antagonize insects. There have been preliminary successes in this area, including the cloning of chitinase genes from Serratia marcescens by John

NATO ASI Series, Vol. H 4
Recognition in Microbe-Plant Symbiotic and
Pathogenic Interactions. Edited by B. Lugtenberg

Bedbrook's group at Advanced Genetic Sciences [EMBO J. 5: 467 (1986)] and the demonstration that their introduction into certain microflora bacteria will antagonize fungi. A similar transformation of the appropriately engineered genes into plant cells may lead to disease resistance to fungal pathogens. Roger Beachy's group at Washington University in St. Louis has also observed a measure of disease control against tobacco mosaic virus by transforming the coat protein gene of the virus into tobacco (Savannah Plant Molecular Biology meeting, October, 1985). Although the mechanism of disease control is not yet known, the exercise demonstrates the power of transforming certain defined DNA sequences into plants and it is likely that other genes will be used in analogous ways.

(iii) Transform genes into plants that diminish the plant environment available to a pathogen. For example, the introduction of genes coding for efficient microbial iron-binding siderophores or efficient permeases which transport glucose and amino acids into plant cells may result in very low levels of these nutrients in plant intercellular spaces for pathogens and therefore engender disease control. It is also possible to utilize the high specificity of vertebrate antibodies against pathogens. One of the more appealing of these approaches is to clone rearranged genes for the light and heavy antibody subunits from a monoclonal hybridoma cell line and then engineer these for expression in plant cells. The use of such antibodies directed to pathogen enzymes important in pathogenesis, to pathogen feeding structures such as haustoria, or to viral replication proteins may result in highly specific disease control.

(iv) More intensively study plant-pathogen interactions with an emphasis on critical tests of hypotheses using the appropriate biochemical and genetic tests. Clone the genes for relevant specificity-imparting mechanisms and express these in the appropriate pathogen or plant genotypes. For instance, race-specific pathogen elicitors such as the Colletotrichum lindemuthianum glycoprotein side chain carbohydrates, the Cladosporium fulvum peptide or victorin from Helminthosporium victoriae are all potential probes to attempt isolation of the putative complementary receptor proteins encoded by plant disease resistance genes. Once such receptor peptides are identified, cloning of their genes should be straightforward. It is indeed possible that the most direct route to cloning of plant disease resistance genes is through the use of specific elicitors.

The above are but a few of the approaches that will be utilized in future attempts to effect disease control approaches permitted by the advent of recombinant DNA technology. It is suspected that they will also result in a wealth of additional knowledge on how plant diseases operate.

PLANT-AGROBACTERIUM INTERACTION

R.A. Schilperoort, Department of Plant Molecular Biology, Biochemical Laboratory, Leiden University, Wassenaarseweg 64, 2333 AL Leiden, The Netherlands

The plant-*Agrobacterium* interaction is a well-studied example of cell to cell signaling. It is giving us important information about factors and processes which play a role in pathogenesis.

Agrobacterium species induce tumors and hairy roots via introduction and integration of T-DNA, carrying oncogenes, into the chromosomal DNA of plant cells (see other papers in this volume). Studies on the function and targets of these genes will lead to a better understanding of the genetic and molecular regulation of plant cell differentiation and regeneration. Since the oncogenes are involved in the synthesis of phytohormones (auxins and cytokinins) and possibly also in the regulation of phytohormone activities, they offer a genetic tool for studying the molecular mechanisms by which phytohormones act. Knowledge of these mechanisms is essential to achieve regeneration of plants from cells in a more controlled manner than is possible nowadays for many important crops.

The capacity of *Agrobacterium* to incorporate foreign DNA more or less at random in the plant genome of many dicots, and possibly also in various monocots, has made this micro-organism a valuable tool for the genetic manipulation of plants. Moreover, studies on plant-*Agrobacterium* interaction have contributed to the technological advancement in plant molecular biology and the development of novel genetic approaches in plant breeding.

Our knowledge about the plant-*Agrobacterium* interaction, however, is still far from complete. We now are at a stage that for further progress in the areas mentioned it is becoming increasingly important to stimulate multidisciplinary collaboration.

NATO ASI Series, Vol. H 4
Recognition in Microbe-Plant Symbiotic and Pathogenic Interactions. Edited by B. Lugtenberg

Various problems especially need a stronger biochemical approach of teams working together on one subject. We have to spend creativity in finding good in vitro systems to understand the function of genes which are directly engaged in the interaction between the bacterium and the plant cell. Although quite a number of agrobacterial genes involved in virulence and oncogenicity have been identified so far, still relatively little is known about their functions. Even worse is the situation for the plant genes involved. Actually none are known. More emphasis should be given to the plant side. This not only relates to the plant genes whic contribute to tumor or hairy root development but also to questions concerning host range and host defense systems. Broad host range strains, for example, appear to induce a hypersensitivity reaction instead of tumors on certain plant species like grapevine, while tumors are induced on the same varieties of grapevine by limited host range strains. The molecular genetics of Agrobacterium is well developed and can be used to understand the genetic basis of the difference between both types of strains. This may lead to the identification of processes and genes in the plant involved in the hypersensitivity reaction. Through this kind of experiments use can be made of the good molecular genetic characterization of Agrobacterium in the unraveling of defense systems in plants. Since studies in this area may contribute to our understanding of pathogen resistance in plants, which is of prime importance as an agricultural trait, they must get more attention. It would be a pity if the attention to the plant-Agrobacterium interaction in future would be restricted to the genetic engineering of plants.

RECOGNITION, ELICITORS AND THE HYPERSENSITIVE REACTION

J.W. Mansfield, Biological Sciences Department, Wye College
Ashford, Kent TN25 5AH, U.K.

In order to understand the biochemical basis underlying control of resistance to plant pathogens it is essential to integrate biological, physiological and molecular studies. During the meeting it became clear that there are disturbing variations in the meaning that different workers attach to terms such as elicitor, recognition, hypersensitive reaction, and even plant cell death. I would like to comment on the use of these terms and attempt to provide a framework to focus attention on the critical processes occurring during infection development and to allow well defined interdisciplinary studies to develop.

Hypersensitive reaction

This may be defined as the rapid and localized death of challenged plant cells associated with the restriction of microbial colonization. The term was first used by Stakman (1915) to describe the rapid death of cells around penetration sites in cereal leaves resistant to the obligate parasite *Puccinia graminis*. More recently, HR has been used to describe not only plant cell death but also the discoloration of cells and their accumulation of phytoalexins. Given our increasing understanding of the biochemical events occurring during incompatible interactions it seems best to continue using the term HR in a broad sense. Thus, plant cell death and restriction of microbial growth are _essential_ features of all hypersensitive reactions but there is scope for variation in other biochemical events associated with the HR *sensu lato*. For example, in some plants phytoalexins may accumulate whereas in others different antimicrobial systems (e.g. lignification, chitinase and hydroxyproline rich glycoprotein synthesis) may be activated.

Elicitor

In my opinion (and also that of Wood, 1984) this is a very

NATO ASI Series, Vol. H 4
Recognition in Microbe-Plant Symbiotic and Pathogenic Interactions. Edited by B. Lugtenberg

confusing word to use if it is not made clear what is being elicited. For example, an elicitor of phytoalexin biosynthesis need not be an elicitor of other components of the hypersensitive reaction. If we are to understand control of the activation of defence reactions it is important to realize that elicitors may have very selective effects. It is also important to attempt to identify the elicitors of phytoalexin biosynthesis that are active during infection.

Plant cell death

In the infected plant, rapid phytoalexin accumulation is invariably associated with the death of plant cells as recognised by various vital staining techniques. Bailey (1983) has argued that the death of plant cells may release endogenous (plant derived) elicitors of phytoalexin biosynthesis. By contrast, Slusarenko et al (this volume) have argued that plant cell death is the end product of elicitation of numerous processes associated with resistance. The contrasting ideas seem to stem, in part, from different interpretations of what is meant by a dead cell. Thus Bailey (1983) equates the time of plant cell death with the time at which compartmentation is lost. Loss of compartmentation is considered by Gahan (1984 and pers. comm.) to be very closely associated with slightly earlier irreversible damage to the plasma membrane. Clearly, loss of compartmentation occurs some time before the affected cell may cease to respire or become discoloured. Slusarenko et al (this volume) have implied that a cell is not dead until it has ceased metabolic activity. The latter is the medically more acceptable usage of the adjective "dead" and draws attention to the numerous processes which may be co-ordinately or separately activated during the plant's response to infection. However, this interpretation rather obscures the possibility that irreversible membrane damage may be the primary event controlling subsequent activation of mechanisms of resistance.

Confusion would be avoided if the meaning attributed to "dead" was always clearly defined as, to give one possible definition, by stating that ...."cells were dead as indicated by their failure to plasmolyse." Incidentally, failure to plasmolyse using solutes of low molecular weight such as KNO_3 probably provides the earliest test for irreversible membrane damage.

Recognition

Recognition can occur at many stages of plant parasite interactions. The term merely implies that one or other partner (not always both) becomes aware of the other's presence. Recognition of the parasite by the host involves the transfer of a biochemical signal and is indicated by the initiation of a response by the challenged plant. The response may not be that which restricts microbial development.

Having discussed the use of terms it is necessary to consider how recognition, loss of compartmentation and elicitation of active processes of resistance may be linked together during the expression of resistance. The schemes outlined below are modified versions of those reported in a previous NATO symposium (Mansfield and Brown 1986); use of the term plant cell death has been avoided here. The end product of each of these schemes would be a hypersensitive reaction in which restriction of the invading microbe is associated with the localized death of plant cells.

a) Loss of compartmentation as the primary cause of induced responses

Avirulent pathogen
↓
Recognition
↓
Loss of compartmentation
↓
Release of endogenous elicitor
↙ ↘
Phytoalexin biosynthesis | Other resistance mechanisms induced

b) Loss of compartmentation as part of an induced response

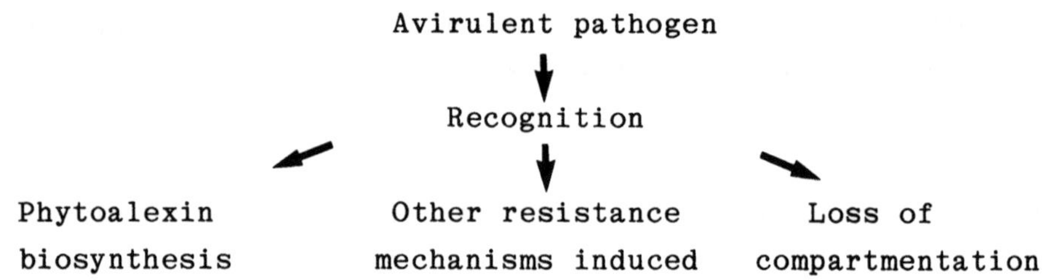

c) Induced responses controlled by separate recognition events

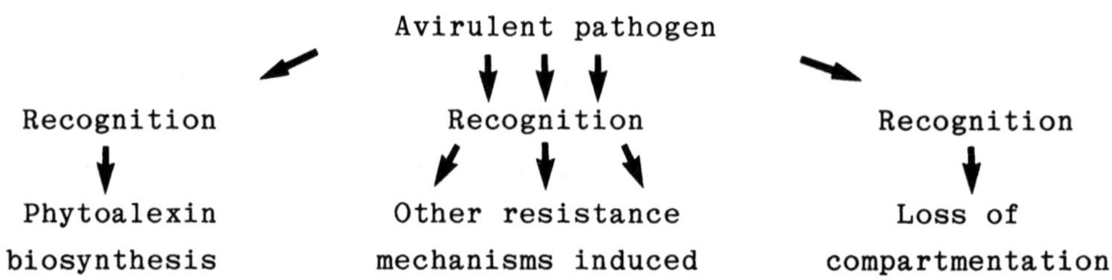

In each scheme there is a requirement for a phase during which the plant recognizes the invading avirulent pathogen. In a) recognition leads directly to loss of compartmentation, and in b) loss of compartmentation is one of a cascade of responses concurrently induced following a single recognition event. Either scheme a) or scheme b) is likely to operate during race specific resistance because this is usually controlled by single genes in the host and parasite. Scheme (c) illustrates how induced responses may be controlled by separate recognition events. Although (c) is unlikely to operate during gene for gene interactions as expressed during race-specific resistance, the scheme emphasizes that in certain interactions activation of single responses should not be overlooked. Currently, we have insufficient knowledge of the biology, physiology or molecular biology of plant-pathogen interactions to be certain which if any of these schemes operates during the expression of resistance.

REFERENCES

Bailey JA (1983) Biological perspectives of host-pathogen interactions. In : Bailey JA, Deverall BJ (eds) The Dynamics of Host Defence. Academic Press, Sydney p.1

Gahan PB (1984) Reversible and irreversible damage in plant cells of different ages. In : Davies I, Sigee DC (eds) Cell ageing and cell death (SEB seminar series; 25). Cambridge University Press p. 155

Mansfield JW, Brown IR (1986) The biology of interactions between plants and bacteria. In : Bailey JA (ed) Biology and Molecular Biology of Plant-Pathogen Interactions. Springer-Verlag (in press)

Stakman EC (1915) Relations between Puccinia graminis and plants highly resistant to its attack. J Agric Res 4:193-200

Wood RKS (1984) Establishment of infection. Plant Pathol 33: 3-12

MOLECULAR ANALYSIS OF PLANT DEFENCE MECHANISMS

H. Grisebach
Biologisches Institut II
Universität Freiburg
D-7800 Freiburg
FRG

1. Elicitors are now being used not only to investigate elicitation of plant defense reactions but also to stimulate the production of secondary metabolites in plant cell suspension cultures.

The demonstration of a high-affinity binding site for a glucan elicitor from P.megasperma on plasma membrane enriched fractions of soybean opens the possibility for further studies on elicitor binding. Such studies should be initiated in other systems as well. Isolation of the putative receptor should be pursued. An important question to be answered is whether an elicitor remains bound to the plasma membrane or whether it can be internalized. This question is closely related to the still unsolved problem of how elicitors trigger activation of defense genes in the nucleus. If the elicitor is internalized it could bind to receptor proteins in the cytoplasm or in the nucleus, as has been found in the case of steroid hormones.

More research is needed on the role of endogenous elicitors and their interaction with exogenous elicitors.

2. Several results suggest that Ca^{2+} ions could be involved in elicitor action. Methods should be developed to measure changes in intracellular Ca^{2+} concentrations in higher plants. Such methods would also be valuable for other areas of research, for example hormone action.

3. More research is needed on the biochemistry of the hypersensitive response which is connected to restriction of microbial development.

NATO ASI Series, Vol. H 4
Recognition in Microbe-Plant Symbiotic and
Pathogenic Interactions. Edited by B. Lugtenberg

4. Phytoalexins seem to play an important role in plant defense against certain pathogens. More genetic evidence for the role of phytoalexins is desirable. Transposon mutagenesis of genes involved in phytoalexin biosynthesis should lead to plants which are unable to synthesize phytoalexins. For example, transposon mutagenesis of stilbene synthase, which catalyses a single-step reaction to phytoalexins, would make it possible to assess the importance of stilbenes in plant defense.

Besides elucidating the possible role of phytoalexins in plant defense, investigations on the induction of phytoalexin pathways have already broadened our understanding of gene activation and should be continued in several host-pathogen systems.

5. Plant defense is a multifactorial process. Investigations on the contributions of cutin, lignin, callose, hydroxyproline rich-glycoproteins, active oxygen species, etc. to resistance should be pursued on a wide scale.

6. Better understanding of the mechanisms of plant defense will help to develop new methods of pest control.

SUBJECT INDEX

(A pagenumber marked* refers to the whole paper)

NATO ASI Series H

Vol. 1: **Biology and Molecular Biology of Plant-Pathogen Interactions.**
Edited by J.A. Bailey. 415 pages. 1986.

Vol. 2: **Glial-Neuronal Communication in Development and Regeneration.**
Edited by H.H. Althaus and W. Seifert. 842 pages. 1987.

Vol. 3: **Nicotinic Acetylcholine Receptor: Structure and Function.**
Edited by A. Maelicke. 489 pages. 1986.

Vol. 4: **Recognition in Microbe-Plant Symbiotic and Pathogenic Interactions.**
Edited by B. Lugtenberg. 449 pages. 1986.

MIX
Papier aus verantwortungsvollen Quellen
Paper from responsible sources
FSC® C105338

If you have any concerns about our products,
you can contact us on
ProductSafety@springernature.com

In case Publisher is established outside the EU,
the EU authorized representative is:
Springer Nature Customer Service Center GmbH
Europaplatz 3, 69115 Heidelberg, Germany

Printed by Libri Plureos GmbH
in Hamburg, Germany